IET RADAR, SONAR AND NAVIGATION SERIES 33

Radar Automatic Target Recognition (ATR) and Non-Cooperative Target Recognition (NCTR)

Other volumes in this series:

Volume 1	**Optimised radar processors** A. Farina (Editor)
Volume 3	**Weibull radar clutter** M. Sekine and Y. Mao
Volume 4	**Advanced radar techniques and systems** G. Galati (Editor)
Volume 7	**Ultra-wideband radar measurements: analysis and processing** L. Yu. Astanin and A.A. Kostylev
Volume 8	**Aviation weather surveillance systems: advanced radar and surface sensors for flight safety and air traffic management** P.R. Mahapatra
Volume 10	**Radar techniques using array antennas** W. Wirth
Volume 11	**Air and spaceborne radar systems: an introduction** P. Lacomme (Editor)
Volume 13	**Introduction to RF stealth** D. Lynch
Volume 14	**Applications of space–time adaptive processing** R. Klemm (Editor)
Volume 15	**Ground penetrating radar, 2nd edition** D. Daniels
Volume 16	**Target detection by marine radar** J. Briggs
Volume 17	**Strapdown inertial navigation technology, 2nd edition** D. Titterton and J. Weston
Volume 18	**Introduction to radar target recognition** P. Tait
Volume 19	**Radar imaging and holography** A. Pasmurov and S. Zinovjev
Volume 20	**Sea clutter: scattering, the *K* distribution and radar performance** K. Ward, R. Tough and S. Watts
Volume 21	**Principles of space–time adaptive processing, 3rd edition** R. Klemm
Volume 22	**Waveform design and diversity for advanced radar systems** F. Gini, A. De Maio and L.K. Patton
Volume 23	**Tracking filter engineering: the Gauss–Newton and polynomial filters** N. Morrison
Volume 25	**Sea clutter: scattering, the *K* distribution and radar performance, 2nd edition** K. Ward, R. Tough and S. Watts
Volume 26	**Radar techniques using array antennas, 2nd edition** W. Wirth
Volume 101	**Introduction to airborne radar, 2nd edition** G.W. Stimson

Radar Automatic Target Recognition (ATR) and Non-Cooperative Target Recognition (NCTR)

Edited by
David Blacknell
Hugh Griffiths

The Institution of Engineering and Technology

Published by The Institution of Engineering and Technology, London, United Kingdom

The Institution of Engineering and Technology is registered as a Charity in England & Wales (no. 211014) and Scotland (no. SC038698).

First published 2013

The Institution of Engineering and Technology

Michael Faraday House
Six Hills Way, Stevenage
Herts, SG1 2AY,
United Kingdom

www.theiet.org

British Library Cataloguing in Publication Data
A catalogue record for this product is available from the British Library

ISBN 978-1-84919-685-7 (hardback)
ISBN 978-1-84919-686-4 (PDF)

Typeset in India by MPS Limited
Printed in the UK by CPI Group (UK) Ltd, Croydon

Contents

Foreword

The spectrum of armed hostilities – conventional/declared warfare, terrorism, police action, armed robbery and contemporary asymmetric armed conflicts – continues to evolve towards less structured hostile engagements. Commanders of military and police units as well as individual soldiers, patrolmen, security personnel, etc., increasingly confront situations with blurred military/civilian targets and with no discernible difference between legitimate combatants and civilians.

Making operational decisions, sometimes with loss-of-life consequences, in such uncertain conditions is framed by extensive training, adherence to pertinent tactics, techniques and procedures (TTP) and rules of engagement (ROE) that provide general guidance in most conceivable situations. Notwithstanding this background material, evaluating a potentially hostile situation and making an appropriate responsive decision, often an instantaneous decision, remain, and will always remain a personal challenge. Mitigating this challenge by developing robust decision aids regarding the disposition of potentially antagonistic targets is the essence of this volume.

This book is a revised and edited compendium of lectures on automatic target recognition (ATR) and non-cooperative target recognition (NCTR) originally commissioned by the NATO Science & Technology Organization (STO) (formerly known as the Research & Technology Organization) Sensors and Electronics Technology Panel (SET) with the title 'Lecture Series on Radar Automatic Recognition (ATR) and Non-Cooperative Target Recognition (NCTR)'. These lectures have been presented recently in many of the NATO member nations.

The 2013 publication of these lectures by the Institution of Engineering and Technology (IET) is timely, as the process of recognising the disposition of a non-cooperative and potentially hostile target is becoming increasingly challenging.

Consider, as a trivial example, the challenge of determining the nature and intent of an unidentified civilian open-bed pick-up truck with a tarp covering some object approaching a restricted zone, vulnerable personnel or critical infrastructure assets. How does one respond in a manner compliant with the rules of engagement? What decision aids facilitate the determination of target intent within time to initiate an appropriate responsive course of action?

Moreover, the trend to network all sensor and weapon platforms, command centres, communication nodes and relays, and personnel within a theatre of operations, while providing unprecedented situational awareness, sometimes produces sensory input overload that interferes with the process of mentally sorting through options in a timely manner. This alone demands an automated capability to sort, evaluate and prioritise options to be presented to a decision maker who can be anywhere along the chain of command. Reduced platform manning, made possible by wideband communication links among networked assets and by automated condition-based platform maintenance, also decreases platform staff available to contribute to decision making processes. In addition, the recent operational capability of unmanned vehicles/platforms forces remote decision makers to rely on distant sensor platforms to provide actionable near real-time, all-weather imagery as the basis for targeting decisions. There is a clear and pressing demand for an expeditious but robust automatic target recognition capability that supports the integration of all types of unmanned platforms into the order of battle – mitigating bandwidth requirements of shared communication systems, prioritising actionable information and providing different users with information appropriate to their level in the chain of command, or as requested on an ad hoc basis.

Putting ATR and NCTR in a more general context, the *NATO AAP-6 Glossary of Terms and Definitions*, as revised in 2008, provides a framework for assessing the nature of a target. *Target acquisition* is defined as '*The detection, identification, and location of a target in sufficient detail to permit the effective employment of weapons*'. Detection and identification are inextricably linked to and depend on recognition. NATO defines *recognition* as 'The determination of the nature of a detected person, object or phenomenon, and possibly its class or type. This may include the determination of an individual within a particular class or type' and *identification* as 'The process of attaining an accurate characterization of a detected entity by any act or means so that high confidence real-time decisions, including weapons engagement, can be made'.

This book on automatic target recognition (ATR) and non-cooperative target recognition (NCTR) describes the challenge of target recognition, presents state-of-the-art resources and indicates directions for future research. There is still much work to be done.

W.J. Miceli
University College London

Chapter 1
Introduction

Hugh Griffiths and David Blacknell

1.1 Motivation

The ability to detect and locate targets by day or night, over wide areas, regardless of weather conditions, has long made radar a key sensor in many military and civil applications. However, the ability to reliably distinguish different targets (e.g. hostile from friendly) has proved much more elusive, and has conventionally relied on skilled operators and on separate devices such as identification friend or foe (IFF). To be able to do this automatically represents a much more difficult challenge, although steady progress has been made over the past couple of decades.

The applications are diverse, and range from the detection and identification of buried landmines and other unexploded ordnance, to the detection and identification of vehicles in high resolution synthetic aperture radar (SAR) images, to the detection of satellites and debris in space.

The problem is perhaps best approached from a physical viewpoint, determining in each case the information contained in the radar echoes and the features that may distinguish the wanted target from competing echoes that at first sight may appear similar, and immediately the value of high resolution and high signal-to-noise ratio becomes apparent. It is also useful to realise that the radar echo is the result of the scattering of the microwave signal, whose wavelength will be of the order of centimetres, from the target, and this is quite different from the scattering of visible light. Furthermore, unlike an optical image, the radar explicitly measures range. Hence, we should not expect radar images to look like optical images – they contain different information.

The way in which targets scatter signals of different polarisations also contains information that can be exploited in target recognition. Furthermore, advanced techniques such as interferometry and coherent change detection provide extraordinarily sensitive means of exploiting the information contained in the radar echoes. The characteristic modulation of the radar echoes due to the rotating parts of jet engines, or the rotating blades of a helicopter, provides distinctive information, which can allow the specific type of aircraft or helicopter to be determined.

In military operations the problem may be compounded by decoy targets of various degrees of sophistication, and it may be extremely difficult for an automatic target recognition system to be able to distinguish between true targets and decoys.

1.2 Definitions and acronyms

The use of automatic techniques to classify radar data gives rise to different acronyms depending on the particular radar domain involved. It is usually known as automatic target recognition (ATR) when dealing with air-to-ground activities, which mostly use synthetic aperture radar (SAR) imaging, while it is usually known as non-cooperative target recognition (NCTR) for ground-to-air or air-to-air activities, which mostly use high-resolution range (HRR) profiles, jet engine modulation (JEM) and inverse SAR (ISAR) imaging.

What is meant by the word 'recognition'? Care is needed in answering this as the ATR 'vocabulary' is still evolving. Indeed, only a few terms have been standardised by NATO but, somewhat confusingly, some of these words actually have more than one formal NATO definition.

Taking the word 'identification' as an example, the NATO AAP-6 Glossary of Terms and Definitions says that identification is the separation of friend and foe. However, in most modern conflicts a third class has to be added to this dual separation of the world to take into account the 'neutral' targets that exist independently of the 'classical' enemies. Moreover, this third class tends to be the focus of most actual identification efforts to avoid collateral damage. It is clear that this identification process will depend on the people involved (countries, coalition forces) and on the context (in both space and time): it should also be taken into account that a civilian 'neutral' vehicle may be easily turned into an enemy weapon. Currently, 'identification' relies mostly on human interpreters or transponders like the identification friend or foe (IFF) system. It is somewhat difficult to characterise this definition purely in terms of scientific criteria and thus very difficult to automate.

In contrast, the word 'recognition' as defined by the NATO AAP-6 Glossary of Terms and Definitions is a little more precise. The process is decomposed into a kind of classification 'tree' in which the targets are categorised into more and more precise sub-classes as progress is made through the tree structure. Six major classification steps are then described:

- detection, separating targets from other objects in the scene,
- classification, giving the target a meta-class such as aircraft, wheeled vehicle, etc.,
- recognition, specifying the class of the targets such as fighter aircraft, truck, etc.,
- identification, giving the sub-class of the target such as MIG29 fighter aircraft, T72 tank, etc.,
- characterisation, taking into account the class variants such as MIG29 PL, T72 tank without fuel barrels, etc.,
- fingerprinting, leading to an even more precise technical analysis such as MIG29 PL with reconnaissance pod.

It can be seen that the boundaries between these decomposition steps cannot be clearly fixed for all problems and targets. Moreover, these definitions lead to the word 'classification' being reserved solely to describe the process of meta-class separation while it is more often used by scientists to describe the whole process of assigning objects to categories irrespective of the status of those categories. This breakdown of definitions is even more obvious with the word 'identification', which has been seen to take two different meanings within a single official glossary.

The main outcome of this discussion is thus to stress the need for a precise problem formulation and description of the operational conditions applying to the particular ATR problem under consideration.

1.3 Scope of book

This book is based on a set of lectures delivered by the authors as part of the NATO SET-172 Lecture Series, in countries including the USA, Portugal, The Netherlands, Greece, Norway, Belgium and Turkey. We express our thanks to the NATO SET organisation for allowing us to publish the material from these lectures in this way.

The book is organised in nine principal chapters, based on the lectures themselves, covering different domains and different aspects of the overall topic. Chapters 2–4 consider the problem as applied to the ground, air and maritime domains, respectively. Chapter 5 describes the impact of image quality (i.e. resolution, signal-to-noise ratio) on the overall target recognition performance. Chapter 6 considers the performance of different approaches to the classifier algorithm. Chapter 7 considers the improvement in performance to be gained when a target can be viewed from more than one perspective, as well as the ways in which natural systems, such as bats, perform target recognition. Such systems have benefitted from millions of years of optimisation through the process of evolution, and perform the target recognition process in an intelligent, adaptive manner. Chapter 8 considers the impact of compressive sensing, which is a relatively novel processing approach, showing that considerable economies may be made in sampling and processing due to the sparseness of the information in the target scene. Chapter 9 describes advances in change detection, including the very powerful techniques of coherent change detection. Finally, Chapter 10 looks at future challenges and directions for future research.

Chapter 2
Automatic target recognition of ground targets

David Blacknell and Luc Vignaud

2.1 Introduction

Since the advent of radar as a sensor for military remote sensing, the potential benefits in terms of reduced operator workload that automated cueing could bring have been clear. In particular, as radar technology has progressed to provide better and better imaging capabilities, the potential to go beyond simple detection and provide a level of target recognition has become increasingly apparent.

One of the earliest ground imaging radars was the H2S system developed by the Telecommunications Research Establishment (TRE) in Malvern, UK, during the Second World War. One explanation for the name of this system (related by R. V. Jones in his fascinating book *Most Secret War* [1]) was that Churchill's scientific advisor, Lord Cherwell, repeatedly declared that 'It stinks!' (a typical British expression of disapproval) when told of delays to the programme that had resulted from a misunderstanding between him and the developers. As a result, the developers gave the project the codename H2S, i.e. the chemical symbol for hydrogen sulphide, which of course 'stinks' like a rotten egg smell. Fig. 2.1 shows a typical image from the H2S system with a map for comparison. It is fair to say that the resolution of this system is quite coarse by today's standards. However, recognition of landmasses is clearly possible, which resulted in the H2S system being crucial in providing improved bombing accuracy in poor visibility.

In modern radar imaging systems, much better resolutions are now achievable. Fig. 2.2 shows an example of a synthetic aperture radar (SAR) image at 10 cm resolution. The image contains a variety of vehicles in a field and the tracks made by the vehicles moving into position can be clearly seen. There are also a number of corner reflectors in the scene, which give rise to distinctive impulse responses. This image illustrates a number of points. It is apparent that some level of vehicle recognition should be possible at such resolutions but the target characteristics that are most evident in radar imagery may not correspond to the most recognisable optical characteristics. It is therefore important to understand the characteristic 'features' of the image. It should also be noted that, because radar imagery is not intuitively interpretable, the role of automated algorithms is even more important to act as an aid for operators. Furthermore, the tracks left by the vehicles provide a

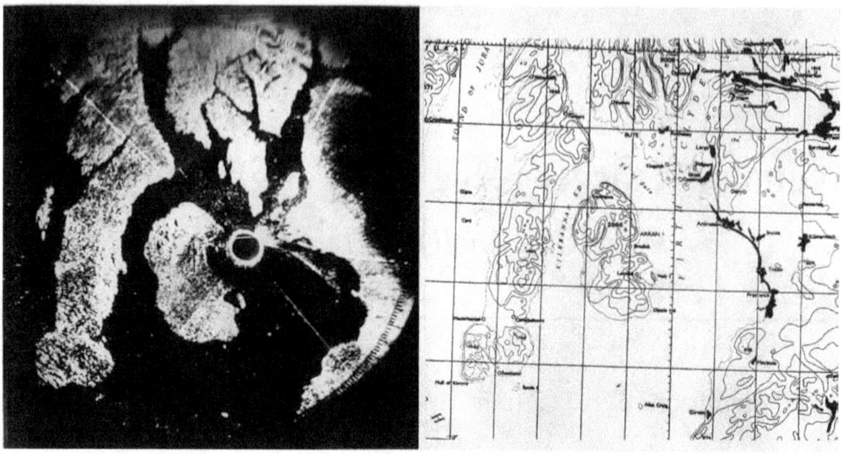

Figure 2.1 The origins of ground imaging radar: an example image from the H2S system

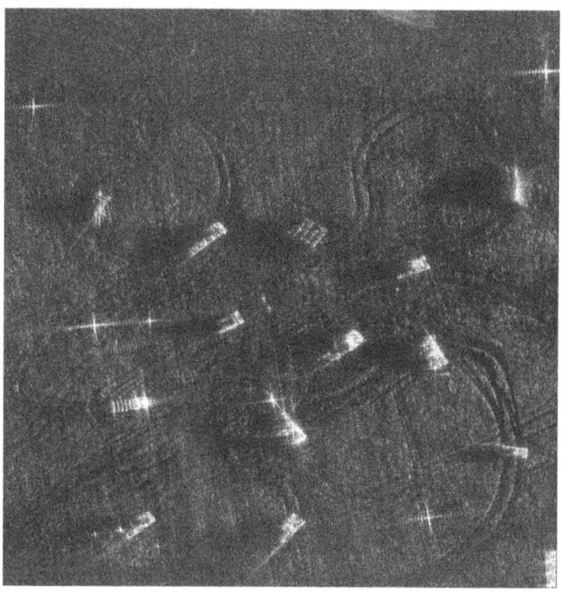

Figure 2.2 SAR image at 10 cm resolution of a variety of vehicles in a field

form of contextual information, which could also be used to aid detection and classification. These points will be discussed at more length in this chapter.

Inspired by the level of information available in high resolution SAR imagery, much research has been undertaken over the past 20 years into automatic recognition of ground targets in SAR imagery [2–5]. The aim of this chapter is to provide an overview of the fundamentals of ground target recognition together with a flavour of some advanced techniques. The discussion will focus exclusively on target recognition in SAR images although many of the techniques discussed are more generally applicable.

The chapter begins with a discussion of the phenomenology of SAR images, i.e. what are the distinctive characteristics of objects when imaged using this radar technique. It should be emphasised that radar images are very different to electro-optic images as a result of the way that the radar signal interacts with the scene and the way in which the returned radar signal is processed and so understanding the phenomenology is a very important part of target recognition. The discussion then proceeds to provide an overview of the principles of target detection and recognition in SAR imagery and how these need to be integrated into an end-to-end system to provide a full ATR capability. A particular unsolved problem is how contextual information can be exploited within the target detection and recognition process and so an outline of a possible approach to the use of contextual information for target detection is included. It will be seen that ATR relies upon having extensive databases of example imagery of the targets of interest. Given the huge degree of variability intrinsic in radar imagery, it is generally impractical to populate such databases entirely with real imagery and so imagery obtained from radar scattering prediction tools applied to target models must also be used. The topic of training databases and target modelling is thus the next topic that is considered. Finally, a crucial aspect of any ATR system intended for military purposes is the ability to assess how well it will perform in given circumstances. Thus the subject of ATR performance assessment is an essential component of any discussion of ground target ATR and forms the last major subject to be addressed. The main points covered are then summarised at the end of the chapter.

2.2 SAR phenomenology

In order for an ATR system to make best use of all the information contained in an SAR image, it is essential that the ATR system formulation must incorporate precise knowledge of radar imaging phenomenology. Such knowledge is gained through understanding of the SAR image formation process, which exploits the relative motion between the sensor and the target. There are many excellent texts discussing SAR image formation (e.g. [6]) and so only a brief overview will be given here.

Two-dimensional SAR imaging is essentially a two-step process:

- *High resolution range profiles (HRRPs) are acquired sequentially*: HRRPs are obtained by transmitting a signal with a large frequency bandwidth, often in the form of a chirp pulse followed by pulse compression (e.g. [7]) to provide a range resolution inversely proportional to the bandwidth. HRRPs are essentially instantaneous 1D projections of the whole scene onto the radar line-of-sight at each radar position from which a pulse has been transmitted.
- *SAR image formation*: HRRPs are integrated coherently over time, exploiting the fact that the variation of geometry with time effectively provides a Doppler

bandwidth and hence resolution in the cross-range direction. Thus cross-range resolution is a consequence of potential targets having been observed over an extended time/angular domain, i.e. a synthetic aperture.

As a result, the SAR image is focused on a projection plane (perpendicular to the apparent target/radar rotation axis), which results in

- sensitivity to the direction and frequency of illumination (i.e. varying scattering properties),
- sensitivity to the 3D geometry of target and ground (i.e. layover and shadowing as will be explained later).

Looking at targets over an extended angular interval, it can be clearly seen that the elementary scatterers from which the target signatures are composed can be highly dependent on the angular directivity of the radar observation, e.g. Fig. 2.3. In this diagram, the illumination angle is shown by an arrow and can be seen to vary by only a small amount over the sequence. However, particular features of the target, e.g. the back edges of the aircraft wings, are seen to appear very differently throughout the sequence. The aspect angle dependency can be so strong that a target may look completely different when seen from directions separated by only few degrees. From an ATR point of view this means that it may be necessary to consider the target images taken at different observation angles to be many 'different' classes.

While there is also some frequency dependence in SAR imagery, the target signature dependency on the radar frequency is probably less important than the angular one. However, within the ATR process, it will be necessary to use reference data taken around the same frequencies as the test data or the signatures may look too different.

Sensitivity to the 3D target geometry is another key point: it is possible to see through very simple examples how the radar energy interacts with the target and ground resulting in layover effects and inner shadowing. This is illustrated in Fig. 2.4 for which the key point to remember is that the radar image is formed taking into account the range of the object from which the radar energy is scattered. In particular point B on the target and point B′ on the ground are at the same range and so

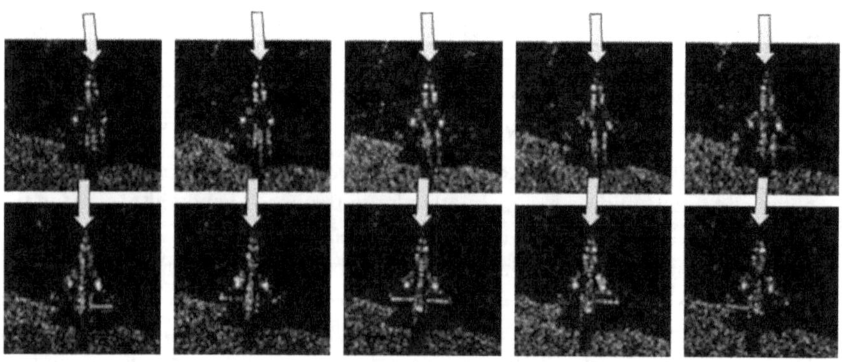

Figure 2.3 Scatterer variability in SAR images formed at 1° intervals

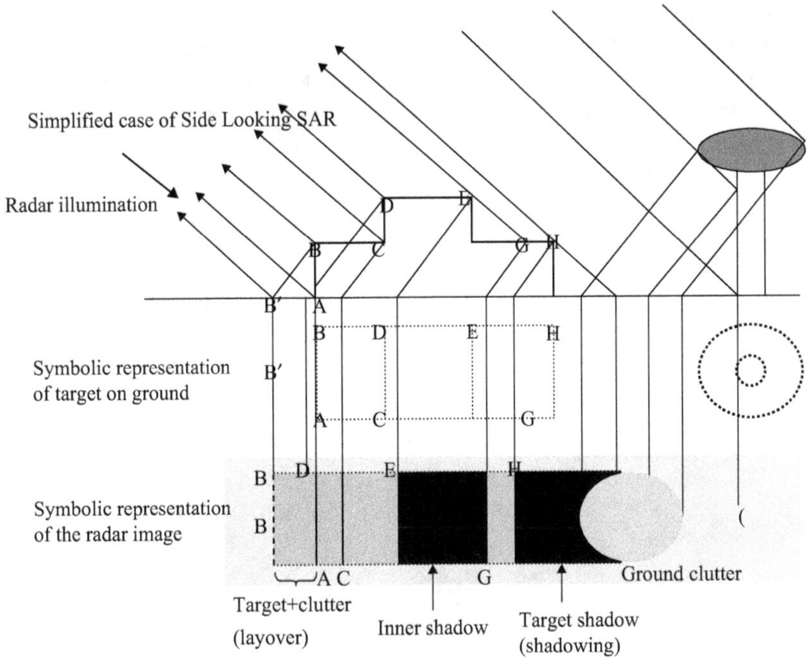

Figure 2.4 Simple illustration of geometrical effects in SAR imaging

reflections from these points will be placed at the same position in the image. In addition, all the reflections from the ground between B′ and A, and from the target between B and A, will occupy the same region in the image. This is known as 'layover'. Another feature of radar imaging is the inner shadow that results from the fact that there is no line-of-sight from the radar to the target between points E and G. This inner shadow will obscure parts of the target itself but the parts obscured will vary depending on the orientation of the target relative to the radar. This is an added complication when trying to design a robust ATR system. There is also a shadow behind the target resulting again from regions that are not in the line-of-sight of the radar. The shadow of the target can provide information about the shape of the target and thus shadow properties can be used in the target recognition process. However, care needs to be taken as objects that are separated from the target on the ground can interact with the target in the image. The illustration shows that the reflections from the tree canopy will prevent the full target shadow being observed, which will corrupt any target shape information being measured from the shape of the shadow. In a more extreme case, the reflections from the tree canopy may actually lay over onto the target returns thus corrupting the target image itself and posing additional difficulty for target recognition.

The specular nature of the radar reflection produces target signatures that can look very 'fuzzy' when compared to their optical equivalent, which may explain the relative difficulty in training human interpreters to work on radar images.

However, these effects may give geometrical 'fingerprints' that are well suited for automated identification. Fig. 2.5 shows an example of shadows and specular reflections from three planes. The shadows provide very well-defined representations of the shapes of the planes, while the actual radar returns from the planes themselves are less well defined because of the specular nature.

The representation of reflected radar energy in the image assumes that the radar signal is reflected once from the scene and returns directly to the radar. However, it is possible for the signal to be reflected multiple times before returning to the radar. Since the signal follows a longer path in this case, the energy from such 'multi-path' events will appear in the image at a further range than direct returns from the object. This is illustrated in Fig. 2.6, which is the image of a tank

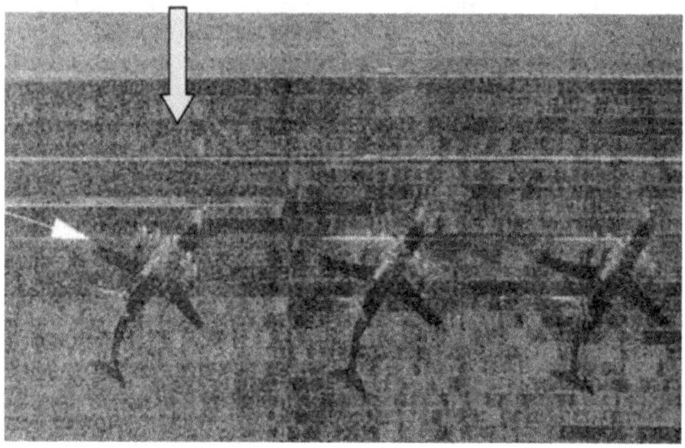

Figure 2.5 Illustration of specular returns and shadows

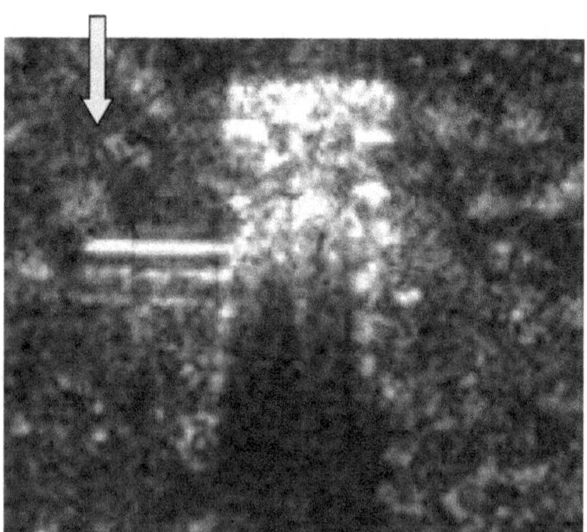

Figure 2.6 Illustration of multi-path reflections from the gun barrel of a tank

with the gun barrel pointing out to the side. It can be seen that the barrel appears three times in the image at three different range positions. The brightest image of the barrel is from the direct reflection and is at the closest range. The next barrel image results from the radar signal reflecting off the ground onto the barrel and then back to the radar (or vice versa). The third barrel image results from the radar signal reflecting off the ground onto the barrel and then back onto the ground before returning to the radar. The energy reduces on each additional reflection and so the barrel images become fainter and appear further down-range as the number of reflections increases. Such multi-path behaviour is yet another SAR phenomenology complexity, which adds to the difficulty of designing a robust ATR system.

In summary, when designing an ATR system it is important to avoid making assumptions based on experience with optical systems. As has been seen, SAR images are the result of complex phenomenology resulting from the imaging geometry, radar parameters and radar scattering mechanisms. It is essential that this phenomenology is taken into account when attempting to characterise target classes in order to achieve a robust ATR system.

2.3 The ATR processing chain

2.3.1 Pre-screening

The first stage in the target recognition process is to automatically detect potential targets in the scene, which can be passed up the processing chain for further analysis. This task in itself can contain a number of stages in which candidate detections are identified and then filtered to reject those that do not meet the criteria for being a potential target. For this reason, this stage of the processing chain is often called 'pre-screening'.

Given a SAR scene, the first stage in the process is to perform a single pixel detection, which flags up pixels that are anomalously bright in comparison to their neighbouring background pixels. The conventional approach to achieve this [8] is to compare the value of the pixel under test with the statistical distribution of the background clutter, which surrounds the pixel. If the pixel value is unlikely to occur as a result of clutter then a target is declared. Pixels within a fixed ring centred on the test pixel are used to estimate the parameters of the clutter distribution. A guard ring (or mask) of pixels immediately neighbouring the test pixel is often excluded from the estimate to avoid including any target information in the estimate. This is illustrated in Fig. 2.7.

Given a statistical model of the target in addition to the clutter it is possible to formulate an optimum statistical test for the presence of a target, i.e. the Neyman–Pearson test [9]. A standard requirement is to maintain a constant false alarm rate, which can be achieved by using an adaptive scheme based on the clutter statistics [10]. Use of a fixed ring to calculate the background clutter parameters has the disadvantage that discrete clutter objects such as trees or hedges may be included in the calculation thus biasing the background estimate. One way to avoid this is to use an adaptive region for clutter estimation based on a segmentation of the scene

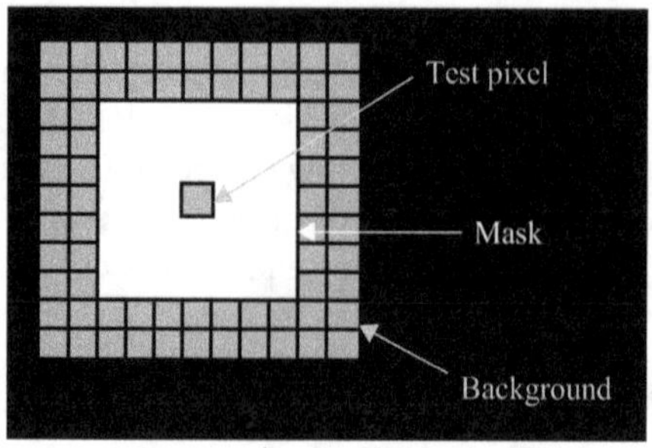

Figure 2.7 Illustration of pixels used for estimating background statistics

so that the largest available homogeneous region is used to estimate the clutter parameters [11].

Target detection involves determining whether a set of image samples belongs to a background class or a target class. Given statistical models for both classes, the likelihoods of observing the test samples are given by the conditional density functions $P_{\mathrm{b}}(x) = P(x|\mathrm{b})$ and $P_{\mathrm{t}}(x) = P(x|\mathrm{t})$ where x is the vector of image samples and b and t indicate background, and target, respectively.

Given the observation x, the maximum likelihood classification for which a target is declared when

$$\frac{P_{\mathrm{t}}(x)}{P_{\mathrm{b}}(x)} > 1$$

will maximise the number of correct classifications given equal probabilities for background and target. If the latter probabilities, $P(\mathrm{b})$ and $P(\mathrm{t})$ (which sum to 1 as a pixel must be either target or background), are unequal then the maximum *a posteriori* (MAP) classification

$$\frac{P(x|\mathrm{t})}{P(x|\mathrm{b})} = \frac{P_{\mathrm{t}}(x)P(\mathrm{t})}{P_{\mathrm{b}}(x)P(\mathrm{b})} > 1$$

will maximise the classification rate. This becomes

$$\frac{P_{\mathrm{t}}(x)}{P_{\mathrm{b}}(x)} > \frac{P(\mathrm{b})}{P(\mathrm{t})} = T$$

which is the likelihood ratio test in which the threshold T can be set to provide a desired probability of false alarm

$$P_{\mathrm{FA}} = \int_{D(T)} P_{\mathrm{b}}(x)dx$$

where $D(T)$ is the domain within which an observation $\boldsymbol{x}$ will be classified as a target. It is thus important to note that setting a threshold to achieve a desired probability of false alarm is equivalent to assigning prior probabilities for background and target. This will be used later in the discussion on use of contextual information.

A number of single pixel detections will be obtained in this way but there is no understanding of which detections may belong to the same potential target. For this reason, a clustering procedure is used as the next stage. One approach to clustering is illustrated in Fig. 2.8. All single pixel detections resulting from the detection stage are shown as circles. The clustering algorithm starts at the top, left-hand corner and proceeds in a raster fashion. When it encounters a detection that does not already belong to a cluster it starts a new cluster. Two parameters are used in growing the new cluster. The first is the number of pixels away from a given cluster member that the algorithm will look to attempt to find a new member of the cluster. In the example here this parameter is set to one pixel (excluding diagonals) so that new cluster members must be neighbours of existing cluster members although the use of larger values allows the clustering algorithm to jump gaps. The second parameter defines how far away from the first cluster member that new cluster members may be located. This value has been taken to be two pixels in the example given here.

Fig. 2.8 shows an intermediate stage of the algorithm. The detections shown in white have already been associated with clusters and four clusters have been identified. The shaded detection at the centre of the square defined by bold lines initiated the generation of the next cluster. All the other shaded detections have been associated with this cluster but the bold square indicates the limit on the size of the cluster has been reached so no more detections can be added. The algorithm will now proceed to the next (black) detection that is not already associated with a cluster and continue. It is clear that there may be some drawbacks with this approach especially in terms of artificially breaking up larger clusters due to the size limitation. The output will also depend on the starting point and the direction through which the image is searched. Alternative approaches are available including those based on morphological filtering using successive dilation and erosion

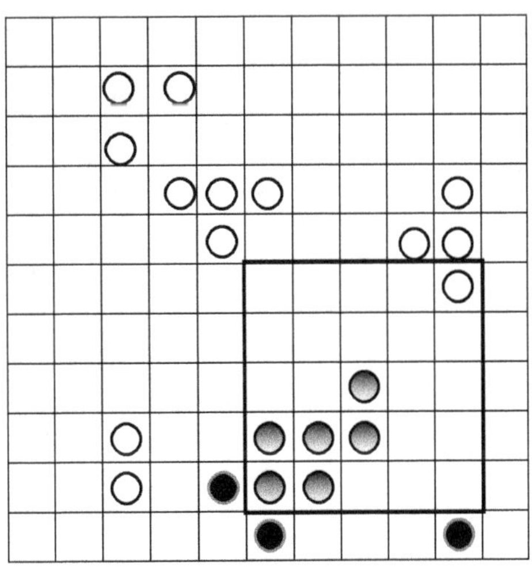

Figure 2.8 Illustration of a possible clustering algorithm

functions [12]. However, the intention here has been to give a flavour of the nature of a clustering algorithm rather than an extensive discussion of available techniques.

The final pre-screening stage is to examine each cluster and to perform a basic assessment of cluster characteristics to eliminate any clusters that clearly do not exhibit the properties of man-made targets. This is a simple discrimination stage and can involve measures such as cluster size and average power. It is possible then to reject a number of candidate targets as being more likely to be discrete clutter objects such as trees rather than man-made objects. The remaining clusters are then passed forward to the classification stage.

2.3.2 Template-matching

Once the candidate targets have been identified by the pre-screening process then the classification process can begin. One conceptually simple approach is to compare the object under test with example images of the various possible targets known to the system. This is illustrated in Fig. 2.9. As discussed with regard to phenomenology, radar images can be highly variable as a function of imaging geometry. So for each possible target class, the database must contain example images of that target at all possible geometries. Fig. 2.9 illustrates a target database containing target images over 360° of aspect angle variation, although in general elevation angle and many other degrees of freedom would need to be taken into account.

A measure of similarity is required to perform the comparison between the object under test and the images in the database. A natural measure is the correlation coefficient between the two images, which has a maximum value of unity for two identical images. Given this maximum, it is reasonable to set a threshold

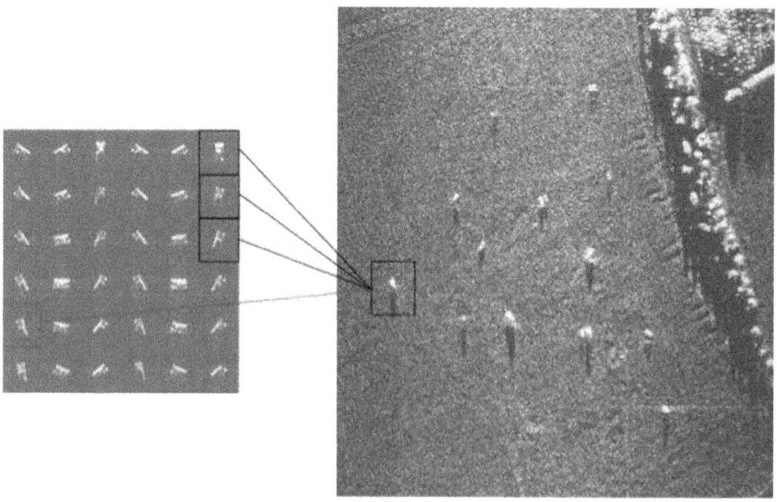

Figure 2.9 Template matching involves comparing the object under test with a number of examples of possible targets

correlation value such that if this threshold is not exceeded then a target classification is not made and the object is declared to be unknown. The option of making an 'unknown' declaration is very important and will be discussed further with regard to performance assessment.

While conceptually simple, the problem with this template-matching approach to classification is that the required databases can be huge when there are many target classes and many potential degrees of freedom. Thus template-matching has an important role for classification problems that are relatively constrained, but an alternative approach is required for less constrained problems.

2.3.3 Feature-based classification

Feature-based classification provides an alternative to template-matching, which solves the issue of a requirement for huge databases of imagery by representing target classes in terms of measured features that are intended to characterise the unique properties of the target class. Before any features can be measured, it is important to establish accurately which pixels belong to the target and which to the surrounding clutter. One way of doing this is to use an active contour or snake algorithm [13] as illustrated in Fig. 2.10.

An initial contour is placed around a point defining the position of the target (e.g. the mean position of the detections comprising the cluster). This is shown as the approximately circular inner contour in the image on the left in Fig. 2.10, which is defined by the number of node positions. This contour is iteratively adapted by randomly moving the positions of the nodes. An objective function is defined as a

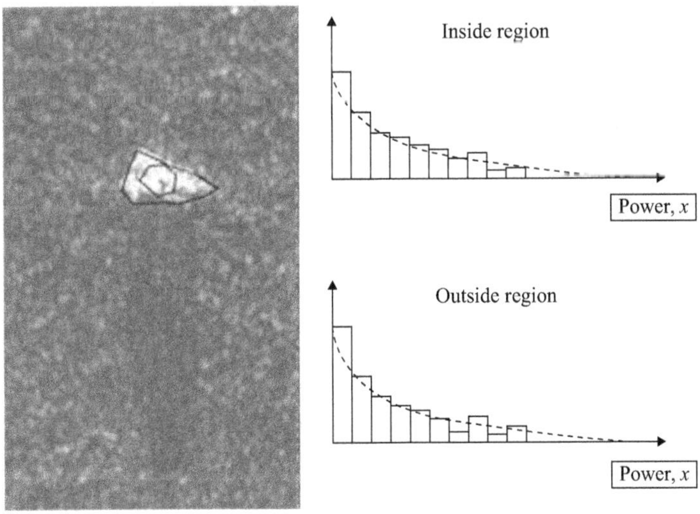

Figure 2.10 An active contour starts with an initial circular contour, which is adapted iteratively until the statistics inside and outside of the region match a given model

function that measures how well the statistics inside and outside of the contour match the assumed statistics for target and background.

An annealing approach [14] is taken such that a change is accepted if the objective function increases but a change is also accepted with some probability if the objective function decreases. This probability decreases as the number of iterations increases. The aim of this process is to avoid the iterative procedure becoming trapped in local maxima instead of finding the global maximum. Once the process has converged, an outline such as the outer contour shown in Fig. 2.10 will be obtained.

The choice of features used to represent the target classes is the key to classification performance. Features can be grouped into three main categories namely geometrical features, texture-based features and contrast-based features. It is desirable to use features from each of the three categories, since the features are then more likely to be approximately independent [15]. Geometrical features include the area (number of pixels over the target extent), ratio of target length to width and the moments of inertia about the centre of mass. Fourier and wavelet coefficients may also be included in this category as they essentially characterise the edges of the target in their high frequency components. Examples of texture-based features are the standard deviation and spatial correlation lengths of the pixel intensities and the entropy, which can characterise the 'randomness' of the pixel intensities of the target. Contrast-based features include the fractal dimension, which measures the spatial distribution of the top scatterers within the target extent and the weighted rank fill ratio, which measures the extent to which the power is concentrated within a few bright scatterers.

A reasonable level of classification can be obtained using such pattern recognition features, but in terms of finding features that provide the greatest robustness to target variability, it may be argued that features that relate specifically to the underlying physical structure of the target are likely to be the most robust as it will be the same underlying structure that is present whatever the imaging geometry or level of obscuration.

It is such scattering events that may hold the key to robust feature-based classification although, on the basis of a single-channel, single-aspect image of a target there may not be sufficient information. However, polarimetric techniques allow more sophisticated characterisation of scatterers while multiple aspect collects or non-straight trajectories allow 3D information to be obtained. The future availability of more advanced collection modes such as these will thus open up additional avenues for defining robust features.

Once a set of features has been established, these features are measured for examples of the different target classes, i.e. the training set. The feature values for an object under test are also measured and compared with those of the known target classes. If there is a sufficiently good match between the test and training values for a particular class then a classification is declared. This basic principle is illustrated in Fig. 2.11 for a two-class problem where two features are being used. The $\times$ symbols mark the positions of feature values obtained from the Class 1 training examples and the $+$ symbols the same for the Class 2 training examples.

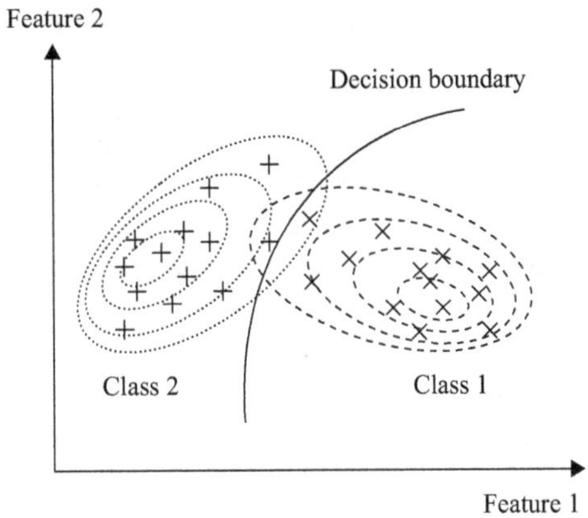

Figure 2.11 Illustration of decision boundary for two-class problem

A decision boundary must be drawn such that a test example with features that lie on one side will be declared as belonging to Class 1 and on the other will be declared as belonging to Class 2. There are many algorithms available for determining the decision boundaries and the appropriate choice is mainly driven by the complexity of the regions of feature space that need to be separated.

In Fig. 2.11, contours have been shown, which arise from approximating the distributions of the feature values by 2D Gaussian distributions. This is easily generalised by the use of multi-variate Gaussian distributions for larger numbers of features. It is then straightforward to set up the decision via statistical considerations, which gives rise to a Bayesian classifier [16]. This approach has the advantage of being based on a rigorous statistical foundation from which confidence values for decisions are easily derived. However, it will perform poorly if the features do not exhibit multi-variate Gaussian behaviour. In this case, it is possible to attempt to fit the feature distributions using other statistical models. This can be problematical for large numbers of features though as the points in feature space become very sparse even for large training sets due to Bellman's curse of dimensionality [17]. It is then very difficult to determine a multi-variate model for the data in a robust manner.

If a statistical model for the feature distributions is not used, then one of the simplest approaches is the nearest neighbour technique [18]. In this approach, the feature vector in the training set that is closest to the feature vector under test is determined and the class of the test vector is declared to be the same as the class of the closest training vector. There can be problems with this approach if the clusters in feature space corresponding to different classes are close together as it is then easy for a test vector from one class to be close to an outlier from a different class. To mitigate this effect, an extension to the approach is to examine the *k* nearest neighbours and declare the class of the test vector to be the dominant class among these *k* nearest neighbours. This approach is simple to implement and it is also easy to introduce another class into the training set. However, it can be computationally expensive for large training sets and it is difficult to define a robust confidence measure for decisions.

An alternative approach is to attempt to determine the boundaries in feature space between the classes. In the simplest case, it is assumed that the boundaries are linear, which gives rise to linear discriminant analysis [18]. This is clearly a good approach when the assumption of linear boundaries holds but will break down for more complex boundaries. Arbitrarily complex boundaries can be determined using the multi-layer perceptron (MLP) neural network approach [18]. An MLP consists of a number of layers of 'neurons'. The first layer will have as many neurons as there are features. The final layer will have as many neurons as there are classes. Examples from the training set are presented to the first layer and the MLP is 'trained' (i.e. the weights governing the flow of information through the MLP are adaptively adjusted) to achieve a final state in which only the output neuron corresponding to the correct class 'fires', i.e. has a value close to 1. The inclusion of one middle layer provides the MLP with the flexibility to learn arbitrarily complex (e.g. disjoint, non-convex) decision boundaries although the number of neurons used will have an effect on the level of complexity that can be represented. All neurons in one layer are connected to all neurons in the next layer and transmit their value via a weighting and a non-linear function. The non-linearity allows curved rather than straight decision boundaries to be formed. The training of the MLP is effectively an optimisation over the weights given the constraint that the outputs must match the training classes. The MLP approach can be computationally expensive in the training phase but this can be done off-line and the testing phase is very simple. A particular drawback is that it is not a simple matter to introduce a new class into the classifier. The whole network would need to be trained again over all the classes. It is also again difficult to define a confidence measure for the decisions made.

An alternative approach to learning the decision boundaries is to use the support vector machine (SVM) methodology [19]. The basic operation of an SVM is to maximise the margin between two classes by determining two linear boundaries between the classes with maximal separation. This assumes that the classes do not overlap but there are generalisations to 'soft' margin approaches to cope with overlapping classes. While the basic SVM is linear, it is possible to learn non-linear decision boundaries using the 'kernel trick' such that the data points are projected into a higher dimensional space in which non-linear boundaries can be approximated by linear boundaries. The basic SVM can only implement a binary decision although decisions between multiple classes can be achieved through a series of binary decisions. In this case, the introduction of a new class to a trained SVM involves learning one more binary decision and so is not too onerous. In common with all approaches not based on an underlying statistical model, it is difficult to assign confidences to decisions.

There are many classification techniques that can be applied to ATR and it has not been the intention here to give a comprehensive or detailed review of these. Rather, the aim has been to provide a flavour of the types of technique that are available and indicate some of their strengths and weaknesses. Ultimately, as will be seen in Chapter 6, the choice of classification scheme is secondary to the choice of features when designing an ATR system.

2.4 Use of contextual information in target detection

2.4.1 Motivation

It has been the standard approach in ATR schemes to extract an image chip containing the target and perform the classification purely on the basis of this chip. However, certainly in the ground target case, there is a substantial amount of contextual information that an operator would use in making a target declaration, which is ignored in this 'tunnel-vision' approach. For example the nearby lines of communication, the type of terrain and the military doctrine, which govern deployment of targets, will influence the probability of a target of a particular type being present.

2.4.2 Statistical formulation

The likelihood of a target being present at a particular image position will be influenced by the image context. For example military vehicles may tend to be parked in groups close to hedges and the edges of woods rather than individually in the middle of open ground. This contextual information can be exploited by allowing the target probability, $P(\mathrm{t})$ to vary as a function of spatial position depending on the context. This will mean that the corresponding threshold used in the likelihood ratio test will vary with spatial position and hence the false alarm rate will not be constant over the scene.

It will normally be the case that the target probability will be small, $P(\mathrm{t}) \ll 1$, so that

$$T = \frac{P(\mathrm{b})}{P(\mathrm{t})} = \frac{1 - P(\mathrm{t})}{P(\mathrm{t})} = \frac{1}{P(\mathrm{t})}\left(1 + O\left(P(\mathrm{t})\right)\right) \approx \frac{1}{P(\mathrm{t})}$$

Also, when only a single sample is being tested, the conditional density functions will normally only cross at a single point so that the equation

$$\frac{P_{\mathrm{t}}(x)}{P_{\mathrm{b}}(x)} = T$$

will only have a single solution at $x = x_0$. Thus targets are declared when $x > x_0$ and the probability of false alarm is given by

$$P_{\mathrm{FA}} = \int_{x_0}^{\infty} P_{\mathrm{b}}(x)dx$$

Frequently the target model $P_{\mathrm{t}}(x)$ will not be known precisely but will need to embody the idea that targets are bright in comparison to the background. Thus it is convenient to assume a target model in which targets with values $x < x_0$ have zero probability, while targets with values $x \geq x_0$ have a uniform probability, c, up to some appropriate cut-off point. Thus $P_{\mathrm{t}}(x_0) = c$ and manipulation of the above equations gives

$$P_{\mathrm{b}}(x_0) \approx cP(\mathrm{t})$$

Now consider a particular image position at which the target probability is $Q(\mathrm{t}) = \alpha P(\mathrm{t})$, where $P(\mathrm{t})$ is the target probability associated with a nominal false alarm probability P_{FA}. Then the value x_α above which targets are declared can be obtained by solving

$$P_{\mathrm{b}}(x_\alpha) \approx cQ(\mathrm{t}) = \alpha cP(\mathrm{t}) = \alpha P_{\mathrm{b}}(x_0)$$

and the probability of false alarm is given by

$$P_{\mathrm{FA}}(\alpha) = \int_{x_\alpha}^{\infty} P_{\mathrm{b}}(x)dx$$

For SAR images, the standard background model is given by the negative exponential distribution, which describes the speckle fluctuations that arise in coherent imaging systems, i.e.

$$P_{\mathrm{b}}(x) = \frac{1}{\mu_{\mathrm{b}}}\exp\left(-\frac{x}{\mu_{\mathrm{b}}}\right)$$

which gives a probability of false alarm of

$$P_{\mathrm{FA}} = \exp\left(-\frac{x_0}{\mu_{\mathrm{b}}}\right)$$

and hence a pixel intensity threshold of

$$x_0 = -\mu_{\mathrm{b}} \ln(P_{\mathrm{FA}})$$

Thus, given a modified target probability, manipulation of the above equations gives the modified pixel intensity threshold

$$x_\alpha = x_0 - \mu_{\mathrm{b}} \ln(\alpha)$$

and the modified probability of false alarm

$$P_{\mathrm{FA}}(\alpha) = \exp\left(-\frac{x_0}{\mu_{\mathrm{b}}} + \ln(\alpha)\right) = \alpha P_{\mathrm{FA}}$$

To recap, target detection can be formulated on a statistical basis, which will normally result in a scheme in which targets are declared if the pixel value exceeds some intensity threshold. There is an associated probability of false alarm, which can be maintained at a constant level by adjusting the intensity threshold as the background level varies. The scheme can be interpreted in terms of prior probabilities for the presence of targets and background, which remain constant over the image. The use of context may then be achieved by allowing the prior probability for the presence of a target to vary over the spatial extent of the image depending on the associated contextual information. The above equations then show how the intensity threshold should be modified in a consistent manner and how the false alarm probability will vary as a consequence.

A spatially varying false alarm rate is unavoidable if context is to be exploited in this manner. However, the performance of the detection scheme can be controlled by maintaining a constant average false alarm rate on a scene-by-scene basis. It must be assumed that within a SAR scene there is a typical mixture of contexts, e.g. similar numbers of hedge and wood boundaries, similar mixtures of terrain types, etc., so that there is a typical distribution of values of the factor α in any scene. The average false alarm rate over the scene is then given by

$$\langle P_{\mathrm{FA}} \rangle = \int P(\alpha)\, P_{\mathrm{FA}}(\alpha) d\alpha$$

By requiring that the average false alarm rate equals the nominal false alarm rate already chosen and by using the results obtained for the speckle background model, it is found that

$$\int \alpha P(\alpha) d\alpha = 1$$

must hold for the average false alarm rate to be maintained. This constraint is effectively a generalised version of the non-contextual case when a specified false alarm rate is associated with a particular choice for the prior target probability. In the general case it is the average value of the prior target probability that is constrained via the multiplicative factor α.

2.4.3 Simulated results

To illustrate the use of contextual information [20], Fig. 2.12 shows the process involved in achieving enhanced detection. There are three types of contextual information. First, it is known that some areas of terrain are easier for vehicles to move over than others. Second, military doctrine dictates that vehicles park up close to hedge boundaries and tree lines rather than in the open. Third, military doctrine dictates that vehicles travel in groups rather than individually. These contextual influences have been encapsulated in a statistical simulation shown in the left-hand image. The simulated image consists of different terrain types,

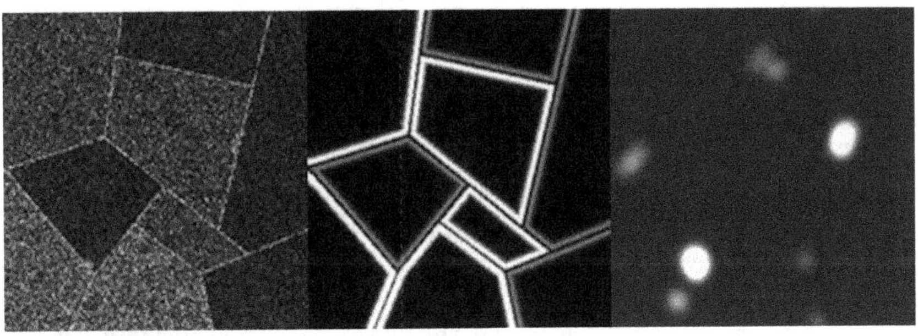

Figure 2.12 Illustration of enhanced detection based on use of contextual information

characterised by different mean pixel values, which are separated by brighter hedge boundaries. Bright target pixels are generated randomly from a target model and positioned randomly according to the contextual influences.

On the basis of terrain type and hedge proximity, the thresholds of a detector are adjusted as illustrated in the centre image in accordance with the analysis above so that regions where it is more likely that targets are present are examined more closely (whiter). Once an initial set of detections has been obtained, an iterative process is followed in which the thresholds are adjusted in the locality of existing detections to accept more detections in these regions on the basis that there are likely to be more targets there as shown in the right-hand image of Fig. 2.12.

This example is discussed in more detail in Reference 20. On comparing the detection rates with and without the use of context, it was found that an improvement in detection performance generally resulted. At worst the performance remained constant. In this example, examining the influences in isolation, the proximity of hedges gave the greatest improvement and hence the strongest contextual information, while the terrain type gave the weakest information. However, it should be emphasised that the target probability weightings due to the various contextual influences were defined fairly arbitrarily for the purposes of the demonstration and so the relative strengths of the various influences should not be assumed to represent the real world situation. It was found that the greatest improvement in detection rate resulted when all the contextual influences were exploited. The proportionate increase in the number of targets detected was fairly constant at about 13.0%, which represented a significant improvement in performance.

2.5 Databases and modelling

2.5.1 Database construction

As will have become apparent in the discussions on classification algorithms, a key requirement is the availability of an extensive database of example images of the targets of interest. Building such an ATR reference data bank for radar presents greater challenges than, e.g. building an EO database when it may be sufficient to use a handheld camera or search over the Internet for a picture of a particular vehicle. To obtain representative SAR imagery of targets in the field, multiple flights are needed over an area where targets are deployed to get enough data for multiple-angle acquisition. The experiment will also be more representative if it is possible to provide variants of the same target classes on the field.

The first large data collection was performed in the 1990s by DARPA and is known as the MSTAR (Moving and Stationary Target Acquisition and Recognition) database [21]. A small part of this database was then released to the public at the beginning of the 2000s and has motivated a substantial body of SAR ATR research, which has been reported in the literature (see, e.g. [2]).

However, when taking into account the totality of real world variability, it is clear that it is not feasible to collect real data that would cover all the operating conditions that may be encountered in modern conflicts. Thus it is essential that reference

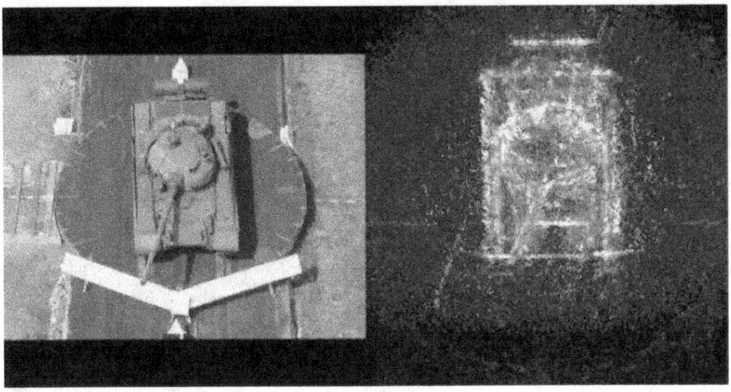

Figure 2.13 An example of a radar image of a tank obtained from turntable measurements

databases are populated to a large extent by other means such as turntable measurements or modelling. Fig. 2.13 shows an example of a radar image of a tank obtained from a turntable. This provides a very controlled situation for image formation in which the radar is stationary and relative motion is introduced by rotating the turntable. Image formation is then performed using inverse SAR techniques. It is important to be aware that such imagery may not be entirely representative of target imagery taken in the field. In particular, the background clutter against which the target appears and hence ground/target interactions may not be representative.

An alternative, as mentioned earlier, is to use modelling techniques, which predict the radar signature from a CAD (computer-aided design) model of the target and thus allow images to be simulated. There exist various simulation tools from sophisticated electromagnetic (EM) codes that solve Maxwell's equations through ray-tracing approaches to more simplistic ones that will only generate the outline of the target's shadow. An example of a ray-tracing approach is given in Reference 22. The compromise is often between accuracy and computation speed. 'Exact' codes may take days to compute a single-aspect angle image and so the simulation tool really has to be fitted to the ATR philosophy.

A key consideration is how the accuracy of the CAD model used affects the accuracy of the simulation and hence the performance of the ATR system. If the ATR performance is critically dependent on having a very exact CAD model representation then this is unlikely to be a robust solution since the actual targets in the field may easily vary from the CAD model representation. Also there is a question as to how the CAD models are to be obtained. If the vehicle is available then it may be possible to use laser scanning techniques to obtain an accurate CAD model. However, in some situations the vehicle will not necessarily be freely available to measure and so it will be necessary to produce a CAD model from possibly a limited number of photographs of the target as illustrated in Fig. 2.14. A continuing challenge is to understand the impact of CAD model fidelity on ATR

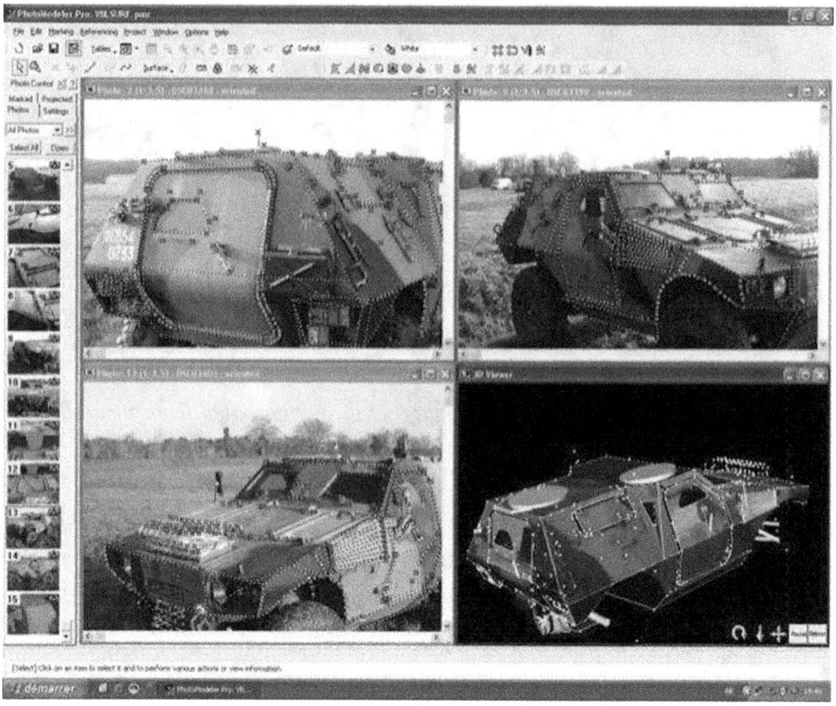

Figure 2.14 Construction of CAD models from photographs

performance and how this relates to the accuracy of CAD models that can be obtained from, e.g. photographs.

In summary, for most ATR applications some form of database of imagery simulated from CAD models is likely to be necessary. However, it needs to be taken into account that the target signature is sensitive to geometry and radar parameters, the targets may have many variants such as articulations or extra equipment attached and targets are not always available for detailed analysis. There is a need to build CAD models either via techniques such as laser scanning or from photographs, but a key question is how the accuracy and complexity of the model in terms of the number of facets, parts, articulations and material properties affect simulation accuracy and hence ATR performance. There is thus a trade-off between accuracy and computation speed, which needs to be taken into account when designing an ATR system.

2.5.2 Case study: model-based ATR using MOCEM

MOCEM [23] is a SAR image simulator that can produce high resolution SAR images from CAD models (VRML, 3DS, etc.) taking into account basic material descriptions. The software has been developed by the Alyotech company under specifications and contract provided by the DGA Maitrise de l'information (the

French MoD Expertise Unit). The process used in MOCEM is not based on the usual principles of radar cross-section (RCS) prediction codes but on the estimation of EM effects based on the CAD representation of the object. The aspects that make it well adapted for model-based ATR purposes are that it combines

- a novel EM formulation based on an analysis of the object geometry to build SAR images in a very short computation time,
- a novel geometrical algorithm, which handles the visibility of the various facets, including the multiple inner reflections and the interaction with the object surroundings like ground.

The analysis is thus obtained in a few seconds, without the computational drawbacks of ray-tracing techniques.

In theory, producing a simulated SAR image of an object requires computation of the complex reflectivity of the scene along the radar trajectory. However, this process is much too time consuming for large or detailed scenes or objects. It also requires a very detailed geometrical and material description of the whole scene, which is rarely available in practice. Moreover, it is not well suited for natural or diffuse surfaces surrounding the object. The MOCEM concept is based on a standard CAD model (not specifically built for radar applications and RCS prediction code constraints) and basic EM material descriptions. After a geometrical analysis, it localises the important EM phenomena for SAR imaging and computes their level and directivity responses. Two main phenomena are considered:

- the coherent response coming from reflective surfaces, typically steel or dielectric plates (including multiple bounces and interaction with surrounding ground),
- the diffuse effect resulting from many incoherent scatterers in the image resolution cell.

SAR images are then produced by applying a 2D SAR radar transfer function that takes into account the main parameters of the SAR system. For each simulated SAR image, it is possible to highlight the individual contribution of the major scattering processes, such as incoherent scatterers, specular scatterers, dihedral or trihedral effects, ground coupling effects, along with their coherence and lobe effects. This allows an exploration of the SAR phenomenology giving rise to various features in the SAR image.

NB: A dedicated version called MOCEM LT has been developed for educational purposes within European and NATO countries. It offers an attractive way to explain the content of SAR images and EM effects encountered to the scientific community, which is more familiar with optical and IR images than radar. A copy of the software can be provided by DGA Maitrise de l'information (please contact Dr Luc Vignaud).

To illustrate the use of MOCEM to support ATR, an example is provided using a large set of SAR images and corresponding CAD models of cruise ships in Oslo harbour (PicoSAR data and CAD models kindly provided by FFI). MOCEM was

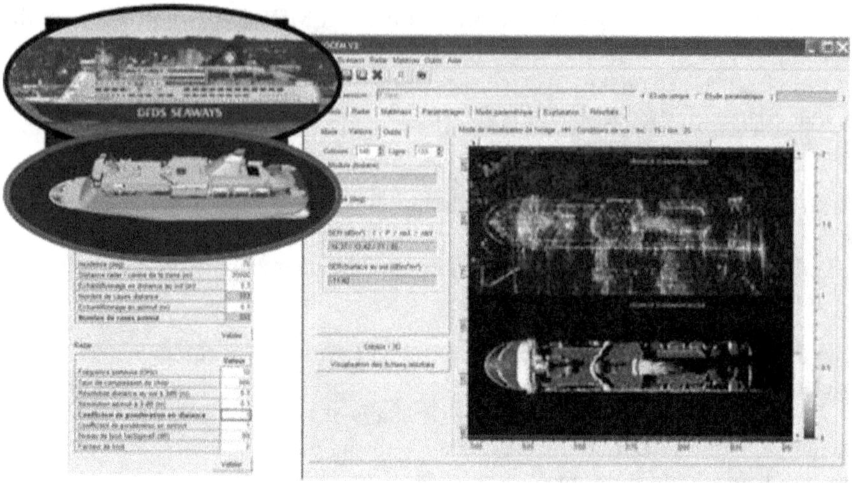

Figure 2.15 SAR image and corresponding MOCEM simulation of the 'Crown of Scandinavia' cruise ship

used to build a fully simulated reference database of these ships, and one example is shown in Fig. 2.15.

Fig. 2.16 shows the classification result of a SAR image (Crown of Scandinavia ship) against the simulated SAR images computed by MOCEM on eight CAD models. The grey scale at the side of each MOCEM simulation shows the degree of similarity found between the real SAR ship signature and the simulated ones where lighter shades indicate greater similarity. It can be seen that the Crown of Scandinavia is found to be the best match by a significant margin.

2.6 Performance assessment

2.6.1 Receiver operating characteristic (ROC) curves

As has been seen previously, the end-to-end ATR processing chain contains a pre-screening stage and a classification stage. The pre-screening stage is essentially a two-class problem, which aims to identify targets and reject clutter. For this type of problem, the receiver operating characteristic (ROC) curve is a convenient measure of performance.

Fig. 2.17 illustrates the underlying principle. It is assumed that some discriminant value is measured to determine whether an object is target or clutter. For many examples of targets this discriminant value will have a PDF and similarly for clutter as shown. A threshold is used to decide whether an object is target (threshold exceeded) or clutter (threshold not exceeded). Given a threshold, it is possible to calculate the probability of detection (PD) by integrating the target PDF from the threshold to infinity and the probability of false alarm by integrating the

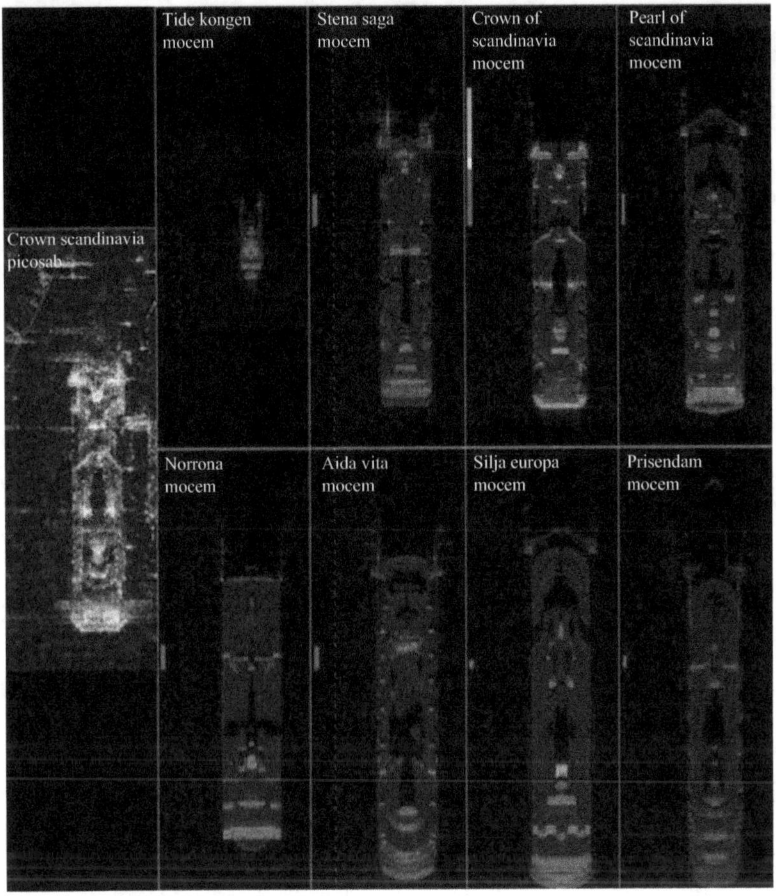

Figure 2.16 Classification result of a SAR image (Crown of Scandinavia ship) against eight simulated SAR images computed by MOCEM on CAD models

clutter PDF from the threshold to infinity as shown. As the threshold is varied, the PD and PFA will vary and this is shown for two values of the threshold.

As the threshold is varied, the resulting values of PD and PFA can be plotted against each other as shown in Fig. 2.18. The ROC curve is thus the locus of PD versus PFA as defined by the threshold parameter. However, care must be taken as the same threshold will not correspond to the same PD or PFA for different missions and so maintaining equivalent performance in different circumstances can be challenging.

Fig. 2.19 provides an interesting example (taken from Reference 24) of the use of ROC curves to show the performance of a target detector. In this example, the targets were either without optical camouflage netting (bare) or with it (cam). Also they were either in the open (open) or in the light vegetation (scrub). In Fig. 2.19 it

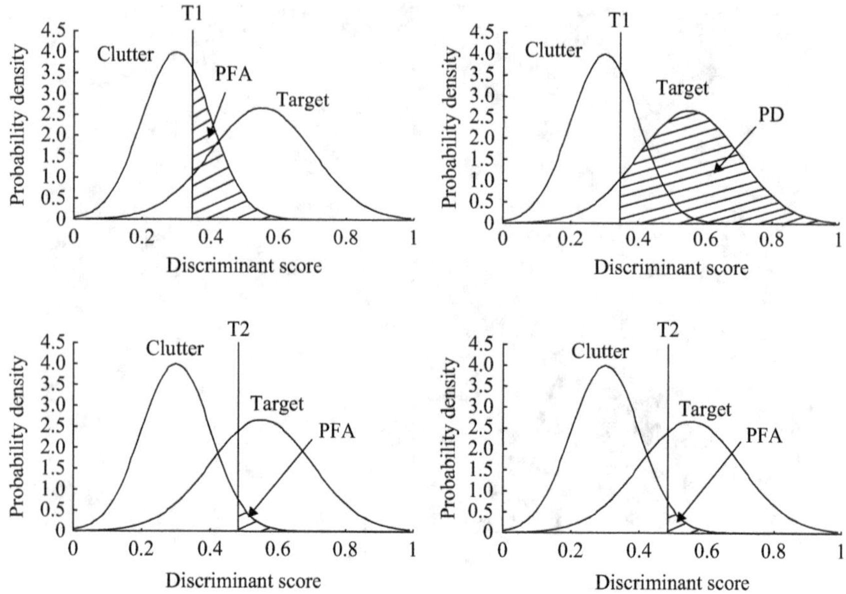

Figure 2.17 The effect on probabilities of detection and false alarm of altering the threshold

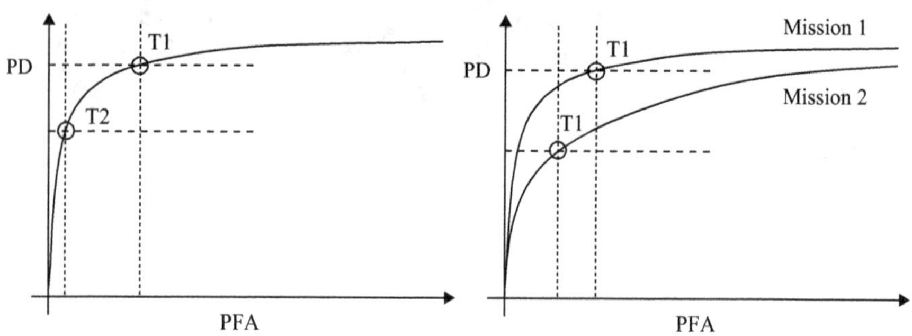

Figure 2.18 ROC curves are the locus of PD versus PFA as the threshold parameter is varied

can be seen that better detection performance is obtained when the target is in the open (top two curves) than when in scrub (bottom two curves). However, in both open and scrub the detection performance is better when optical camouflage netting is used than when it is not. The reason for this has not been determined conclusively but it is speculated that the netting was wet and it was a windy day. Hence in the resulting radar imagery, the netting caused a bright and smeared return, which was more easily detected. The use of ROC curves was thus important in understanding what was being seen in this example.

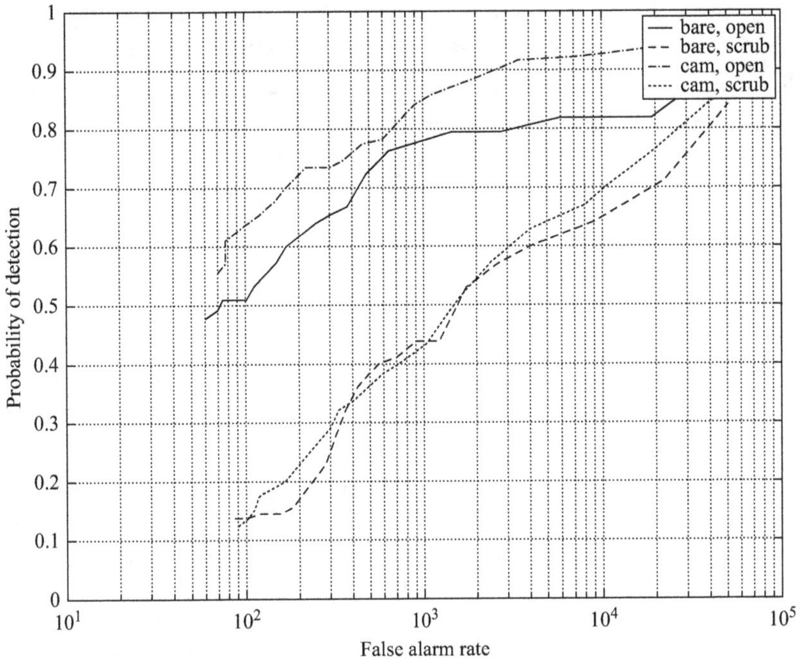

Figure 2.19 An example of the use of ROC curves

2.6.2 Confusion matrices

At the classification stage of the ATR chain, there will normally be multiple possible classes to which a target may belong and so the concept of an ROC curve is less useful. In this case, the performance of the system can be more fully understood through the use of a confusion matrix. Table 2.1 shows a basic example of a confusion matrix for a four-class problem. In this example there are two classes of friendly vehicles (blue) and two classes of enemy vehicles (red). The numerical

Table 2.1 A basic confusion matrix for the performance of a four-class classifier

	ATR system output				
	Declaration				
Truth	Blue1	Blue2	Red1	Red2	PCC
Blue1	0.78	0.03	0.06	0.13	0.78
Blue2	0.01	0.95	0.02	0.02	0.95
Red1	0.10	0.03	0.83	0.04	0.83
Red2	0.02	0.01	0.03	0.95	0.95
PCL	0.86	0.93	0.88	0.83	

Table 2.2 A confusion matrix for the performance of a four-class classifier including confuser targets

	ATR system output				
	Declaration				
Truth	Blue1	Blue2	Red1	Red2	PCC
Blue1	0.78	0.03	0.06	0.13	0.78
Blue2	0.01	0.95	0.02	0.02	0.95
Red1	0.10	0.03	0.83	0.04	0.83
Red2	0.02	0.01	0.03	0.95	0.95
Conf1	0.24	0.28	0.22	0.27	N/A
Conf2	0.28	0.16	0.31	0.25	N/A
PCL	0.55	0.65	0.56	0.57	

entries in the main body show the proportion of targets of a given true class that has been declared as a given declaration class. The PCC is the probability that a target of a given class is declared as that class and this is shown in the right-hand column. However, of more relevance to the war-fighter is the probability of correct label (PCL), which is the probability that a target that is declared to be of a given class is actually a member of that class. The PCL is shown in the bottom row and clearly tells a different story. For example, targets from class Red2 are declared as such 95% of the time. However, if a target is declared as Red2, it is actually only a member of that class on 83% of the time.

This distinction between PCC and PCL becomes more evident when the more realistic situation is considered in which the targets under test contain vehicles of two class types that are not known to the classifier, i.e. confuser classes. Typical results in this case are shown in Table 2.2. It can be seen that the PCC is unaltered in this case but the PCL is dramatically affected. Now a target declared as Red2 will only actually belong to the Red2 class 57% of the time and the effect is similar for all classes.

This illustrates a problem with the classifier whose performance is being assessed by these confusion matrices. Basically it is being forced to classify every target under test as one of the four classes it knows about even though it is being exposed to targets outside its database as is virtually inevitable in ground target recognition situations. It is thus important, as has been mentioned earlier, to include an unknown class in the classification procedure. Table 2.3 then shows an example output from the classifier. It can be seen that the PCC values are reduced as some of the previous correct declarations were not actually sufficiently confident and have now been declared as unknown. However, the PCL is now significantly improved over the case when a forced decision was made. This emphasises the importance of including an unknown class.

Another factor that needs to be taken into account is that some declaration errors can have more serious consequences than others. For example if Blue is declared but the target is Red then an enemy attack may not be averted, while if Red is declared but the target is Blue then fratricide may occur. Thus it is important

Table 2.3 A confusion matrix for the performance of a four-class classifier including confuser targets and an unknown class

	ATR system output					
	Declaration					
Truth	Blue1	Blue2	Red1	Red2	Unknown	PCC
Blue1	0.70	0.01	0.00	0.12	0.17	0.70
Blue2	0.00	0.95	0.01	0.00	0.04	0.95
Red1	0.09	0.01	0.79	0.02	0.09	0.79
Red2	0.01	0.01	0.02	0.95	0.02	0.95
Conf1	0.04	0.01	0.00	0.03	0.93	N/A
Conf2	0.04	0.00	0.07	0.04	0.85	N/A
PCL	0.80	0.96	0.89	0.82	N/A	

Table 2.4 A confusion matrix for the performance of a four-class classifier including confuser targets, an unknown class and showing probability of critical error

	ATR system output					
	Declaration					
Truth	Blue1	Blue2	Red1	Red2	Unknown	PCC
Blue1	0.70	0.01	0.00	0.12	0.17	0.70
Blue2	0.00	0.95	0.01	0.00	0.04	0.95
Red1	0.09	0.01	0.79	0.02	0.09	0.79
Red2	0.01	0.01	0.02	0.95	0.02	0.95
Conf1	0.04	0.01	0.00	0.03	0.93	N/A
Conf2	0.04	0.00	0.07	0.04	0.85	N/A
PCL	0.80	0.96	0.89	0.82	N/A	
PCE	0.11	0.02	0.09	0.16	N/A	

to include the probability of critical error (PCE) as has been done in Table 2.4 to understand the significance of a false declaration.

Yet another factor that needs to be considered is the 'order of battle', i.e. the probable number of units present. In this example it is assumed that there are ten times as many Red units as Blue units and confusers. The confusion matrix declarations in Table 2.5 have been adjusted to give the number of declarations if there were 1000 of each Red unit and 100 of each other unit. This does not change the PCC but has a drastic impact on the PCL and PCE. In this situation, the war-fighter would have a lot more confidence acting on a Red declaration but there would be little confidence in a Blue declaration.

Thus it has been seen that the confusion matrix provides a powerful means of representing classifier performance, but it is important to take all factors into

Table 2.5 A confusion matrix for the performance of a four-class classifier including confuser targets, an unknown class and showing PCE

	ATR system output					
	Declaration					
Truth	Blue1	Blue2	Red1	Red2	Unknown	PCC
Blue1	70	1	0	12	17	0.70
Blue2	0	95	1	0	4	0.95
Red1	90	10	790	20	90	0.79
Red2	10	10	20	950	20	0.95
Conf1	4	1	0	3	93	N/A
Conf2	4	0	7	4	85	N/A
PCL	0.40	0.81	0.97	0.96	N/A	
PCE	0.57	0.17	0.01	0.02	N/A	

account when interpreting the results. A few key considerations have been introduced here but this has by no means been exhaustive.

2.6.3 Operational assessment

There are many, many sets of ATR performance results published in the literature in the form of confusion matrices. Almost every paper presents a new technique and shows that it gives better performance than existing techniques. How can this be? Fig. 2.20 illustrates a few of the ways in which reported performance results may not truly reflect the operational performance of the system.

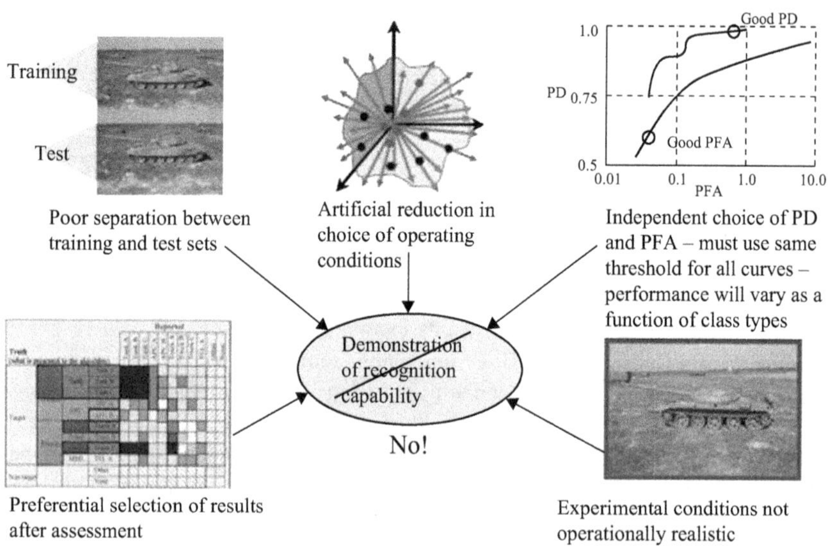

Figure 2.20 Some common mistakes made when assessing performance

A common problem is the limited data that is available on which to train and test the algorithm. This is an inevitable consequence of the expense of conducting experiments with an airborne SAR system. However, it does mean that the data on which the algorithm is tested may not be truly independent of the data on which the algorithm was trained. For example, an experiment in which targets from various classes are imaged at one degree intervals may be used to provide the data. A training set may be formed by taking alternate images (i.e. at 1°, 3°, etc.) and the test set formed from the remaining images (i.e. at 2°, 4°, etc.). While the training and test sets contain different images, they will not be independent and so the performance results will be biased.

Another problem is unintentional preferential selection of results. This can be as simple as choosing different declaration thresholds for different assessments and then comparing the results. The difficulty is that for any ATR implementation the declaration threshold must be fixed beforehand and so assessments must be made using the same threshold. Another aspect of this problem is that, in the development of an ATR system, it is inevitable that the algorithm will be assessed at various stages and the development of the algorithm will be driven by how well it performs on the data being used for assessment. The danger is that the algorithm will become tuned to the test dataset being used and so its performance will be preferentially assessed. One way to avoid this problem is to have a sequestered independent test dataset, which is only ever used for the final assessment of the ATR algorithm.

Other performance assessment problems are related to how well the diversity of operational conditions is represented in the test data. Inevitably, it will not be feasible to represent all possible operational configurations when assessing an algorithm but it is important to consider the problem carefully to attempt to span the diversity space as fully as possible using both real data gathered from an airborne platform or turntable and augmenting this with appropriate simulated data.

The problem of robustly assessing the potential operational performance of an ATR system is a significant issue and emphasises the need for a strategy for ATR performance assessment at a national or international level in which some independent body maintains a set of test data, which is used only for performance assessment and rigorously dictates the assessment procedure. Without this, many reported ATR performance figures are of limited value in terms of guiding military procurements.

Performance assessment should, of course, reflect the performance that will be achieved operationally. However, a major issue for operational assessment is that algorithm development and performance assessment take place in circumstances that are not necessarily representative of the real operational environment. This is illustrated in Fig. 2.21. In this illustration, the reference conditions relate to data that was gathered to support algorithm development; the model-based conditions relate to data that has been used to populate databases; the test conditions relate to data collected to support algorithm performance assessment; and the operational conditions relate to data to which the system is to be applied operationally. As has been illustrated conceptually, the intersection of all these conditions is very small and, in particular, there is a large portion of operational space, which is not represented. The challenge for the future is to develop assessment methodologies,

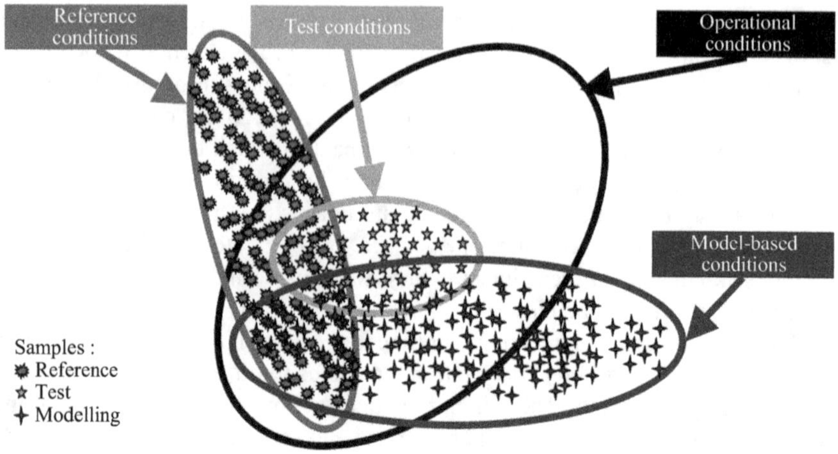

Figure 2.21 Illustration of performance assessment issues (courtesy of SET111)

which provide the procurer of such systems greater confidence that they will indeed perform as intended when used operationally.

2.7 Conclusions

An overview of the fundamentals of ground target recognition using SAR has been given. There is a tendency, when discussing ground target ATR, to consider only the most complex problems consisting of very many target classes and challenging clutter environments. However, it should be borne in mind that radar ATR is not a single problem that can or cannot be solved but that it is a continuum of problems of varying degrees of difficulty and complexity all of which provide useful military capability. This was essentially the viewpoint articulated by the NATO SET111 Task Group on ground target recognition at the conclusion of that activity. Operational radar ATR systems already exist and the overarching challenge is to push forward the solution space to achieve successful operation in more difficult and complex circumstances. To achieve this, a number of key specific challenges must be overcome, some of which are identified and discussed in Chapter 10, and to a large extent it is these challenges that current researchers are tackling. Hopefully this book will inspire those reading to contribute to the effort to solve these challenges.

Acknowledgements

The material contained in this chapter is based substantially on material presented in the NATO SET172 lecture series on Radar Automatic Target Recognition and Non-Cooperative Target Recognition. Much of the ground target recognition aspects of SET 172 arose in turn from collaborative activities and discussions under a sequence of NATO task groups, SET053, SET111 and SET163, concerned primarily with target recognition in SAR images. The authors gratefully acknowledge the contribution made by members of these various NATO groups in terms of both material provided and extensive in-depth technical discussions over the years.

Aspects of the material presented in this chapter (and Chapter 10) were produced under various contracts for the UK MOD while one of the authors, David Blacknell, was employed by QinetiQ. This author gratefully acknowledges the MOD's role in funding these activities.

This author also gratefully acknowledges QinetiQ's role in the work and especially the many former colleagues at QinetiQ who contributed to the SAR ATR activities from which some of this material is derived. In particular, these activities gave rise to Figs. 2.2, 2.8, 2.9, 2.10, 2.12 and 10.1, which are reproduced courtesy of QinetiQ.

References

[1] R. V. Jones, *Most Secret War*, Hamish Hamilton, London, 1978.

[2] E. G. Zelnio and F. Garber (editors), *Proceedings of SPIE Conference on Algorithms for SAR*, Orlando, FL, 1993.

[3] P. Tait, *Introduction to Radar Target Recognition*, The Institution of Engineering and Technology, London, 2005.

[4] L. M. Novak 'State-of-the-Art of SAR Automatic Target Recognition', *IEEE International Radar Conference*, Alexandria, VA, pp. 836–843, 2000.

[5] B. Bhanu, D. E. Dudgeon, E. G. Zelnio, A. Rosenfeld, D. Casasent and I. S. Reed, *Special Issue on Automatic Target Detection and Recognition, IEEE Trans. Image Process.*, 6, 1997.

[6] M. Soumekh, *Synthetic Aperture Radar Signal Processing with Matlab Algorithms*, John Wiley & Sons, United States, 1999.

[7] M. I. Skolnik, *Introduction to Radar Systems*, McGraw-Hill, New York, 2001.

[8] P. M. Woodward, *Probability and Information Theory with Applications to Radar*, McGraw-Hill, New York, 1953.

[9] J. Neyman and E. S. Pearson, 'The Problem of the Most Efficient Tests of Statistical Hypothesis', *Phil. Trans. Roy. Soc. (Lond.)*, Vol. A231, pp. 289–333, 1933.

[10] G. B. Goldstein, 'False-Alarm Regulation in Lognormal and Weibull Clutter', *IEEE Trans.*, Vol. AES-9, pp. 84–92, New York, 1973.

[11] I. McConnell and C. J. Oliver, 'Segmentation-Based Target Detection in SAR', *Proceedings of SPIE Conference on SAR Image Analysis, Modelling and Techniques IV*, Florence, Italy, 1999.

[12] J. P. Serra, *Image Analysis and Mathematical Morphology*, Academic Press Inc., 1982.

[13] C. Chesnaud, P. Refregier and V. Blolet, 'Statistical Region Snake-Based Segmentation Adapted to different Physical Noise Models', *IEEE Trans. Pattern Anal. Mach. Intell.*, Vol. 21, No. 11, pp. 1145–1157, November 1999.

[14] N. Metropolis, A. W. Rosenbluth, M. N. Rosenbluth, A. H. Teller and E. Teller, 'Equation of State Calculations by Fast Computing Machines', *J. Chem. Phys.*, Vol. 21, No. 6, pp. 1087–1092, 1953.

[15] C. J. Oliver and S. Quegan, *Understanding Synthetic Aperture Radar Images*, SciTech Publishing Inc., 2004.

[16] K. Fukunaga, *Introduction to Statistical Pattern Recognition* (2nd edition), Academic Press Inc., 1992.

[17] R. E. Bellman, *Dynamic programming*, Princeton University Press, 1957.

[18] R. O. Duda, P. E. Hart and D. G. Stork, *Pattern Classification* (2nd edition), Wiley-Blackwell, 2000.

[19] V. N. Vapnik, *The Nature of Statistical Learning Theory* (2nd edition), Springer, 2000.

[20] D. Blacknell, 'Contextual Information in SAR Target Detection', *IEE Proc. Radar, Sonar Navigat.*, Vol. 148, No. 1, pp. 41–47, 2001.

[21] R. Hummel, 'Model-Based ATR Using Synthetic Aperture Radar', *IEEE Radar Conference*, Alexandria, VA, pp. 856–861, 2000.

[22] S. D. Turner, 'RESPECT: Rapid Electromagnetic Scattering Predictor for Extremely Complex Targets', *IEE Proc.*, Vol. 137, Part F, No. 4, August 1990.

[23] C. Cochin, D. Le Hellard, F. Aubineau and P. Gosselin, 'MOCEM: An All in One Tool to Simulate SAR Images', *EUSAR 2008*, Friedrichshafen, Germany, 2–5 June, 2008.

[24] G. Blucher, D. Blacknell, N. J. Redding and D. Vagg, 'Prescreening Algorithm Assessment within the Analysts' Detection Support System', *Proceedings of RADAR*, Adelaide, Australia, 3–5 September, 2003.

Chapter 3

Automatic Recognition of Air Targets

Peter D. F. Tait

3.1 Introduction

Two topics are addressed in this chapter on the non-cooperative target recognition (NCTR) of air targets using radar: the fundamentals of the automatic target recognition process and the techniques that specifically address the recognition of jet aircraft and helicopters.

The fundamentals of how targets can be differentiated are discussed. A summary of current radar NCTR techniques is presented. The processes involved in measuring the signature of the target, the conditioning of this data for feature extraction and the recognition of the target from the measured features are presented. The employment and assembly of a database representing the target features and the application of the classifier are discussed. The utilisation of the confusion matrix to assess classifier performance is presented.

Jet engine modulation (JEM) is one type of NCTR technique employed and provides an estimate of the number of blades located on the first and often subsequent stages of the compressor stage of jet engines. The JEM technique is based upon analysing the spectrum of the reflected radar signal from the jet engine.

An understanding of the basic mechanics of the jet engine and the various levels of symmetry associated with the whole engine and the rotor stages is used to determine the JEM spectral components generated. The recognition of the engine type is based upon extracting key spectral attributes in order to establish the numbers of blades present. By using a database of blade numbers for jet engines, the actual engine type is then recognised.

The recognition of helicopters is also discussed followed by range-Doppler imaging (RDI), which provides a 2D signature of targets. The helicopter's physical attributes, particularly those associated with the propulsion system, are discussed. The interaction of these moving parts with the radar signal forms the basis of how helicopters are recognised using radar. The design of the radar signal needed to measure the helicopter's features is discussed. The method for using the features to identify the helicopter type is presented.

RDI provides a 2D map of the target with high range resolution and high Doppler resolution data simultaneously presented. Relatively short dwell periods are utilised, which are shorter than are normally required to generate an inverse synthetic aperture radar (ISAR) image. These techniques have not been studied very widely, are relatively new and have the potential to contribute to the recognition of air targets. RDI signatures of jet aircraft, a helicopter and a propeller aircraft are presented. However, these techniques tend to be limited by the ability to provide high range and high frequency resolution simultaneously.

3.2 Fundamentals of the target recognition process

3.2.1 Introduction

Whatever radar technique is used to recognise a target from its signature, the fundamentals of the process employed are the same. The actual methods employed can vary considerably, but the essence of the process is fundamental. The main elements involved are the measurement of the aircraft's signature, the conditioning and extraction of the features, the assembly of the database, which represents the target's features, and the application of the classifier to the features and database to recognise the target. In many military situations in which an NCTR function is employed, there is no time for an operator to study a radar signature and try to determine the target type or even to study any features that have been extracted, so it is necessary to automate the whole process. The challenge for radar and other engineers working in the field of NCTR is to make the process sufficiently reliable and effective that military commanders and operators have full confidence in a machine making a vital decision, which could have a severe impact on the mission and on people's lives.

3.2.2 Target features

In order to recognise a target and differentiate it from others, it has to possess a feature or features that can be used to discriminate it from other targets. The feature could be a time, spectral or distance measure. Fig. 3.1 shows some typical military aircraft targets, which would be required to be identified.

In order to discriminate targets, they must have features
that radar can measure to differentiate them

Figure 3.1 Typical aircraft targets to be recognised

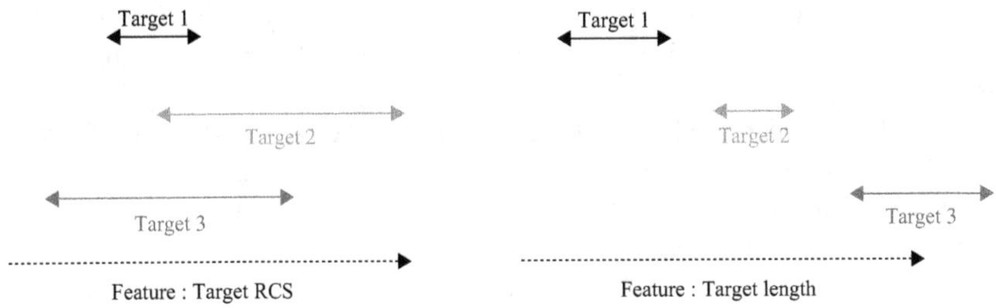

(a) Radar cross-sections of targets 1, 2 and 3 overlap, so RCS is not a good recognition discriminant

(b) Lengths of targets 1, 2 and 3 do not overlap, so length is a good recognition discriminant

Figure 3.2 Feature spaces used for discriminating attributes between targets should not overlap

The numerical values of any features employed have to differ to be useful as recognition attributes. The less that the overlap is in feature space between targets, the better is the ultimate classification performance that can be achieved. If the feature spaces of targets do significantly overlap, it is necessary to use probabilistic methods and several features to differentiate them.

In Fig. 3.2 a simple example of two features is shown with the respective variations in value for three targets. In Fig. 3.2(a) the feature is target radar cross-section (RCS) and the feature spaces of all three targets overlap. They cannot be differentiated by the RCS measurement unless probability density functions are available for the features and would also need to be combined with other feature measurements to provide confidence in the performance. The outputs would be ID probabilities for the three respective targets.

In Fig. 3.2(b) the feature used is target length with no overlap in the lengths of the targets. In this case length is a robust feature as there is no ambiguity in target ID if a target length is measured. In order to really understand target features and to make judgements about which are robust it is necessary to make measurements on representative types of targets. The statistics of the robustness of any potential feature can then be made.

3.2.3 Aircraft recognition techniques and waveform design

The main NCTR techniques used for recognising aircraft targets are summarised in Table 3.1. Jet engine modulation (JEM), high resolution range profiling (HRRP), helicopter rotor modulation (HERM), inverse synthetic aperture radar (ISAR) and range-Doppler imaging (RDI) are shown.

A radar waveform needs to be designed to extract the feature of interest and if necessary the radar planned to perform the measurement may need to be modified or specially programmed to support the waveform required. Normally a high frequency or high range resolution waveform would be needed.

For jet engine recognition, spectral analysis is performed to extract target features, which are then used to determine the engine blade count numbers. JEM is a frequency domain technique (Fig. 3.3). A high frequency resolution waveform with a high pulse repetition frequency (PRF) is needed to obtain a high integrity JEM spectrum.

For HRRP, which produces a 1D set of range data, the information of interest is in the time domain (Fig. 3.4). The range profile of the target is compared to entries in the database to find a good match. A wideband waveform with sub-metre range resolution would normally be required to obtain a high resolution range profile. Many pulses would be coherently integrated to provide the necessary sensitivity.

For helicopter blade flash or helicopter rotor modulation (HERM), detection occurs in the time domain on a pulse-to-pulse basis (Fig. 3.5). The periodicity of the main rotor blade flashes is determined as the recognition discriminant. A high PRF waveform is needed in order to ensure that the blade flashes are detected and are not missed between radar pulses. The dwell time has to exceed the periodicity of the blade flashes in order to ensure at least two blade flashes are detected.

ISAR and RDI use high range resolution data to obtain high down-range resolution and high Doppler resolution to provide high cross-range resolution. The 2D target image data is compared to database entries for matching. ISAR has a relatively long dwell time and an ISAR image of an aircraft is shown in Fig. 3.6.

Table 3.1 Features extracted for various aircraft recognition techniques

Target recognition technique	Signature data	Features extracted
JEM	Spectral	Engine blade count
HRRP	Time domain (range gate-to-range gate)	1D range profile
HERM	Time domain (pulse-to-pulse)	Helicopter blade flash rate
ISAR	Time and frequency domain	2D target image
RDI	Time and frequency domain	2D target image

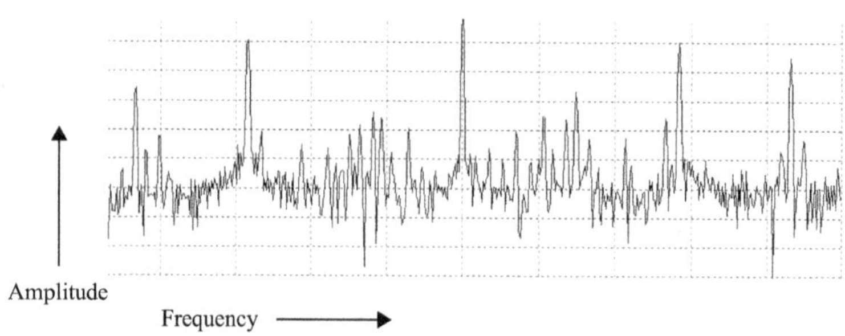

Figure 3.3 Jet engine modulation spectrum: frequency domain signature data

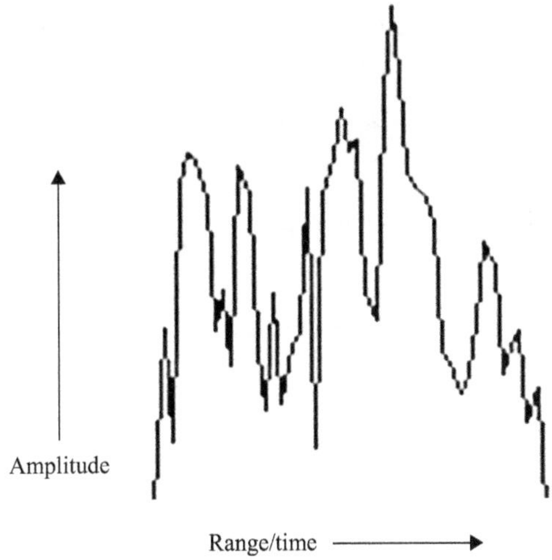

Figure 3.4 High resolution range profile of aircraft target – time domain signature data (range gate-to-range gate)

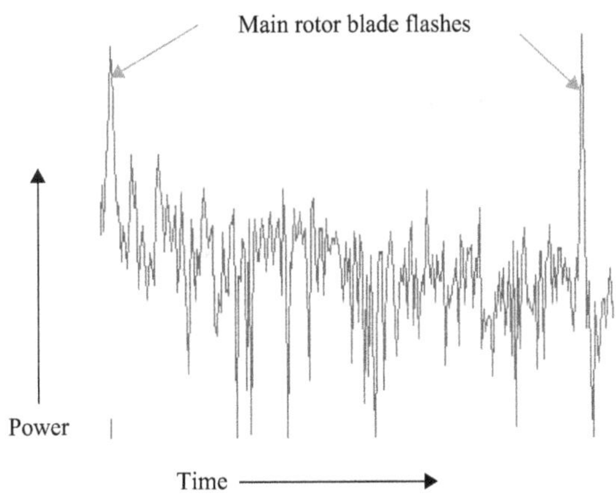

Figure 3.5 Helicopter main rotor blade flashes – time domain data (pulse-to-pulse)

RDI uses relatively short dwell periods and ISAR effects are not normally detected, but the locations of the propulsion systems can be detected and localised in range.

3.2.4 Target signature measurement

It is clearly necessary to design the radar appropriate to the type of measurement, which has to be performed, to provide the type of target signature required. In order to obtain signatures of high integrity, the waveform must be carefully designed and the radar must support the transmission and reception of the signal without

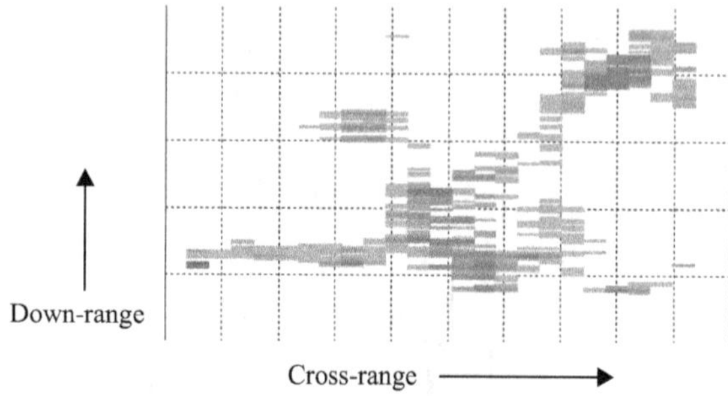

Figure 3.6 Aircraft ISAR image

distortion. The waveform, the associated signal processing, radar phase noise and dynamic range performance also have to be designed to minimise the effects of clutter.

3.2.5 Radar range equation for radar target recognition

There must also be sufficient energy illuminating the target to ensure that the smallest contributions to the target signature, which are needed for the recognition process, are detected reliably.

A *radar range equation* can actually be defined for radar target recognition and is clearly dependent upon the target features used. For conventional radar detection functions the RCS of the whole target are normally employed, taking into account any RCS fluctuation statistics. For target recognition a similar approach can be followed, but the effective RCS of the target is much smaller. For target recognition the RCS that is inserted into the *radar range equation* is the RCS of the smallest element of the target, which is used by the classifier.

All of the other parameters in the equation are generally similar to those used in the standard radar modes. Essentially the difference in range for the application of a target recognition mode for the same dwell period, antenna aperture and transmitter power compared to detection is the fourth root of the ratio of the full target RCS to that of the smallest RCS element used by the classifier. The *radar range equation* for target recognition is

$$R^4 = \frac{PA^2 \sigma N \tau}{4\pi \lambda^2 LSkT_0 NF} \tag{3.1}$$

where:

R = Radar range
P = Peak transmitter power
A = Effective antenna aperture
Σ = RCS of smallest recognition discriminant
N = Number of coherently integrated pulses
t = Duration of pulse

Λ = Radar wavelength
L = System losses
S = Threshold above noise
k = Boltzmann's constant
T_0 = Reference noise temperature (290 K)
N_F = Noise figure

3.2.6 Main classification functions

The target recognition process is summarised in Fig. 3.7. After the radar has transmitted the high resolution waveform, detected and processed the return from the target of interest, the signature is conditioned ready for the recognition process. The signal processing algorithms provide the signature data, which then has to be localised in range appropriate to its range extent and any clutter has to be filtered out. If range profile data is needed from the signature, any frequency domain data is also filtered out.

In order to perform target recognition, a good deal needs to be known about the targets which are to be recognised. High integrity high resolution reference signatures of all targets of potential interest, at aspect angles of interest and possibly with different stores, flight or assembly configurations, are needed. It is necessary

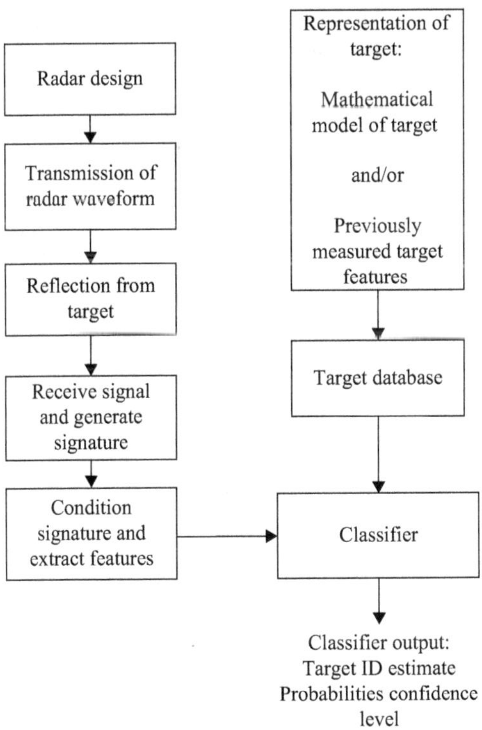

Figure 3.7 Target recognition process

to mathematically predict the signatures of the targets, which are to be encountered, or to physically measure the high resolution signatures of the targets. A combination of these techniques can also be used. When many measurements or signature modelling computer runs have been performed, it is necessary to assemble them into a database. The type of target recognition technique being utilised and the type of recognition algorithm or classifier employed determine how this data is subsequently used. It can be averaged in various ways and it is then translated into some form of mathematical model representing the targets' characteristics.

3.2.7 Database

A representation of a database of targets is shown in Fig. 3.8. Database entries are required for all possible targets which are likely to be encountered. The number of parameters to be represented in the database is dependent upon the target recognition technique being employed. The parameters stored in the database represent the features that are extracted from the target signature measured. For each parameter its value could have a tolerance or a probability distribution and an angular dependence, so the database could be extremely large. For example suppose a parameter's value varied with aspect angle such as to be required at every 5° in azimuth over 360° and for elevation angles from plus to minus 30° also in 5° steps. About 500 database entries would be needed for each target for each parameter stored. Hence, if there were 100 targets in the database, there would be a requirement for 50,000 database entries per parameter.

3.2.8 Classifier

Classifiers are mathematical techniques, algorithms or formulae that compare the signatures which have been measured with the mathematical models or representations of targets which have been assembled in the database. The classifiers are designed to perform this function and are required to assess how close a match a particular signature is to its mathematical model reference and if any other targets

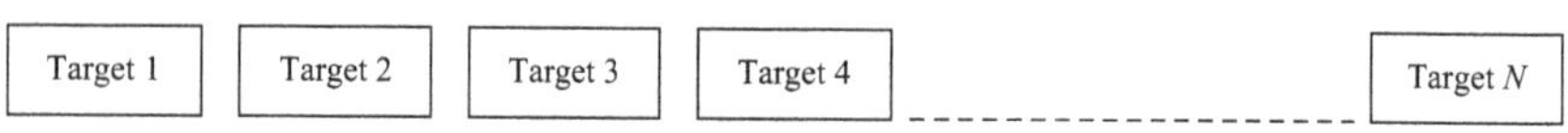

For each target in database:

	Parameter 1	**Parameter 2**	**Parameter 3**	………	**Parameter *N***
Value					
Tolerance/ Probability Distribution					
Azimuth and Elevation Angular Dependence					

Figure 3.8 Representation of database

also have a likelihood of being the same target. This is illustrated in Fig. 3.9 for a measured range profile of an aircraft target.

The signature shown is a typical range profile of an aircraft. The classifier compares the signature obtained with the five target database entries at the required target aspect angle and provides an estimate of the best candidate for the unknown target's identity. In this example target '3' is the best match as can be seen by comparing the shapes and sizes of the references with the measurement.

When several target parameters are used, their matches have to be combined. Although many different techniques can be used to combine the various features, the outcome is always a measure of how well a particular set of measured features match the database entries for the target. The outputs can be presented probabilistically for the various candidate targets.

The recognition of targets has a probability of correct assignment and a probability of wrong assignment to target type. These probabilities depend upon the quality of the radar measurement, the similarity of the two targets, the quality of the mathematical model of the targets and the effectiveness of the recognition algorithm. The quality of the radar measurement depends on the signal-to-noise ratio, whether clutter, interference or jamming is present and in some cases, whether the

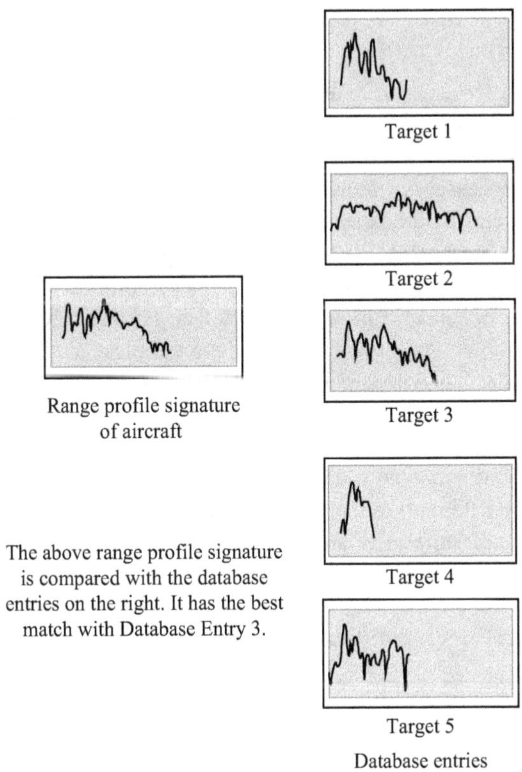

Figure 3.9 Matching of measured range profile signature with database entries

target is changing aspect angle or accelerating during the measurement. For example in certain circumstances, when making high range resolution measurements, a manoeuvring or accelerating target can blur the high resolution image and thereby degrade the recognition process.

If possible, the classifier should be designed to reliably recognise targets with a very low error rate.

The type of output obtained from the operation of the classifier is shown in Table 3.2. Assuming that three different targets in the database have some level of match with the measured signature data, the resulting level of matches can be translated into probabilities for each of the respective candidate targets. In addition a column is provided, which gives a level of confidence in the decision. For example if the measurements were performed at a low signal-to-noise ratio and the level of matches obtained was low, then there would not have been much confidence in the decision. However, if at least one very good match is obtained, there would be a high level of confidence in the decision. Criteria would have to be developed for various probability levels for candidate targets and the confidence level required for any resulting operational decisions to be made. Clearly a single candidate target with a high probability value and a high level of confidence is required for a positive confident operational decision to be made.

3.2.9 Assembly of database

The assembly of a representative and up-to-date database for all targets of interest is one of the most important and challenging aspects of NCTR. In addition to friendly targets, there are neutrals and potential adversaries. As well as military aircraft it is also necessary to include civil aircraft in the database.

There are several methods available for the assembly of the database: mathematical modelling; measurements of real target signatures; measurements of scale models; and combinations of all three of these. The complexity of the compilation of the database is dependent upon the type of target attribute or attributes being employed by the classifier. Probably one of the simplest databases comprises the blade count numbers for various compressor stages of the jet engine for performing JEM. Only one or two numbers would normally be needed and as the blade count is independent of target aspect angle, i.e. there is no angular term. This is an example of a very simple mathematical model-based approach.

In contrast HRRP involves the generation of range profile representations at all aspect angles for all stores fits and configurations for each type of target. The assembly of the database entries can be performed by measuring the various targets

Table 3.2 Output of target classification function

Target	Probability	Confidence of decision
Target 1	$A\%$	$X\%$
Target 2	$B\%$	
Target 3	$C\%$	

at all aspect angles, which is theoretically possible for friendly aircraft, but is extremely costly. It becomes progressively more difficult to measure the signatures of neutrals' and then adversaries' assets, so another approach is needed. If the dimensions and general materials making up the neutral's or adversary's target are known, a mathematical model based on electromagnetic theory can be used to model each target at each aspect angle of interest. This is clearly logistically much easier than trying to measure the signatures of adversaries' targets, but tends to still be a very costly exercise. The construction of the mathematical model and coding it in software tends to be very expensive and running the software to provide HRRP data requires a lot of computer time.

The other method is to build scale models from knowledge of the target's dimensions and materials and use a wideband millimetre or sub-millimetre wave radar to generate the range profiles by performing measurements on the models in an anechoic chamber. This approach avoids the time and cost associated with the mathematical modelling or making real target measurements, but is highly dependent upon the scale model being a high integrity representation of the target.

One solution to this difficult problem is to use all three methods on a sub-set of the targets and aspect angles. If agreement can be reached between the three methods, use the scale modelling to complete the process of assembling the database.

The inclusion of target signature data from potential adversaries remains a major challenge and also has to rely on radar measurements and mathematical modelling.

3.2.10 Classifier performance

When developing the capability to recognise targets, it is necessary to assess the performance obtained. The key parameters are the levels of correct and incorrect decisions which are being made. If the performance is good and is showing potential, the operational application can be addressed. If the performance is only likely to provide a marginal benefit, it may be difficult to justify its operational employment.

Classifier performance is conventionally assessed using a confusion matrix as shown in Table 3.3. Normally a confusion matrix has $N \times N$ cells for N target types stored in the database. The actual target types stored in the database are located in the left-hand column of the matrix. Along the top row of the matrix are the potential target types, which the classifier can declare. In a 'closed world' of N targets with no chance of any other target being detected all of the ID decisions can be included in the $N \times N$ matrix cells.

The ideal performance of a classifier with five targets in the database is shown in Table 3.3. The values of the diagonal are all 100% and mean that all the targets are correctly identified all of the time. In reality this does not happen, as some incorrect ID decisions are made, which are manifested as off diagonal entries in the matrix. Any off diagonal values that are not 0 represent incorrect target ID decisions being made. The objective is to maximise the values of the diagonal entries. For an operationally useful system, entries in the high 90% would be expected.

Table 3.3 Ideal confusion matrix with values expressed as percentages

Estimated Target ID **Actual Target ID**	**Target A**	**Target B**	**Target C**	**Target D**	**Target E**
Target A	100	0	0	0	0
Target B	0	100	0	0	0
Target C	0	0	100	0	0
Target D	0	0	0	100	0
Target E	0	0	0	0	100

In order to determine the values in the confusion matrix, the extracted features of many signatures of each target would be applied to the classifier. The percentages of occasions in which the respective target type decisions are made are collated. Clearly for targets with similar values of attributes there would be more chance of incorrect decisions being made and correspondingly higher values in the off diagonal matrix entries than for targets that have dissimilar characteristics.

3.2.11 Conclusions

The reliable recognition of air targets is achievable provided all the requirements on the elements in the target recognition chain presented above are met. It is a very complex task measuring and analysing large numbers of aircraft signatures and understanding and identifying robust target features. The provision of the radar and the development of the signal processing and feature extraction algorithms, the classifier and the assembly of the database are also very significant challenges. The whole process takes many years to achieve a high performance NCTR function.

3.3 Jet engine recognition

3.3.1 Introduction

The range profiling techniques summarised above are dependent upon the stability of the aircraft target's signature. For JEM the opposite approach is adopted. The physically stable parts of the target are not used and it is the moving parts of the target that are utilised in the target recognition process. The moving parts of the engine, which are normally of interest, are associated with the compression stages of the jet engine, which are at the front of the engine. At head-on aspect angles these are visible and are of particular interest for any approaching threatening aircraft target. The radar signal interacts with the rotating engine compressor stages and generates the JEM spectrum. It is the understanding of how this process occurs and of the resulting spectrum that enables the first, and possibly second, stage blade counts to be determined. This information is then used to identify the engine type and enables an estimate of the actual aircraft type identity or options to be determined.

A fairly simple but representative explanation of the JEM spectrum is presented. The key spectral components, which are utilised to extract the blade counts at the front of the compressor stage, are presented.

3.3.2 Jet engine mechanics

The JEM spectrum results from the illuminating electromagnetic radar signal interacting with the jet engine. The mechanics of the front of the jet engine are now discussed.

The front of a typical jet engine is shown in Fig. 3.10. There are many variations on the basic design of the jet engine, but the principles as far as the radar engineer is concerned for measuring the JEM spectrum are quite similar. The air flowing into the front of a jet engine is compressed by a series of blades attached to a common shaft, which normally rotates at a few thousand revolutions per minute. The individual compression stages at the front of the engine normally have increasing numbers of rotor blades attached to the shaft with depth into the engine. The blade sizes also tend to reduce with distance into the compression stage, as the air is progressively compressed at higher pressure into a smaller volume.

3.3.3 Interaction of radar signal with engine blades

A simple representation of three compression blade stages at the front of a jet engine is shown in Fig. 3.11. As the shaft is normally common to the first few stages of the compressor, each rotor stage rotates at the same rate.

Initially the radar signal is reflected from the first rotor stage of the engine. Each of the rotor blades individually contributes to the net signal returns. As well as being reflected, some of the energy propagates between the blades through to the second rotor stage. Again some of the energy is reflected back towards the radar as well as some of it continuing to propagate further into the engine. As the signal gets deeper into the engine, the smaller sizes of the blades and the associated smaller gaps between them allow progressively less energy to propagate to the deeper rotor

Figure 3.10 Jet engine: the front of the engine contains the compression stage rotor blades, which generate the jet engine modulation spectrum when illuminated by radar

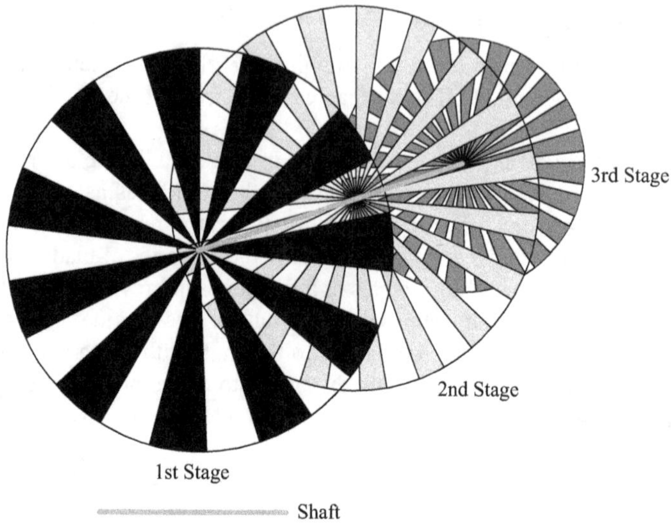

Figure 3.11 Representation of rotor blade stages at front of jet engine

stages. Most of the energy of interest at wavelengths at which JEM is employed results from reflections from the first two or three rotor stages of the compressor, which are the stages of interest for the JEM recognition function.

Radar signals having the shorter wavelengths penetrate further into the engine than at the longer wavelengths as the signal can propagate more effectively through the smaller gaps between the blades. Radars operating at higher frequencies penetrate deeper into the engine, so the signals received have a far higher content of returns from the deeper compressor stages of the engine than radars operating at longer wavelengths.

3.3.4 Jet engine modulation spectrum: engine rotational rate

The scattering effects, which occur as the electromagnetic wave interacts with the engine, are extremely complex and are probably far too complex to model in detail. However, by understanding the physics of the situation and the mechanics of the jet engine and using the symmetry of the effects, progress can be made in understanding the resulting spectrum and in devising techniques to determine the blade count on the first couple of compressor blade stages.

The rotation of the whole of the compressor stage on its shaft is now considered. The rotation rate of the compressor is called the shaft rate or spool rate and can be in units of Hz or RPM. It is assumed that the engine is running at a constant speed, the range to the radar is not changing significantly and the aspect angle of the engine to the radar is not changing over the measurement period. The signal received by the radar over each rotation cycle of the engine is then the same. Each time the engine rotates a full cycle it returns to the same point. Over a cycle of rotation the amplitude and phase of the radar signal vary considerably, but over the next cycle these characteristics are repeated. This is illustrated in Fig. 3.12.

Fig. 3.12 shows the periodicity of the engine rotation, which is repeated with reasonable integrity on an engine cycle-to-cycle basis. The fine detail comes from higher frequency modulation effects, which are discussed below.

The characteristics are due to the net effect of all of the contributions of all the blades reflecting the signal back in the direction of the radar. Each blade can be considered to be a reflector and for the incident electromagnetic wave generates a reflected wave with amplitude and phase components, which are dependent upon the look angle.

Rotor blades are pitched for aerodynamic reasons so that the plane of the blade is not the same as the plane of rotation of the blade about the spool. The result is that as the blade rotates on the spool, it is continuously presenting a different aspect angle to the radar. The amplitude measured at the radar varies over a cycle of the engine and repeats again next cycle. This occurs for all of the blades, so the net amplitude at the radar is the sum of all the amplitude and phase contributions from all the blades. As the blade is moving, during certain parts of the rotation cycle it is approaching the radar and it is receding at other times, so that there is a Doppler effect which is continually changing. This is manifested in a continuously varying phase change for each blade during the rotation cycle. Hence, both the amplitude and phase contributions from each blade are varying continuously and the net effect at the radar is the vector summation of all the contributions.

Although the time domain signal received by the radar from the jet engine is exceedingly complex and is not known, it is known to be periodically repeating every cycle of the engine. Hence, the spectrum that the jet engine is generating has a fundamental frequency, which is offset from the radar's transmission frequency by the rotational frequency of the engine. All possible frequencies generated are at harmonics of the rotational rate and also modulate the radar signal. The lowest

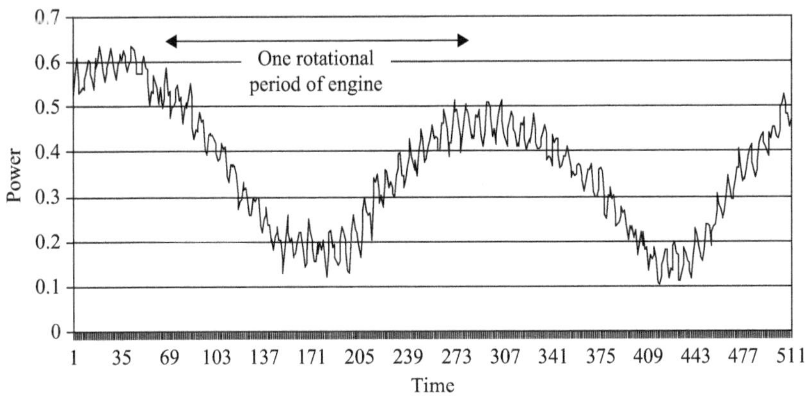

Figure 3.12 Power versus time characteristic of radar signal reflected from jet engine

possible frequency that can contribute to the JEM spectrum is the rotational rate of the engine. The next lowest frequency components are its second and third positive and negative frequency harmonics. For a rotational period T, the spectral components f_n have values of $\pm n/T$, where n is a whole number.

$$f_n = \frac{n}{T} \tag{3.2}$$

Hence, if the JEM spectrum can be analysed and the fundamental frequency of the spectral lines can be determined, the engine rotational rate can be found. Fig. 3.13 is the frequency domain equivalent of the time domain characteristics of Fig. 3.12.

3.3.5 Jet engine modulation spectrum: rotor stage spectrum

In the previous section the rate of rotation of the engine and the generation of the lowest frequency components of the JEM spectrum were discussed. In this section the generation of the higher frequency spectral components is presented.

The characteristics of the signal reflected from the first rotor blade stage are now considered. Consider a single rotor stage as shown in Fig. 3.14. The engine has N blades on a single rotor stage each of which is assumed to be the same. The engine is symmetric about the spool and every time that it undergoes a rotation of $1/N$ of a cycle, the mechanical action is repeated. As the radar illuminates the engine, at any particular time, energy is reflected from all the blades simultaneously. The amplitude and phase measured by the radar receiver result from the contributions from all the blades. The analytical calculation of the returned signal is dependent upon the detailed mechanics of each blade's dimensions and positions

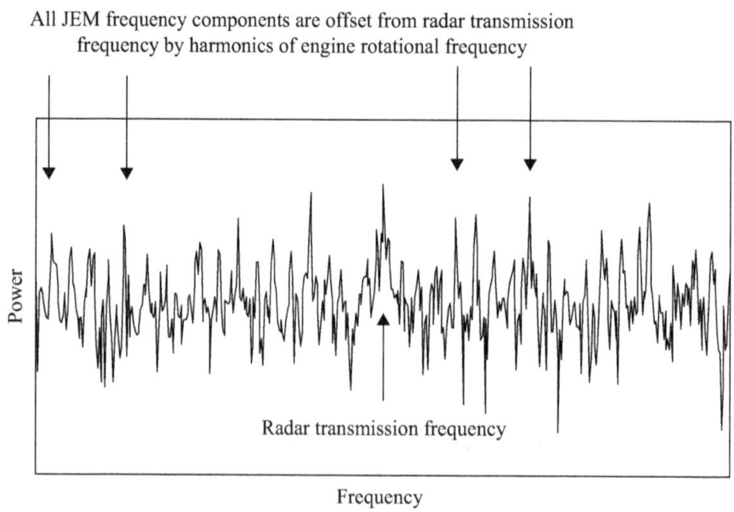

Figure 3.13 All frequencies in JEM spectrum are at harmonics of the engine rotational frequency

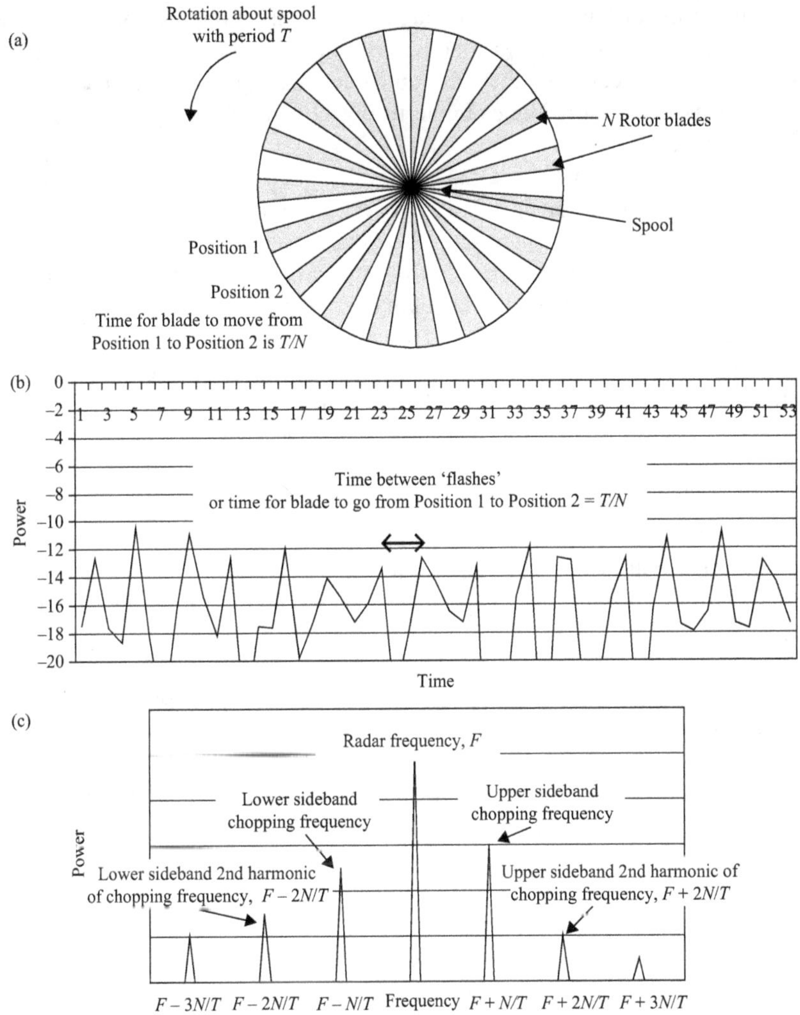

Figure 3.14 Time domain characteristics and typical JEM spectrum obtained from single rotor stage of engine with N blades. (a) Single rotor stage with N blades. (b) Power versus time history for single rotor stage showing blade flashes. (c) Spectrum from single rotor stage

and the radar wavelength. As the detailed dimensions of the blades and engine are not normally known to the radar engineer and the calculations are extremely complex it is not practical to obtain a unique analytical electromagnetic model for an engine.

In Fig. 3.14(b) the time history of the reflection of a radar signal from a single rotor stage of a jet engine is shown. Ideally the time history should repeat exactly for every 'blade period', T/N, but the reality of the practical measurement did not quite match the theory. Although the actual amplitude and phase characteristics are very complex in practice, the important fact is that they repeat at a rate which is equal to the time taken for a single blade to move to the position of its neighbour. The example in Fig. 3.14(a) shows adjacent blades at positions 1 and 2.

The radar signal repeats at a time interval, T/N. It is the Fourier transform of this time domain data that is the JEM spectrum for a single rotor stage with N blades. This spectrum is characterised by lines spaced at frequency intervals of N/T from the radar signal transmission frequency as illustrated in Fig. 3.14(c). The actual spectrum obtained and the amount of energy in each line are dependent upon the detailed time domain amplitude and phase characteristics received by the radar. The dimensions, shape and orientation of the blades determine the details of the spectrum, due to a single rotor stage. The spectral components F_n for a single rotor blade stage have values of $\pm Nn/T$, where n is a whole number.

$$F_n = \frac{Nn}{T} \tag{3.3}$$

Experimental results from real engines at microwave radar frequencies strongly confirm that most of the JEM energy from a single rotor stage resides in the main 'chopping' sidebands, which are offset from the radar signal frequency F at harmonic intervals of N/T. Although the details of the signal received by the radar for a single rotor are not generally feasible to predict, the corresponding frequency domain signal measured is composed of discrete spectral components, which have values that are related to the mechanics of the engine.

It should also be noted that the chopping sidebands identified at harmonics of N/T are all individually harmonics of the spool rate $1/T$, which was discussed in section 3.2.4. For each individual rotor stage taken in isolation, the spectrum is of the form shown in Fig. 3.14(c) with the actual chopping frequencies being proportional to the blade count on the rotor stage.

Hence, for each rotor stage of the engine a family of spectral lines is generated consisting of a set of harmonics resulting from the product of the engine rotational rate with the number of blades on the rotor, each line of which is offset from the radar frequency.

3.3.6 Jet engine modulation spectrum: mixing products from rotor stages

In addition to the chopping lines, spectral lines which are mixing products from the different rotor stages are generated. At the second rotor stage some energy is reflected back towards the first stage, of which a proportion is reflected from the first blade stage and some propagates through the first stage blades out of the engine. These reflection and 're-reflection' effects occur as the radar signal penetrates through the engine.

The signal that is reflected back to the radar from the first rotor stage consists of the transmission frequency of the radar and modulation sidebands at the chopping rate harmonics. Signals with this same frequency content, but with differing amplitudes, are also propagated past the first rotor stage and encounter the second rotor stage.

In Table 3.4 typical components of the JEM spectrum are shown due to the mixing of the chopping frequencies from the front three rotor stages. The first four harmonics from each stage and the second- and third-order intermodulation products are shown.

3.3.7 *Determination of blade count*

A typical JEM spectrum is shown in Fig. 3.15 and includes the main chopping lines, some inter-modulation products and spool lines. In order to determine the blade count the spectral components must be extracted and analysed. There are several methods available and one technique is to determine the first-stage chopping frequency, N/T from (3.3), and the engine rotational period frequency, $1/T$ from (3.2). The ratio of these two frequencies is N, the first stage blade count. As an example, if the engine is rotating at 3000 RPM or 50 Hz and the first stage chopping frequency is 1500 Hz, then the first stage blade count is 30.

Similarly, if the second-stage chopping frequency, M/T, can be determined, the second stage blade count, M, can be established. There are several ways available for determining the engine rotational rate based upon extracting the periodicity of the spool lines. If the first and second stage blade counts can be determined, they provide the robust attributes needed to recognise a jet engine.

3.3.8 *JEM waveform*

The key parameters for a JEM waveform are high frequency resolution and unambiguous frequency measurement. Normally the JEM waveform is low bandwidth as high frequency resolution performance is more important than high range resolution. A dwell period from a few tens to several hundred milliseconds

Table 3.4 Typical JEM frequencies generated from mixing products from front three compressor stages of engine

Rotor stage	1 (Blade count N)	2 (Blade count M)	3 (Blade count P)
1 (Blade count N)	$\pm N/T$	$\pm(N+M)/T$	$\pm(N+P)/T$
	$\pm 2N/T$	$\pm(N-M)/T$	$\pm(N-P)/T$
	$\pm 3N/T$	$\pm(2N+M)/T$	$\pm(2N+P)/T$
	$\pm 4N/T$	$\pm(N+2M)/T$	$\pm(N+2P)/T$
2 (Blade count M)		$\pm M/T$	$\pm(M+P)/T$
		$\pm 2M/T$	$\pm(M-P)/T$
		$\pm 3M/T$	$\pm(2M+P)/T$
		$\pm 4M/T$	$\pm(M+2P)/T$
3 (Blade count P)			$\pm P/T$
			$\pm 2P/T$
			$\pm 3P/T$
			$\pm 4P/T$

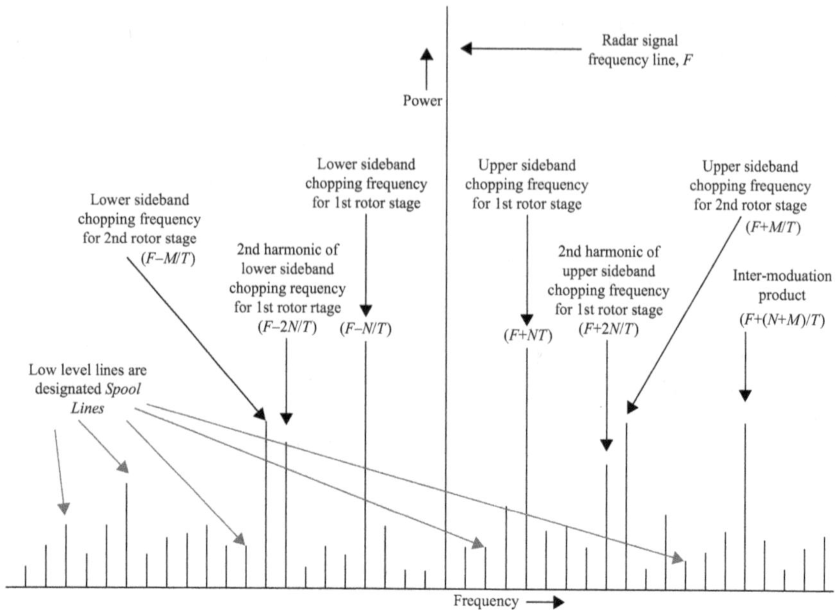

Figure 3.15 Typical JEM spectrum

and a PRF of up to a few hundred kHz would normally ensure that these criteria are met.

3.3.9 System requirements

The system requirements for supporting JEM are not much more demanding than for the surveillance and tracking functions of the radar, provided the dwell time is available and the radar is coherent. One key parameter is the close-to-carrier phase noise to ensure that any clutter present does not degrade the JEM spectrum. The spurious free dynamic range of the digitiser is also important as any spurious signals present could be confused with JEM lines of interest.

3.3.10 Conclusions

The basic principles involved in the generation of the JEM spectrum and its application in the recognition of jet engines have been presented. The key physical factors that determine the JEM spectral content are the radar wavelength, the rotational symmetry of the whole engine and the symmetry of individual rotor stages. These spectral components are multiples of the rotational speed of the engine.

The algorithms required to extract the engine rotational rate and the chopping frequencies of the front stages of the compressor are based upon the physical principles and mechanics of the engine. They tend to be quite complex and have to also include clutter rejection functions and to be able to deal with the presence of multiple engines on the aircraft.

3.4 Helicopter recognition

3.4.1 Introduction

For recognising helicopters, the radar signatures of the parts of the aircraft that move relative to the main body of the helicopter are normally used. Any mechanical parts of the helicopter that rotate and are illuminated by the radar signal are potential candidates for providing attributes for target recognition purposes. The rotor hubs and rotor blades can interact with the radar signal to provide reflections that are periodic or have spectral recognition discriminants and can be extracted as target features.

The main helicopter features tend to be impulses from the blades, known as blade flash, and are normally detected in the time domain. In order to detect these, a waveform with good time resolution is required, but there is no necessity to provide high range resolution, so a low range resolution waveform can be used.

A helicopter has rotary wings which provide the lift, which keeps the aircraft aloft and generate the force required to move it in the required direction. Helicopters are normally powered by one and sometimes two jet engines, which cause the main and rear rotor hubs to rotate. They typically have one main rotor supporting relatively long blades mounted near the front of the helicopter and the plane of the blades is horizontal. The second rotor is mounted near the tail of the helicopter and the relatively small blades move in the vertical plane. These helicopter features are shown in Fig. 3.16.

3.4.2 Main rotor blade flash

The main rotor blades of a helicopter rotate at typically a few hundred revolutions per minute and are limited to a maximum air speed, which is the speed of sound and is about 330 m/s at sea level. The rotor blades are long and thin and can be represented in a very simple model as a long flat plate to demonstrate its interaction with the radar signal. In Fig. 3.17(a) a rotor blade is represented in a simple model by a plate of length L having a narrow height dimension H rotating about a hub in the horizontal plane.

Figure 3.16 Helicopter with main and tail rotor blades

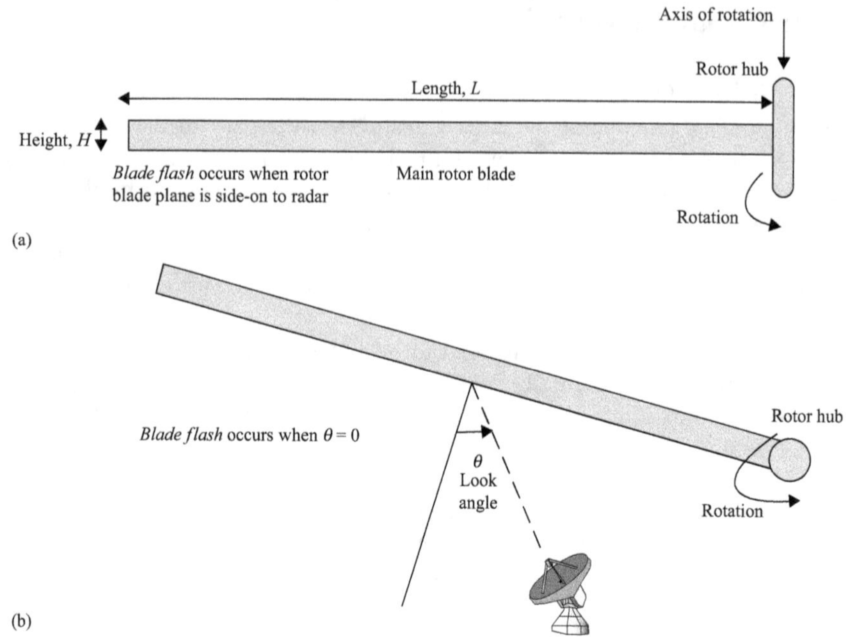

Figure 3.17 Main rotor blade and look angle of radar, θ. (a) Viewed in horizontal plane. (b) Viewed from above

For a rectangular plate of length L and height H the radar cross-section measured in the same horizontal plane of the blade is

$$\sigma = \frac{4\pi L^2 H^2}{\lambda^2}\left[\frac{\sin((2\pi/\lambda)L\,\sin\theta)}{((2\pi/\lambda)L\,\sin\theta)}\right]^2 \cos^2\theta \tag{3.4}$$

where:

$\theta =$ Look angle

For simplicity and with no loss of generality it is assumed that the plane swept through by the blade as it rotates and the radar are the same horizontal plane. As the plate is narrow, there is far less angular dependence in the vertical plane; elevation angle effects are neglected in this simple model. It can be seen that the angular dependence in the horizontal plane follows a $(\sin x/x)^2$ function multiplied by $\cos^2$ factor.

The peak occurs when the radar is orthogonal to the blade ($\theta = 0°$) and reduces either side of this peak by the sin x/x function shown. This means that a sharp peak signal is detected when the blade is orthogonal to the radar, with relatively low level signals measured at all the other angles. The ability to detect this signal, which is popularly known by radar engineers as 'blade flash', is therefore dependent upon the sharpness of the peak and the pulse parameters of the radar. It is necessary for the radar to be transmitting a pulse when the blade is near this

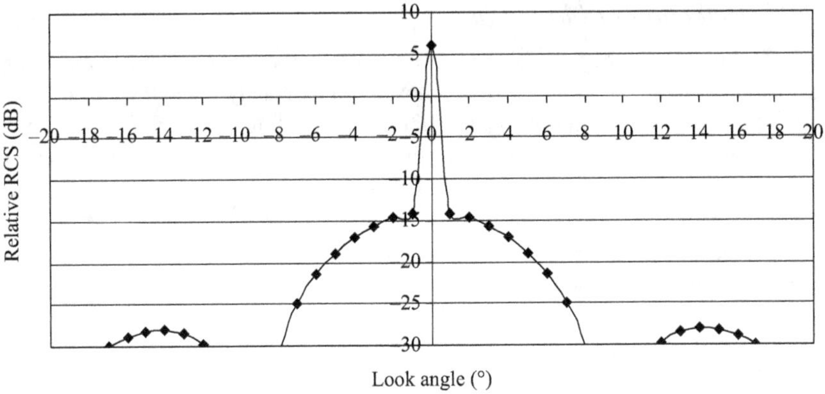

Figure 3.18 Modelling variation of radar cross-section of rotor blade with look angle – blade flash peak occurs at RCS peak

orthogonal position. The angular width of this blade flash is approximately λ/L, which is analogous to the beamwidth obtained from an aperture of the same length from antenna theory. For a typical helicopter the main rotor blade length is several metres.

The value of the peak blade flash signal reduces as the wavelength of the radar signal increases. Hence, in order to detect blade flash there is a trade-off between the value of the peak signal obtained and the angular width of the peak, which are both dependent upon the wavelength used. The modelled RCS of the rotor blade look angle θ is shown in Fig. 3.18 and radar measurements of blade flashes are shown in Fig. 3.19.

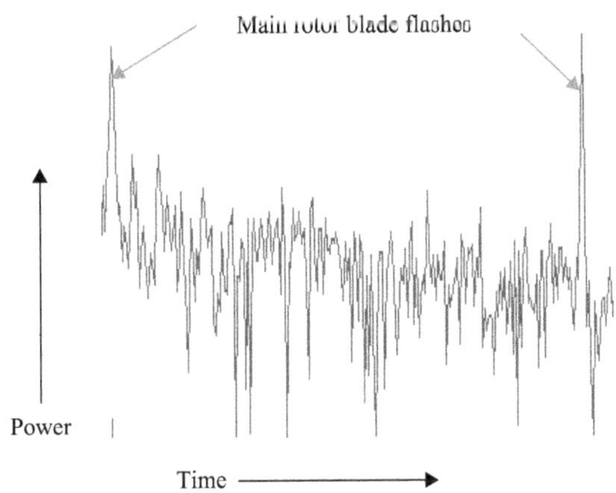

Figure 3.19 Helicopter main rotor blade flashes

3.4.3 Detection of blade flash

Blade flashes occur when any blade is orthogonal to the radar's radial vector. For a four-bladed helicopter, flashes occur almost simultaneously when the two opposite rotors (e.g. blade nos. 2 and 4 in Fig. 3.20(a)) are orthogonal to the radar's radial vector. One blade is receding while the opposite blade is approaching the radar. The next blade flash occurs when the blades 1 and 3 are orthogonal to the radar. Hence, for a four-bladed helicopter the main rotor blade flash rate is one quarter of the rotational period *T*/4. In order to measure the blade flash period, a minimum of two blade flashes have to be detected, which means that the radar integration period for a four-bladed helicopter must exceed *T*/2 Although there are sometimes slight variations due to aerodynamic effects, most even-bladed helicopters exhibit this behaviour. For some helicopter designs the angular velocity of the blades changes slightly over the rotational period.

In contrast for an odd-bladed helicopter, as in Fig. 3.20(b), the receding and approaching blade flashes do not coincide and result in a blade flash rate of *T*/10 for a five-bladed helicopter. In the diagram blade flash from approaching blade 2 is occurring and the next blade flash comes from the receding blade 4 in position 4(a) shown. Again, due to aerodynamic effects, the period between blade flashes may not be quite constant, but for most helicopters this is a negligible effect. The minimum radar integration period for detecting two sequential blade flashes for a five-bladed helicopter is *T*/5.

For odd-bladed helicopters the flashes occur alternately between the approaching and receding blades and tend to have different refection properties as the blades are not symmetric with respect to the two aspect angles. This means that there is a potential amplitude discriminant in the time sequence of pulses for differentiating odd- and even-bladed helicopters. The Doppler effect can also be

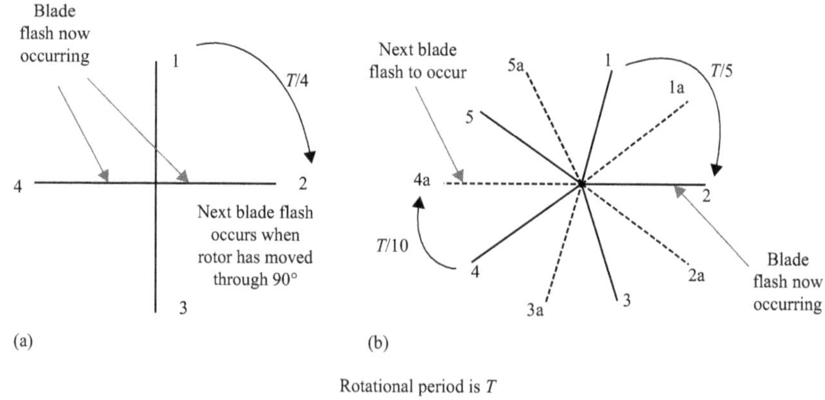

Figure 3.20 Blade flash periods for even and odd numbers of blades. (a) Four-bladed helicopter: blade flash period is T/4. *(b) Five-bladed helicopter: blade flash period is* T/10

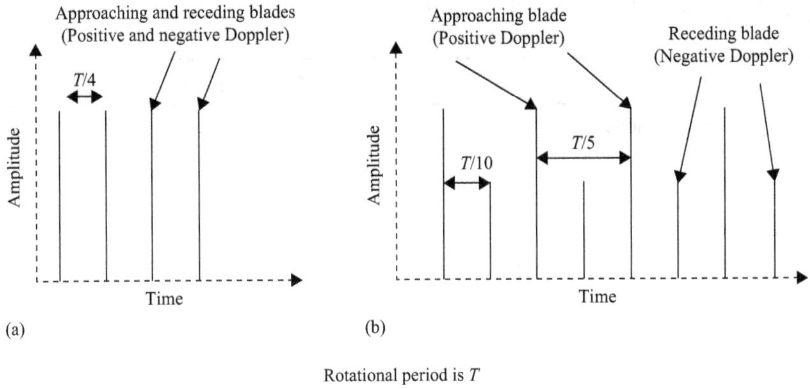

Figure 3.21 Blade flashes versus time for even- and odd-bladed helicopters with Doppler from blades. (a) Four-bladed helicopter: blade flashes of similar amplitude. (b) Five-bladed helicopter: blade flash amplitudes alternate

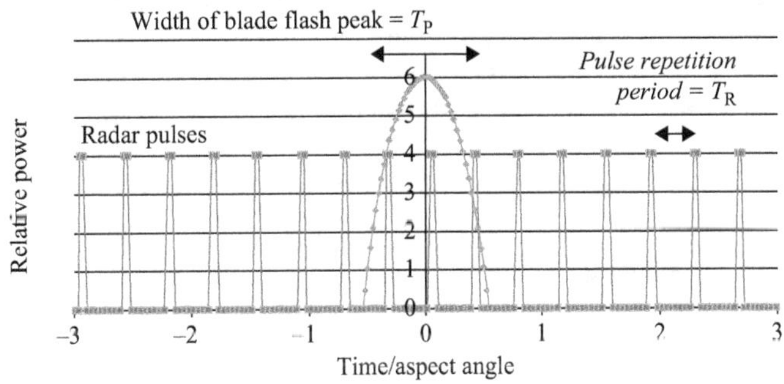

Figure 3.22 Selection of radar pulse parameters to detect blade flashes

used to determine if the blade is receding or approaching the radar as shown in Fig. 3.21.

In order to detect at least a single blade flash, it is necessary to use a PRF which is sufficiently high so that the flash is not missed between radar pulses. With reference to Fig. 3.22, it can be seen that the pulse repetition period T_R must be shorter than the flash duration T_P in order to guarantee at least part of one pulse hitting the blade during the flash. If the pulse repetition interval between transmitted pulses is longer than the flash duration, the detection of the flash is not reliable.

For a rotational period of T, the duration of the blade flash, T_P, is the angular width of the blade flash, λ/L, divided by 2π and multiplied by T, i.e. $\lambda T/2\pi L$.

If the Doppler frequency of the blade flash is to be detected, at least two pulses have to detect the blade flash and be coherently integrated and the corresponding pulse repetition period must be shorter than the pulse duration.

Table 3.5 Blade flash parameters and waveform requirements

	Even bladed	**Odd bladed**
No. of main rotor blades	N	N
Rotational period	T	T
Blade flash period	T/N	$T/2N$
Blade flash frequency	N/T	$2N/T$
Blade flash duration	$\lambda T/2\Pi L$	$\lambda T/2\Pi L$
Minimum radar integration period for blade flash detection	T/N	$T/2N$
Minimum radar integration period to determine blade flash period	$2T/N$	T/N
Minimum PRF for blade flash detection	$2\Pi L/\lambda T$	$2\Pi L/\lambda T$

The blade flash parameters and waveform requirements for blade flash detection are summarised in Table 3.5. For example the main rotor of a helicopter with four blades of length 5 m would rotate at about 8 Hz. The blade flash frequency would be 32 Hz and the corresponding blade flash period would be about 31.3 ms. In order to ensure that two blade flashes are detected, the radar integration period must be at least 63 ms. The blade flash duration for a wavelength of 6 cm would be about 240 ms, so a PRF of at least 5 kHz would be required to detect it and up to 20 kHz for determining the blade flash Doppler frequency polarity.

3.4.4 Waveform and system requirements for blade flash detection

The waveform for blade flash detection uses the information in Table 3.5 for the dwell and PRF parameters. It is necessary to use values appropriate to the extremes of helicopter parameters, which are likely to be detected. The minimum integration time is the highest value of the main rotor rotation period divided by the number of blades for odd-bladed and twice this value for even-bladed helicopters. The minimum PRF is determined by the highest value of the length of the main rotor blade divided by the wavelength of the radar signal and the rotational period. These are illustrated in Fig. 3.23. The range resolution of the waveform is normally sufficient to enable all of the returns from the helicopter to be in a single range gate and equates to a bandwidth of a few MHz. If the Doppler frequency of the returns from the blades is being used for even or odd blade count determination, a high PRF waveform, which is unambiguous in Doppler, should be used.

3.4.5 Blade flash detection

The blade flash periods for helicopters are typically several tens of milliseconds, so radar surveillance and tracking waveforms, which have dwell times of a few milliseconds, are not sufficient to reliably detect blade flash and a longer dwell period is needed. When a radar detects a helicopter, it would normally be tracked with its range, velocity, bearing and possibly height being determined. The radar would employ its pulse-Doppler processing to extract these parameters for all targets of interest.

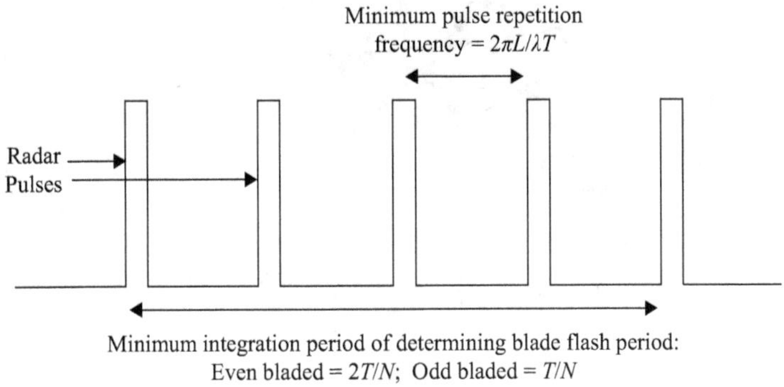

Figure 3.23 Absolute minimum values of integration period and PRF to extract blade flash period

In Fig. 3.24, on the left, the detection and tracking of an aircraft target are shown using standard techniques. Any targets of particular interest, perhaps a closing target with an unknown identity that could be a threat, would then have the helicopter recognition waveform for blade flash detection applied. One method employed to extract blade flash parameters is summarised in the figure. The mode is applied at the appropriate bearing and signature data from the helicopter is acquired at the range determined from the track.

The waveform for helicopter recognition would be pulse-Doppler enabling coherent pulse integration and clutter rejection functions to be performed. The range/time data is transformed into the frequency domain to enable clutter to be filtered out. The clutter frequency is known from the radar's own platform motion, if any. Methods are also available to filter out the helicopter skin return from the data if the skin returns are relatively large, which might make blade flash detection more difficult. The data can then be Fourier transformed back to the time domain. A threshold is set, which would normally be determined by the peak time domain signal detected and the receiver noise floor. Any high level impulses or blade flashes exceeding the threshold would then be declared. The relative levels of the flashes or any associated Doppler from the flashes can then be used to determine whether it is even or odd bladed.

3.4.6 Helicopter classification using blade flash

The helicopter would normally be classified on the basis of the blade flash rate. The assumption is made that the blade flash rates of helicopters are known sufficiently well for them to be differentiated. The database for the helicopters using blade flash parameters is of the form shown in Table 3.6.

3.4.7 Main rotor hub spectrum

The rotor hub spectrum is a recognition discriminant for determining whether the target is a helicopter as shown in Fig. 3.25. The maximum velocity V of the hub is

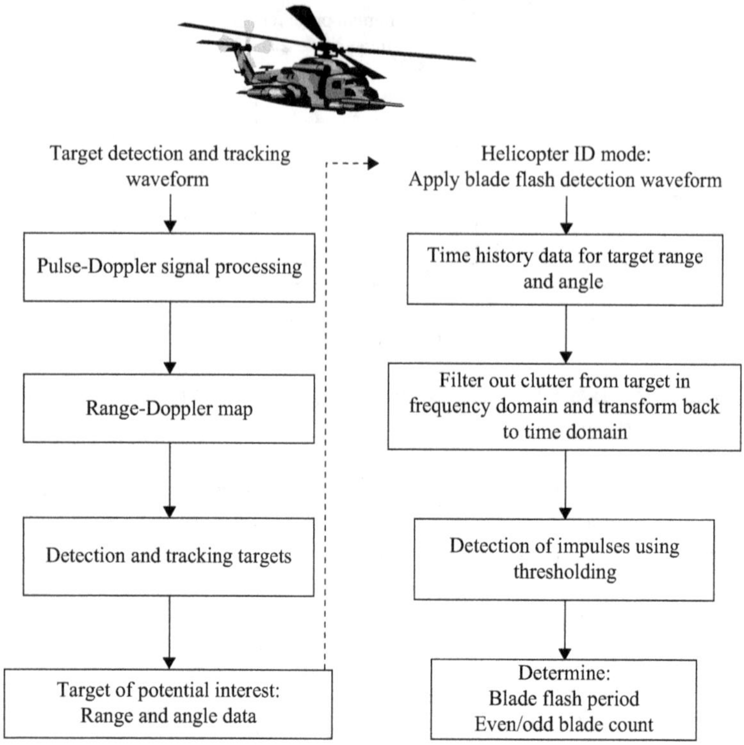

Figure 3.24 Signal processing chains for target detection and recognition of helicopters from detection of blade flash. (a) Main rotor hub velocity. (b) Main rotor hub Doppler frequency variation

governed by the rotational period T and its radius R, and is $2\pi R/T$. All velocities between the positive and negative values of V are potentially detectable from the hub and are characteristic of the frequency domain signature of a helicopter.

The spectrum that was measured from a helicopter is shown in Fig. 3.26. The positive and negative Doppler frequency returns from the rotor hub located around the skin frequency of the helicopter can be seen. Provided the radar has sufficient sensitivity, it does not need the long dwell time associated with blade flash to detect this rotor spectral characteristic.

Table 3.6 Database for helicopter recognition using blade flash

	Blade flash rate (Hz)	**No. of blades: even/odd**
Helicopter A		
Helicopter B		
Helicopter C		

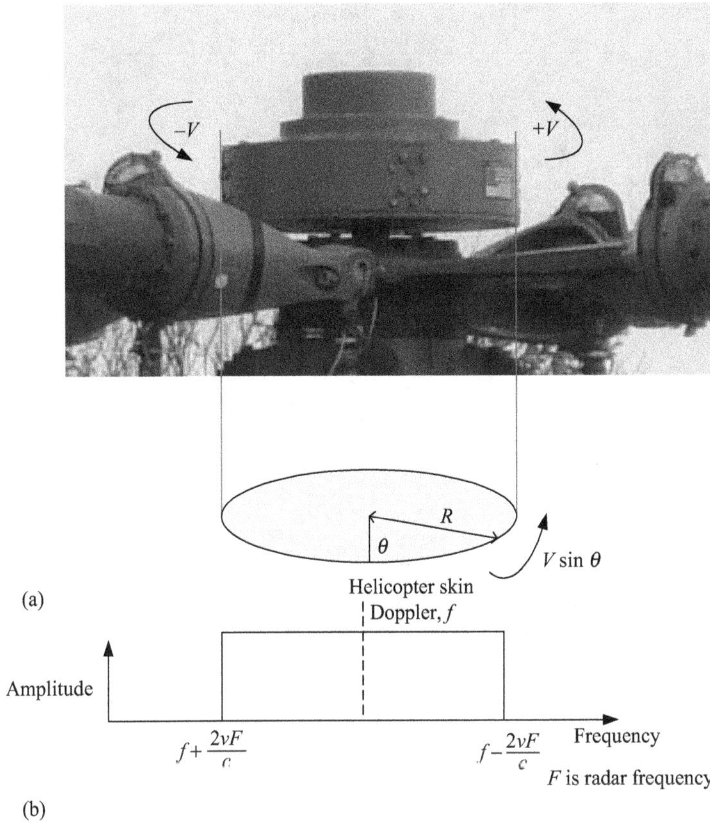

Figure 3.25 Main rotor hub spectrum

Although this technique cannot determine the type of helicopter, it can be used by radars in surveillance and tracking modes with short dwell times to recognise the target as a helicopter. As the hub radius is relatively small and the rotation speeds are quite low, the maximum hub velocity and associated Doppler frequency are typically less than an order of magnitude of the speed of sound, so a relatively low PRF can be used.

3.4.8 Rear rotor blades

The rear rotor blades can also provide discriminants, which can be used for recognising and identifying helicopters. With reference to Fig. 3.16, the main and tail rotors are shown. The main rotor is much larger than the tail rotor and the two are normally driven together via a mechanical gearing system. The gear ratio is typically below 10, with the rear rotor rotating faster. As the rear rotor blades are shorter, they can rotate much faster than the main rotor blades, while still maintaining their maximum speed at the blade tips below the speed of sound. The blade flash from the rear rotor occurs at the blade chopping frequency, which is the

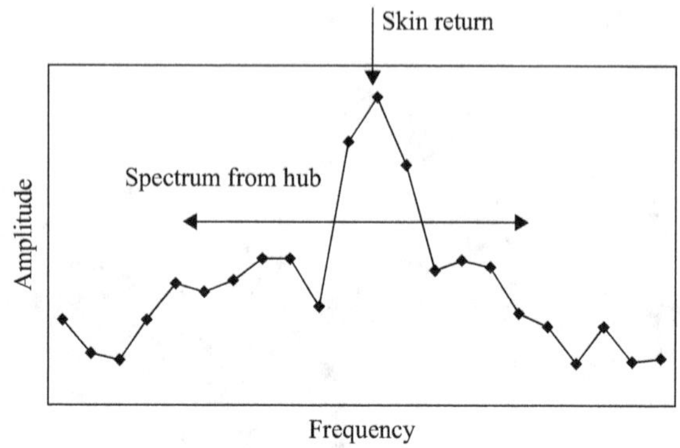

Figure 3.26 Measured spectrum from main rotor hub of helicopter

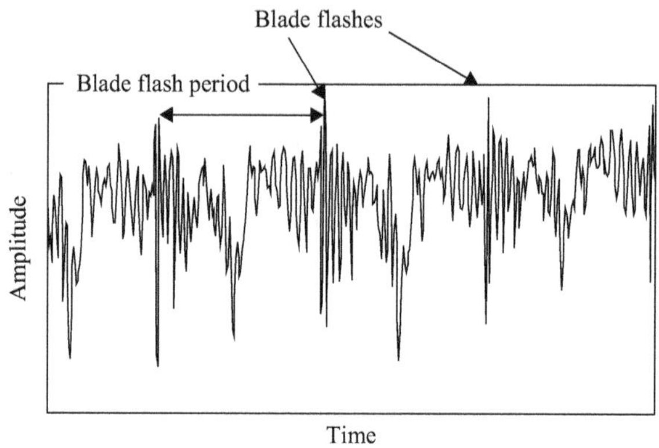

Figure 3.27 Rear rotor blade flashes

rotation rate multiplied by the number of blades on the rotor for even numbers of blades as shown in Fig. 3.27.

As the tail rotor blades are much smaller than the main rotor blades, their corresponding radar cross-sections are much less, so the signal received by the radar is relatively low. They are not readily measurable, particularly in the presence of the main rotor blade flashes, the main rotor hub Doppler returns and the skin return from the helicopter. More complex waveforms and techniques are required to isolate these effects than have been discussed within this section. In the next part of the chapter on RDI techniques the isolation of the rear rotor returns is addressed.

3.4.9 Radar range equation for helicopter recognition

A *radar range equation* can be defined for helicopter main rotor blade flash detection and is dependent upon the effective RCS of the blade flash. The simple model of (3.1) gives an approximate RCS of the blade flash. This value is then the 'σ' which is inserted into the radar range equation. As blade flash detection usually involves a single detected pulse, the transmitter energy and receiver bandwidth inserted into the equation correspond to the single pulse parameter values.

$$R^4 = \frac{PA^2\sigma_F\tau}{4\pi\lambda^2 LSkT_0N_F} \tag{3.5}$$

where:

R = Radar range
P = Peak transmitter power
A = Effective antenna aperture
σ_F = RCS of blade flash
τ = Duration of pulse
λ = Radar wavelength
L = System losses
S = Threshold above noise
k = Boltzmann's constant
T_0 = Reference noise temperature (290 K)
N_F = Noise figure

Most of the radar parameters are self-explanatory and are the same as for detecting any target. No fluctuation statistics have been included as an absolute value of RCS has been defined. The main difference is that it is basically a single pulse detection equation with no pulse integration.

3.4.10 Helicopter recognition summary

The main readily extractable recognition discriminant that is used for helicopter recognition is the period between main rotor blade flashes. The tolerance on this time measure is typically a few per cent and can differentiate many helicopters. In addition, various techniques can be used to determine from a sequence of blade flashes whether the helicopter has an even or odd number of blades. The rear rotor can also provide recognition discriminants, but they are more difficult to extract and have a lower effective RCS than the blade flash so have a shorter application range.

3.5 Range-Doppler imaging

3.5.1 Introduction

Range-Doppler imaging (RDI) involves generating a 2D range-Doppler image of the target and extracting features to identify it. High range resolution and high Doppler resolution are simultaneously employed to provide the 2D signature data. The technique is based upon using a coherent high range resolution waveform to provide the target's range profile and any frequency domain data that is also present in the signature. The first step of dividing the target into high down-range resolution cells is shown in Fig. 3.28.

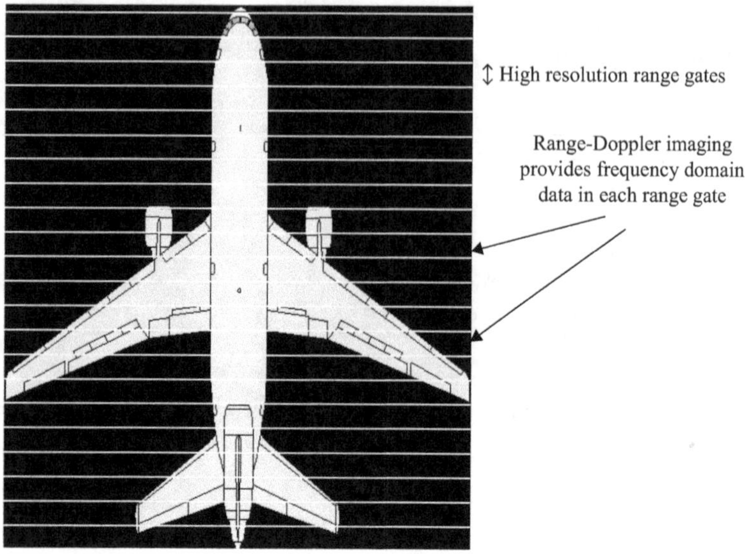

Figure 3.28 Range-Doppler imaging – each high range resolution range gate is Doppler processed for detecting any frequency data present

RDI is generally known for generating 2D ISAR signatures and involves relatively long dwell times to provide the cross-range resolution as illustrated in Fig. 3.29. This chapter is addressing shorter dwell RDI techniques than ISAR, although the latter is first briefly reviewed to put it into context. The cross-range resolution for ISAR is generated by the target's motion, which provides the large synthetic aperture required. Another way of viewing this effect is that the different parts of the target are at slightly different aspect angles from the radar and hence have different values of Doppler frequency. By applying the high resolution Doppler waveform, the different parts of the target at the same range fall into different Doppler cells, which correspond to different cross-range cells. Hence, if high down-range resolution is also employed, a 2D target image is obtained.

Two key points on ISAR are that any modulation frequencies such as JEM must be filtered from the target data and the actual values of the differential Doppler frequencies across the aircraft skin vary typically from a few Hz to a few tens of Hz.

Fig. 3.30 illustrates the Doppler content theoretically available from a jet aircraft using RDI. The ISAR frequency domain data from the skin return of the aircraft cover up to a few tens of Hz of bandwidth while the JEM lines can spread out to several tens of kHz. ISAR uses the data shown in the grey spectral returns and short dwell RDI uses the darker spectral returns. Short dwell RDI does not need the very high frequency resolution needed by ISAR.

Hence, to generate an ISAR signature it is necessary to filter out any JEM returns and use the signature's low frequency Doppler content. In contrast short

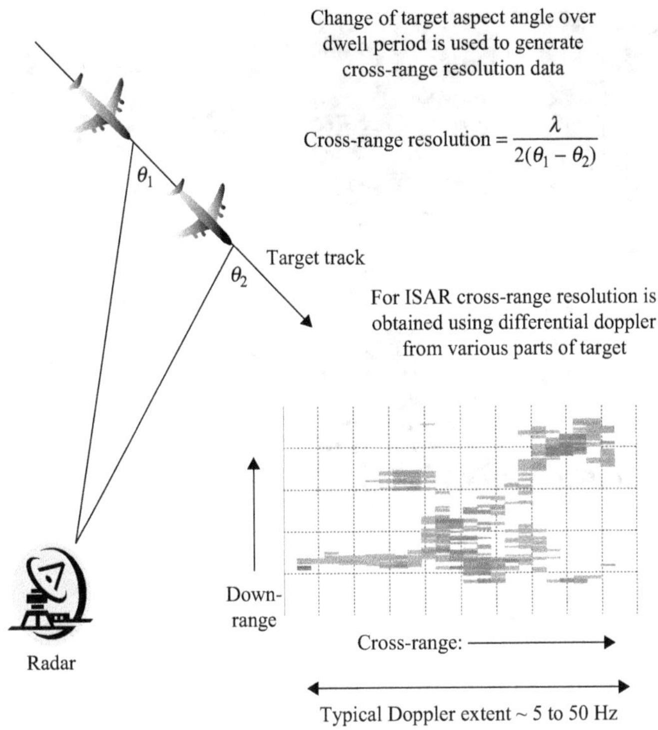

Figure 3.29 RDI is traditionally used to perform ISAR for aircraft targets

dwell RDI uses the high frequency Doppler returns and attempts to avoid ISAR effects by confining all the skin returns to very few Doppler cells.

These short dwell RDI methods are aimed to simultaneously provide high range resolution and modulation information, particularly location information, from the targets' propulsion systems such as jet engines or rotors. RDI discussed in this chapter tends to have dwell times of a few tens of milliseconds.

The short dwell RDI techniques have not been very widely explored and there have not been very many technical papers on the subject to date. RDI has not generally been considered to be a mainstream NCTR technique for aircraft recognition.

Within this chapter some examples of aircraft signatures are provided.

3.5.2 Helicopter signature

The RDI signature of a helicopter is shown in Fig. 3.31. The range and frequency axes are shown and the helicopter velocity has been motion compensated to 0 Hz. A sub-metre range resolution has been used and for a helicopter flying at 80 m/s over a dwell period of 30 ms, it travels a few metres. Hence, in order to gain the full benefit of the range resolution available, motion compensation for range gate walk over the period is required. It can be seen that the signature is reasonably well

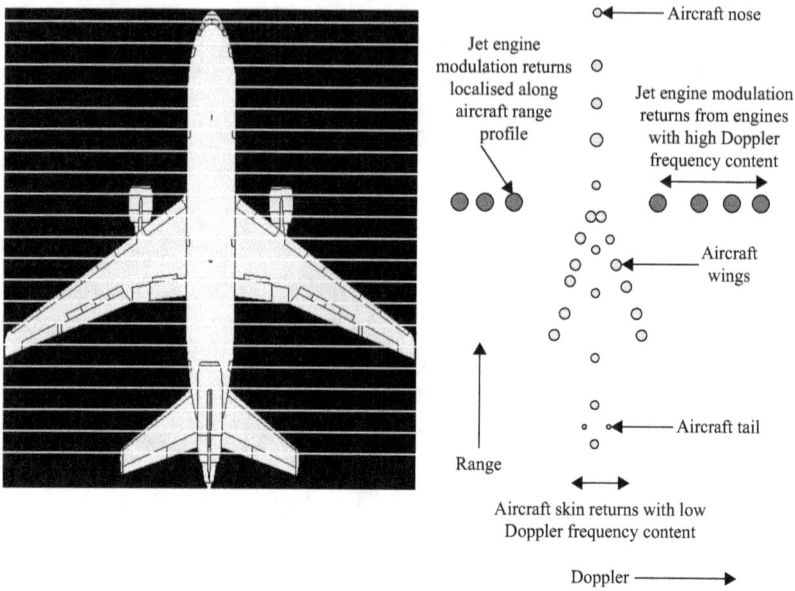

Figure 3.30 Frequency domain data theoretically available from jet aircraft range-Doppler image

focused, so the range gate walk motion compensation has been successfully implemented.

The main features that are visible are the skin return along the 0 Hz axis and the main rotor and tail rotor modulations. Over a coherent dwell period the main rotor blade flashes appear at about the same range cell as the main rotor hub. They are impulse-like in the time domain and are transformed into the frequency domain to cover the majority of the Doppler filters. In addition there are Doppler frequency returns in the same range cells due to the rotating hub. There are also hub modulation and blade flashes from the tail rotor.

Other Doppler returns can be seen in the signature, some of which are considered to be due to the JEM from the jet engines that power the helicopter.

3.5.3 Jet airliner signature

A signature of a civil airliner is shown in Fig. 3.32 with the two dimensions of range and frequency. The signature of this jet aircraft has been measured tail-on and is reasonably well focused and has been motion compensated such that the skin velocity is at 0 Hz. The locations of the JEM from the engines can be seen and actually cover several metres down-range. The distance from the first scatterer, which is in the vicinity of the tail to the nearest JEM return, can be determined, so it gives an approximate tail-to-nearest engine length. The nearest JEM returns detected at this aspect angle are from the last turbine stage. Although the blades are smaller than those of the compressor, they can be readily detected. The extent of the JEM lines in range provides information on the location of the engines along the range profile.

The nose of the aircraft is not detected at this aspect angle, so it is not possible to provide a length estimate of the aircraft.

3.5.4 Business jet signature

A signature of an approaching business jet is shown in Fig. 3.33 and the skin return has been motion compensated to 0 Hz. The return nearest to the radar comes from the nose region of the aircraft. The JEM lines can also be seen, which come from the two fuselage-mounted jet engines located near the tail. There is considerable JEM multi-path in the signature, so localising and counting the number of engines present is challenging. This is due to multiple reflections from the engines, wing and fuselage.

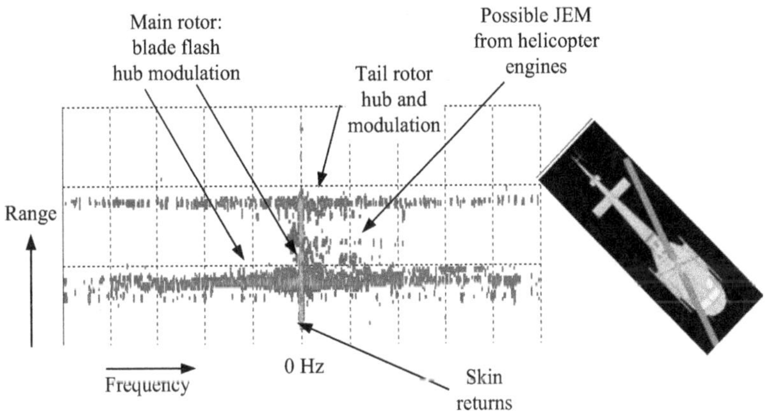

Figure 3.31 Range-Doppler image of a helicopter

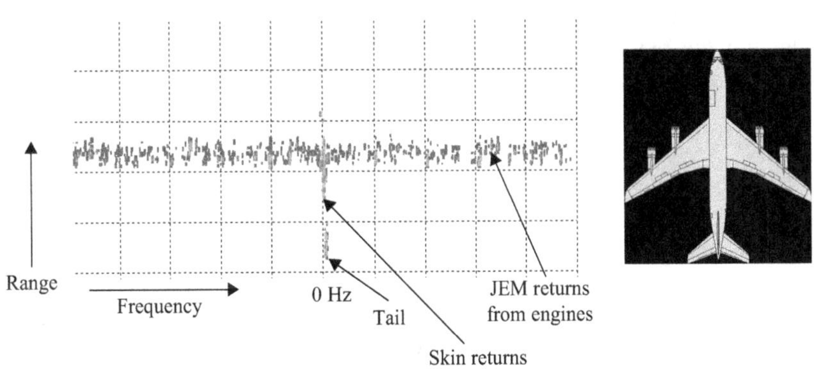

Figure 3.32 Range-Doppler image of a civil airliner

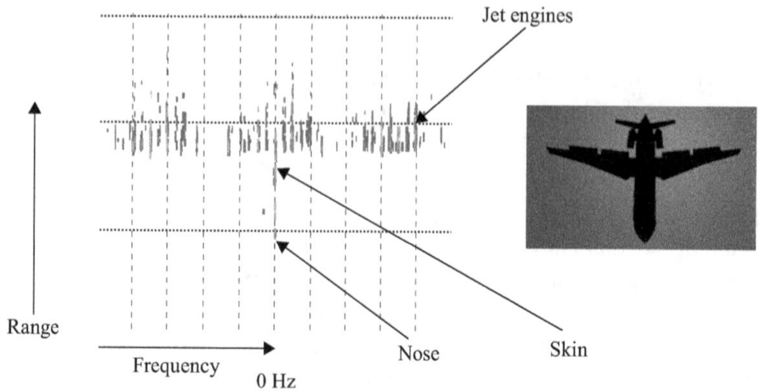

Figure 3.33 Range-Doppler signature of a business jet

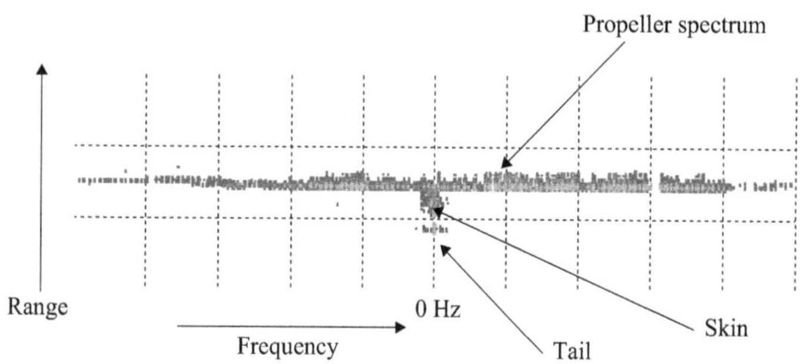

Figure 3.34 Range-Doppler signature of a propeller aircraft

3.5.5 *Propeller aircraft signature*

A signature of a receding propeller aircraft is shown in Fig. 3.34 and the skin return has been motion compensated to 0 Hz. The first scatterer detected is the tail. The modulation due to the single propeller at the front of the aircraft can be seen and the distance to the tail can be determined.

3.5.6 *Waveforms and system requirements for supporting RDI*

The waveform needed to support RDI has to be wideband to provide high range resolution and be of the coherent pulse-Doppler type to obtain high resolution frequency domain data. The waveform should be the same every pulse.

The waveforms that are recommended are repeated pulses, which could be wideband coded waveforms provided they are Doppler tolerant and have good time sidelobe characteristics. They could also be chirps or impulses. These wideband pulses provide high range resolution and their repetition over the radar integration period enables high frequency resolution to be achieved. The PRF and dwell period can also be designed to acquire unambiguous frequency data to support the frequency domain recognition techniques such as JEM.

Provided it meets the sensitivity and bandwidth requirements, any coherent pulse-Doppler radar should be able to support RDI functions. The key issue is the bandwidth, which many radar components cannot readily support without major upgrades. For example, 100 MHz is needed to provide 1.5 m range resolution, which is much a higher bandwidth than most radars utilise in a single pulse.

3.5.7 Conclusions

RDI is an aircraft target recognition technique, which has not been widely explored, but has potential for contributing to automatic air target recognition. The use of simultaneous range and frequency data has benefits in being able to localise aircraft propulsion systems along the range profile, which is not generally possible with other NCTR techniques. For jet aircraft it can theoretically acquire HRRP and JEM data in a single dwell with the appropriate waveform, but in practice it is not normally achievable due to difficulties in designing a radar with the appropriate characteristics. RDI can measure the signatures of jet and propeller aircraft and helicopters.

3.6 Aircraft target recognition conclusions

The main techniques used for the automatic non-cooperative recognition of air targets have been described. The recognition of air targets, as is the case with other targets, is dependent upon identifying target features, which can be extracted using appropriate radar designs and waveforms. In order to provide a fully automated target recognition process, it is also necessary to automatically extract the target features, develop a classifier and assemble a target signature or feature database.

The most popular recognition techniques are based upon the utilisation of the physical attributes of the aircraft, particularly with respect to the reflection of the radar signal from the propulsion systems, which can differentiate the various target types.

The JEM technique analyses the spectrum produced from the interaction of the electro magnetic wave, the radar signal, with the compressor stages at the front of the engine. The reflected signal contains embedded information on the numbers of blades on the compressor stages. The spectrum has to be effectively analysed and decoded to extract these values for recognising the aircraft engine. From the knowledge of which engines are fitted to which aircraft, the aircraft type can be determined.

Helicopter recognition is based primarily upon extracting the rate at which the main rotor blades reflect the incident radar signal. In particular the strongest return, or blade flash, occurs when a blade is perpendicular to the vector joining the radar to the target. More in-depth analysis of the reflected signal can determine whether the helicopter has an even or odd numbers of blades. These features can then be used to differentiate helicopters.

Range profiling and ISAR techniques can also be employed for automatic aircraft target recognition, but these are not specific to aircraft targets, as are JEM and helicopter blade flash detection techniques.

RDI may also be able to contribute to aircraft target recognition, but is limited by the ability of radars to provide high quality range profile and spectral information simultaneously.

Acknowledgements

The author would like to thank his colleagues in BAE Systems for providing support and material for this chapter and to the United Kingdom Ministry of Defence who supported some of the measurements described.

References

[1] Y. Zhou, X. Yu, M. Cui and X. Wang, 'Radar Target Recognition Based on Multiple Features Fusion with Dempster-Shafer Theory', *10th International Conference on Electronic Measurement & Instruments (ICEMI)*, Vol. 1, pp. 243–247, 2011.

[2] P. O. Molchanov, J. T. Astola, K. O. Egiazarian and A. V. Totsky, 'Target Classification by Using Pattern Features Extracted from Bispectrum-Based Radar Doppler Signatures', pp. 791–796.

[3] P. van Genderen and V. Kovalenko, 'Evaluation of Some Features for Extended Target Extraction in Polarimetric Radar', *Tyrrhenian International Workshop on Digital Communications – Enhanced Surveillance of Aircraft and Vehicles*, TIWDC/ESAV2008, pp. 1–6, 2008.

[4] A. W. Rihaczek and S. J. Herschowitz, *Theory and Practice of Radar Target Identification*, Artech House: Norwood MA, USA, 2000.

[5] P. Tait, *Introduction to Radar Target Recognition*, UK Institution of Engineering and Technology, 2006.

[6] L. Savy, 'Model-Based Classification of Aircraft Range Profiles Using Data Association Algorithms', *IEEE Radar Conference*, Boston, MA, 2007.

[7] J. Chadwick and G. L. Williams, 'Air Target Identification – Concept to Reality,' *IET International Radar Conference*, Edinburgh, UK, 2007.

[8] Q. Zhao, 'A New Near Field Radar Target and JEM Modelling and Implementation of Radar Target HWIL Simulation System', *IET International Radar Conference*, Guillin, China, 2009.

[9] J. Martin and B. Mulgrew, 'Analysis of the Effect of Blade Pitch on the Return Signal from Rotating Aircraft Blades', *IEE International Radar Conference Radar 92*, Edinburgh, 1992.

[10] M. R. Bell and R. A. Grubbs, 'JEM Modelling and Measurement for Radar Target Identification', *IEEE Trans. Aerosp. Electron. Syst.*, Vol. 29, No. 1, January 1993.

[11] S. Pardini, P. F. Pellegrini and P. Piccini, 'Use of Target Spectrum for Detection Enhancement and Identification', *AGARD Conference*, Ottawa, Canada, 1994.

[12] M. Burgos-Garcia, 'Radar Signature of a Helicopter Illuminated by a Long LFM Signal', *IEEE Trans. Aerosp. Electron. Syst.*, Vol. 45 July 2009.

[13] H. Essen, 'Non-Cooperative Classification of Helicopters Using Millimetre Wave Radar and ISAR Processing', *Digital Communications – Tyrrhenian International Workshop on Enhanced Surveillance of Aircraft and Vehicles*, TIWDC/ESAV, 2008.

[14] P. Pouliguen, L. Lucas, F. Muller, S. Quete and C. Terret, 'Calculation and Analysis of Electromagnetic Scattering by Helicopter Blades', *IEEE Trans. Antennas Propagat.*, Vol. 50, No. 10, October 2002.

[15] J. Misiurewick, K. Kulpa and Z. Czekala, *Analysis of Recorded Helicopter Echo, Radar 97* IEE Radar 97 International Conference, October 14th–16th 1997, Edinburgh, UK.

[16] S.- H. Yoon, B. Kim and Y. S. Kim, 'Helicopter Classification Using Time-Frequency Analysis', *Electron. Lett.*, Vol. 36, No. 22, 26 October 2000.

[17] C. Ozdemir, *Inverse Synthetic Aperture Radar Imaging with MATLAB Algorithms, Series in Microwave and Optical Engineering*, Wiley, Hoboken, NJ, USA, March 2012.

[18] G. Thomas, J. S. Son and B. Flores, *Range-Doppler Radar Imaging and Motion Compensation*, Artech House: Norwood MA, USA, 2001.

Chapter 4
Radar ATR of maritime targets

Hartmut Schimpf

4.1 Introduction

For commercial ship traffic the automatic identification system (AIS) is obligatory for all vessels above a certain size and thus warrants their cooperative classification and even identification. However, for smaller craft and especially for non-cooperative objects with hostile intent, the classification has to rely on classical approaches of automatic target recognition (ATR), which mostly are based on radar due to its day/night and all-weather capabilities. This leads to applications such as coastal surveillance for border control, the protection of harbour installations, ship self-defence or the suppression of drug trafficking, where the classification of ships by means of ATR schemes becomes more and more important. This is especially true in times of asymmetric (terrorist) threat and piracy.

With a modern high resolution radar one has the choice of two different ways of target imaging. The first is 2D imaging, either from an airborne (SAR) or from a ground-based platform (ISAR). The latter depends on the relative motion of the target itself and therefore may be difficult in the case of non-cooperative targets. The desired axis of rotation should be vertical, which may not be the case when high sea states cause strong roll and pitch motion for smaller ships. Moreover, when the hostile ship is approaching or receding on a straight course, there is no relative rotation that lends itself to ISAR exploitation.

Typical 2D approaches start with the imaging process [1], followed by a detection and segmentation stage [2]. From the target image proper, features are extracted for the subsequent target classification. Valuable information is obtained from polarimetry in connection with SAR imaging (PolSAR) [2–4], which is used for target/background separation in the detection stage, and for the characterisation of special scattering mechanisms that support classification. If available, even interferometry [5] can be added (PolInSAR). Bon et al. [6] reconstruct the 3D structure of dominant scatterers by means of Pauli decomposition and classify the ship via pattern matching against a reference database. Ship ATR based on ISAR imaging and features that are derived therefrom is described by Pastina and Spina [7,8] and by Musman [9].

While 2D imaging is the most promising approach in the case of airborne and spaceborne radar sensors with SAR capabilities, the second possible approach is 1D imaging obtained by means of high range resolution (HRR) profiles. This circumvents all the difficulties of determining an axis of rotation for ISAR imaging or the need of a moving sensor platform for SAR imaging. Therefore, its use is more realistic for ground-based radar. Also, it seems that HRR profiles are more robust against aspect angle variations than ISAR images as was found by Menon et al. [10]. The authors start their approach with ISAR images, but then sum all intensities within each range bin to end up with HRR profiles for further use.

The range resolution that determines the HRR profiles depends on the radar bandwidth. The shape of the HRR profile is a function of aspect angle, but also of range. The latter is due to an increasing clutter contribution in a real-beam situation, and to the influence of multipath at shallow depression angles. This will have severe consequences for the computation of features that are used for the classification of a ship. Important features that can be used for classification are those that use the geometry of ships like length and width. Also, the ATR algorithm can use features that describe the structure of dominant scattering centres on the ship, their relative locations and amplitudes as they appear in the HRR profile [11].

Close inspection of all possible scenarios leads to the conclusion that the problem of maritime ATR falls into two main classes. On one side, there are airborne or spaceborne sensor platforms that are moving and therefore are able to perform 2D imaging. They observe under steep depression angles and thus do not suffer from multipath. Consequently, they can generate high fidelity images that can be used for pattern matching with objects from a library, or for classical similarity tests in some feature space. Hence, essentially all the methods developed for land targets can be applied with the difference that the surrounding clutter is non-stationary and highly variable depending on wind conditions and the sea state. On the other side, there are stationary ground-based (including ship-based) radars that depend on real-beam cross-range resolution and therefore will evaluate HRR profiles. Due to the shallow depression angles and the presence of the marine boundary layer there will be multipath and atmospheric duct phenomena that cause strongly range-dependent signatures. The challenge consists in defining appropriate training and testing vectors because a reference to existing libraries is extremely difficult using local estimates. This latter topic, because it is unique to the maritime domain, will be the subject of the following discussion.

4.2 The use of high range resolution (HRR) profiles for ATR

HRR profiles are naturally applicable to the situation of a hostile ship that is approaching directly. In that case, there is no change of aspect angle that would support the creation of an ISAR image. Rather, the full HRR profile of the ship is seen without any projection effects. However, several problems make the exploitation of HRR profiles rather difficult:

- Shadowing (self-masking) effects due to a very low depression angle δ. This is especially the case in a maritime scenario where δ is usually close to 0°. The highest parts on a ship's body are mostly close to its centre so that either the

bow or the stern region may be occluded. Consequently, this problem is based on the shape of the target and on the aspect angle and cannot be influenced by an appropriate choice of radar parameters.
- Low signal-to-noise ratio due to a very large slant range.
- Low signal-to-clutter ratio due to the large angular extent of the footprint in a real-beam situation: while the scatterers on the ship keep their radar cross-section (RCS), the RCS of the clutter cell increases proportional to the distance R.
- Multiple target situations due to the large angular extent of the footprint in a real-beam situation: the more distant the ship, the more likely it becomes that a second ship at a similar range may enter the beam as well. For example, a 2° beam has a substantial cross-range extent of 350 m at a modest 10 km range. This can be overcome only by constant monitoring of the surveyed area.
- The occurrence of spray near the bow and wakes near the stern of the ship, which will contribute to its signature.

As was seen in Chapter 3 the computation of ATR features from HRR profiles is based on a 'minimum bounding range interval' (MBRI) that is defined by the range extent of a certain number N_{st} of dominant scatterers. The question is how to determine these. As can be seen from Fig. 4.1, in the case of HRR (the profile shown was taken at 17 GHz with 300 MHz bandwidth corresponding to 0.5 m range resolution) most dominant scatterers create peaks of a certain width, which may well be more than one range cell. Thus it might happen that 'the second strongest RCS' corresponds to the flank of the peak with the strongest RCS and physically would not represent a scatterer of its own. Therefore, the original HRR

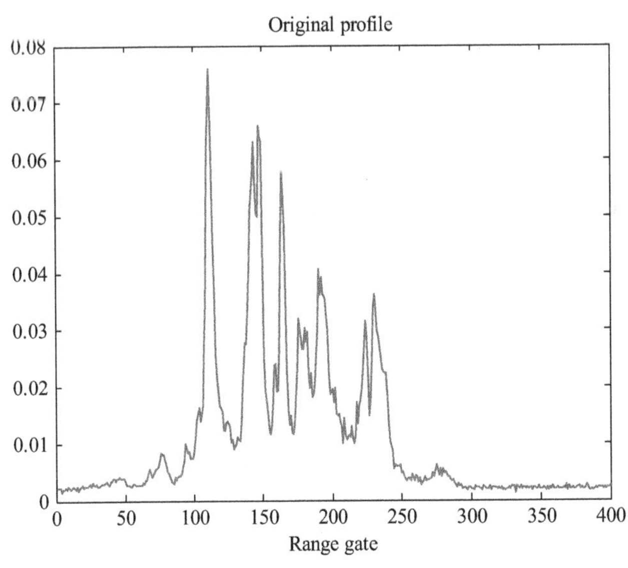

Figure 4.1 Typical HRR profile measured at 17 GHz (RCS/m^2), 0.5 m resolution

profiles have to be transformed into 'ensembles of peaks' based on the following conditions:

- A 'peak' must have at least two values of lower RCS on both sides.
- The peak RCS must exceed a threshold that is given by the average sea clutter level.

Fig. 4.2 shows which of the HRR lines are recognised as individual peaks (marked by asterisks). This leads to an equivalent 'peak profile' (Fig. 4.3) from which the ATR feastures are calculated.

4.3 The derivation of ATR features from HRR profiles

4.3.1 Length estimate

Out of all recognition features based on scatterer geometry that one might think of, 'length' seems to be the simplest and most straightforward. However, the estimation of the length of an extended radar target – be it a land vehicle, an aircraft, or a ship – is a non-trivial task [12].

Several problems like shadowing, and low signal-to-noise and signal-to-clutter ratios make the exploitation of HRR profiles rather difficult as described in Chapter 2. Information about the target can only come from scatterers that contribute to the backscatter signal. The latter is not the case for those parts of the target that are masked by others, and it is not the case for scatterers that although physically belonging to the target and illuminated by the radar cause specular scattering away from the radar antenna and hence are invisible. Even estimating the

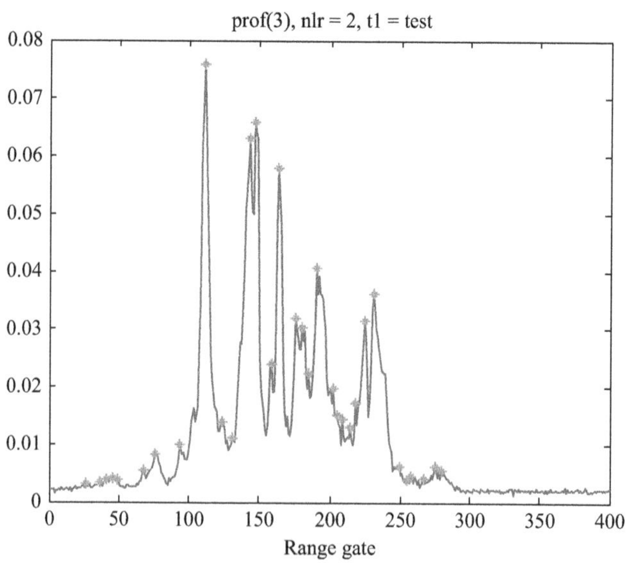

Figure 4.2 Same profile as in Fig. 4.3 with peaks marked by asterisks

length of a ship will be subject to error when characteristic scatterers near the bow or the stern fall below a certain detection threshold due to multipath (Figs. 4.4 and 4.11).

Based on HRR profiles, 'length' can be defined in several ways [12]. Each one has its advantages and disadvantages, and the results normally are not identical. In the case of ships, two different methods were applied [13]. The first one (L_a) equals the extent of the smallest window containing $x < 100\%$ of the total energy of the profile and uses an adaptive threshold that is based on the mean and standard deviation of the sea clutter RCS. This takes into account that the sea clutter RCS increases proportional to R due to the real beam situation. The mean m_{clut} and standard deviation s_{clut} are determined from the n_{clut} lowest values of the sorted HRR profile, N_{Thr} is a 'CFAR-factor', and then the threshold is defined as

$$Thr = m_{clut} + N_{Thr} \times s_{clut}$$

Good results were obtained with $n_{clut} = 140$ and $N_{Thr} = 10$ in profiles of length $N_{HRR} = 200$, and with $\Delta r = 1.5$ m, i.e. a maximum range extent of 90 m of a ship could be accommodated. The 'Planet', a research vessel used for several measurement campaigns, had a length of 73 m corresponding to 49 range cells.

The second method (L_b) is quite different. One starts with a window containing a small number of the N_{st} strongest peaks of the HRR profile (usually $N_{st} = 3$ or 4). Step by step one increases the window size by one cell by incorporating the next adjacent cell (before or behind the present window) on the side with the higher RCS estimate. This estimate is based on a power sum of three range cells in order to

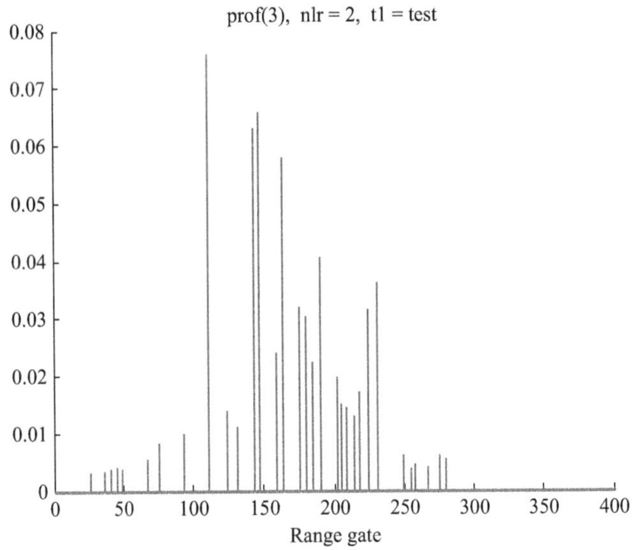

Figure 4.3 Profile from Fig. 4.2 transformed into peak profile as used for feature computation

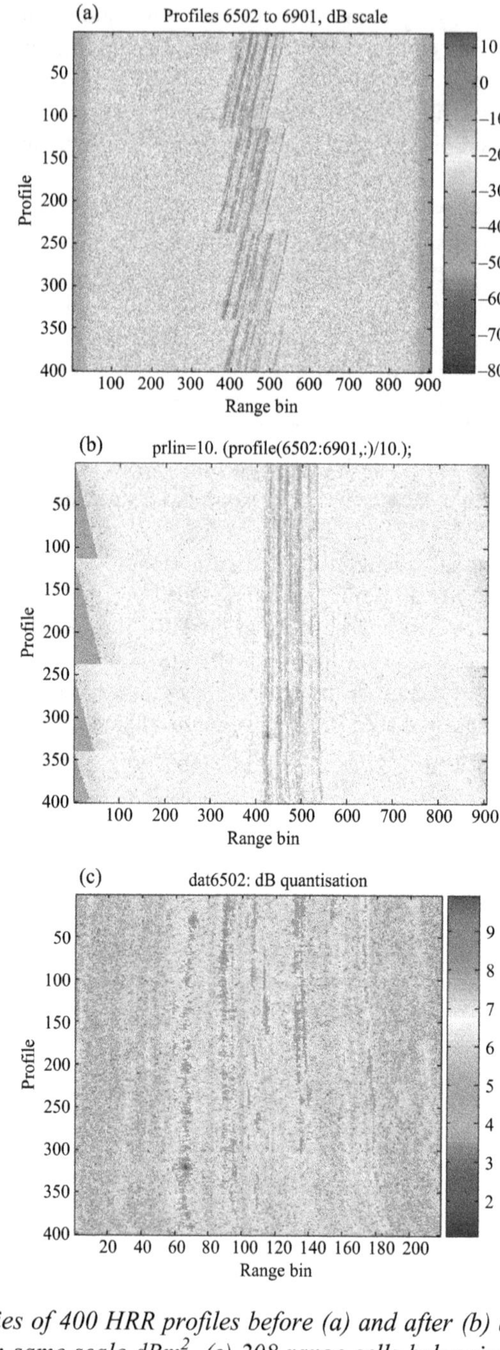

Figure 4.4 *Series of 400 HRR profiles before (a) and after (b) alignment, both with same scale dBm². (c) 208 range cells belonging to the ship, quantised into ten levels*

avoid extreme local minima. At each step the relative power increase $\Delta_i = p_i/\sum p_i$ is determined. The window size has reached the final 'length' L_b as soon as Δ_i stays permanently below a certain threshold. A value of 0.01 was found to be effective.

4.3.2 Position specific matrices (PSMs)

The concept of PSMs was developed for the description of gene sequences in DNA analysis (cf. Cotuk et al. [11]). This technique does not derive single features from the HRR profile but rather treats HRR profiles as an entity. In the same way as a DNA molecule can be represented by a sequence of letters, where each position in the sequence is equal to one of four possible letters, an HRR profile can be represented by a sequence of RCS amplitude values. Working with PSMs comprises the following key steps:

- determination of the length N_p of the ship profile,
- alignment of HRR profiles,
- quantisation into N_q levels,
- creation of reference PSMs (one per class) from a series of N_t profiles,
- comparison of the quantised test profile to the reference PSMs,
- determination of a figure of merit (FoM) to quantify the degree of similarity,
- *classification*: the test profile is assigned to the class that obtained the highest FoM.

4.3.2.1 Determination of length

In a long HRR profile, most of the range cells contain only clutter. The cells that belong to the target ship have to be identified, because only these are important for the subsequent processing.

4.3.2.2 Alignment

HRR profiles in a time sequence can only be compared to each other on a range cell by range cell basis which requires careful alignment. Also, creation of the PSM depends crucially on correct alignment.

4.3.2.3 Quantisation

In the case of HRR profiles, there are not only four possible values as in the DNA case, but rather a number of N_q levels that are created artificially by quantising the more or less continuous range of RCS values. Cotuk chose a value of $N_q = 30$ without giving a rationale or claiming this to be an optimised value. Applying this method to ships [14], it was found that the choice of N_q depends on whether the RCS data are used in linear or in dB space. In the first case, due to the high dynamic range, N_q has to be quite large (several hundred or even 1000) in order to accommodate single dominant scatterers. Therefore, it is advantageous to work in dB space. In that case, a value of $N_q = 10$ is sufficient as can be seen by comparing Fig. 4.4(b) (full dynamic range) and (c) (quantisation to 10 levels). The whole structure of the HRR profiles necessary for classification has been preserved. A small value of N_q has the advantage of needing only limited computing power by handling PSMs of moderate size.

4.3.2.4 Creation of reference PSMs

After having determined the length N_p of the HRR profiles, and after having performed the alignment, a PSM with N_q rows and N_p columns is created. For this purpose one takes a training set of N_t profiles and inserts the frequency of occurrence of each quantised RCS level into the PSM. Stated otherwise, when RCS level m (out of N_q possible values) occurs in range cell n (out of N_p possible values) then the counter in element PSM(m, n) is increased by 1. This provides us with a reference PSM in which the sum of all entries within each column is equal to N_t. This means that the highest value of any entry is less than or equal to N_t. This maximum value N_t is reached only when all RCS values of the respective range cell have the same quantisation value.

The example shown in Fig. 4.5 was constructed on the basis of a series of 400 consecutive HRR profiles of length $N_p = 208$. They were subdivided into groups of 20 profiles each so that 20 different PSMs could be created with the following parameters: $N_q = 10$, $N_t = 20$ and $N_p = 208$. In Fig. 4.5(a) and (b) we see the first and second PSM, respectively. Fig. 4.5(a) shows an average of all 20 PSMs. There is a high degree of similarity, but none are identical. Also, the fact that only few range cells out of 208 reach a count close to $N_t = 20$ (the possible maximum) gives a hint that the HRR profiles show considerable variability.

4.3.2.5 Compare the quantised test profile to the reference PSMs

How can these PSMs be used for classification? For each target class we need one reference PSM. The target under test is represented by its HRR profile (out of a series of profiles), which is quantised in the same way as the training profiles. This quantised HRR test profile (Fig. 4.6(a)) corresponds to a test PSM that contains only ones in the respective positions (Fig. 4.6(b)) and that has now to be compared to the reference PSM of each target class. This comparison is done cell by cell for every matrix element. A high score in PSM_{ref} at a position given by PSM_{test} means high similarity, and a low or zero score means dissimilarity.

4.3.2.6 Determine a figure of merit

In order to quantify the degree of similarity, some 'likelihood measure' or 'figure of merit' (FoM) has to be defined. Cotuk et al. use a 'likelihood score' (LS) by identifying a single row for each column of the reference PSM for the hypothesised class. This row corresponds to the quantised amplitude level in the observed profile at that spatial location as indicated by the test PSM. They then take the product of all these matrix entries because they argue that the entries in a PSM are proportional to the estimated probability of observing a particular RCS level at that position. However, a product is 0 as soon as any single one of its factors is 0. Therefore, the likelihood score is 0 as soon as the test PSM hits an empty entry anywhere in the reference PSM although all the remaining positions may be in perfect agreement. For this reason the LS as defined by Cotuk is not well suited for comparing different degrees of similarity, especially for rather long HRR profiles (in our example $N_p = 208$).

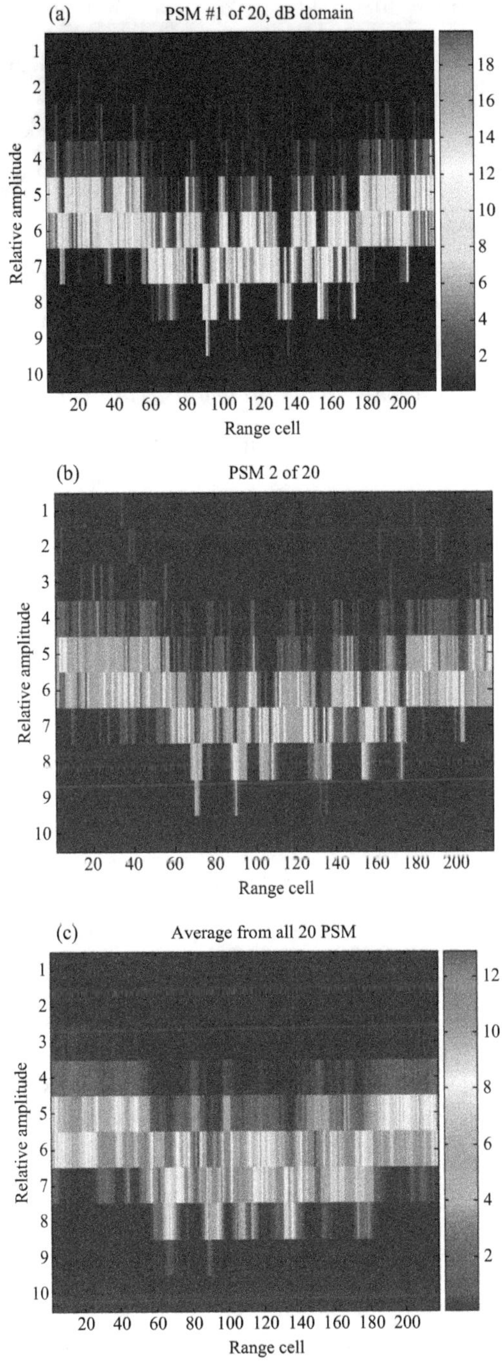

Figure 4.5 PSMs 1 (a) and 2 (b), scale from 0 to 20. The averaged PSM (c) has a scale from 0 to 13, and hence has considerably lower values

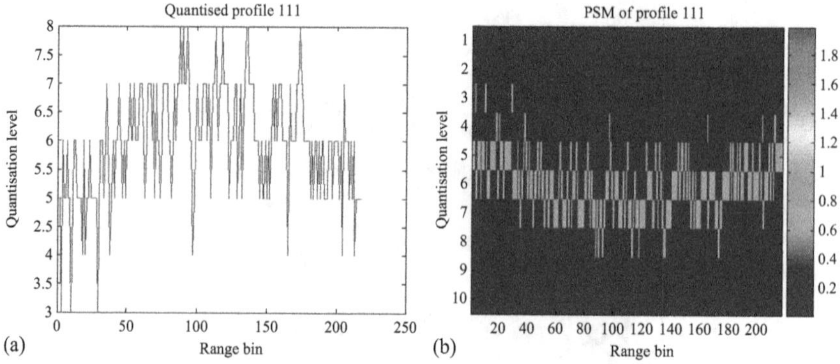

Figure 4.6 Quantised HRR profile (a) and its representation by a PSM (b)

We, therefore, propose an FoM to quantify the degree of similarity that is given by

$$FoM = \sum_{i=1}^{N_q} \sum_{j=1}^{N_p} PSM_{test}(i,j) \cdot PSM_{ref}(i,j)$$

which is the element-wise product of the test PSM (PSM_{test}) and the reference PSM (PSM_{ref}). The highest possible value for FoM is $N_t \cdot N_p$; therefore, a normalised FoM may be defined as $FoM/(N_t \cdot N_p)$.

4.3.2.7 Classification

If reference PSMs are available for several different classes, then the profile under test is assigned to the class with the highest FoM.

4.3.3 Other examples of ATR features

Based on geometrical relations and amplitude ratios of dominant scatterers in the MBRI one can define ATR features of different types, mainly geometrical, statistical or structural. The following set of features [15] that will be analysed in more detail in section 4.5.3 are not meant to be 'best' features, but rather 'generic' features that may be typical for any HRR profile-based ATR scheme. They are all based on HRR profiles of the ships, measured with a single polarisation (*H* on transmit and receive). The main criteria for feature construction are

- Features must not depend on absolute RCS, i.e. no radiometric calibration is required.
- Features should cover different properties of the HRR profile, e.g. geometry, statistics and structure of the ensemble of dominant scatterers.

The features that are used in the present work are Ft1 to Ft14 as described below. Some of them were already assessed in former analysis [13,16]:

- Ft1 = range extent of N_{st} strongest scatterers in the HRR profile. This range extent at the same time defines the MBRI on which all other features are computed.

- Ft2 = mean/std. dev. of RCS values within MBRI.
- Ft3 = power sum of N_{pwr} strongest scatterers/power sum of all scatterers within MBRI; it describes the contribution of a few dominant scatterers to the MBRI.
- Ft4 = log 10(strongest RCS/N_{pos} – strongest RCS) describing the 'decay' of RCS values.
- Ft5 = log 10(highest RCS/lowest RCS) out of N_{st} strongest scatterers.
- Ft6 = relative distance between the two strongest scatterers with respect to range extent of MBRI.
- Ft7 = slope of straight line fit of RCS (in dB) of *n*th strongest scatterer (normalised to highest RCS) versus range extent of *n* strongest scatterers ($n = 2, \ldots, N_{st}$).
- Ft8 = shift of same fit as in Ft7.
- Ft9, Ft10 = slope and shift of straight line fit to normalised RCS of N_{st} strongest scatterers (in dB) in descending order.
- Ft11 to Ft14 = coefficients of cubical fit to normalised power sum of *n* strongest scatterers ($n = 1, \ldots, N_{st}$).

For the present analysis (section 4.5.3) the following parameters were chosen: $N_{st} = 10$, $N_{pwr} = 5$ and $N_{pos} = 5$. Ft1 and Ft6 are of type geometrical, Ft2 statistical, all others of type structural, describing the relative RCS values and RCS ratios of dominant scatterers.

4.3.4 Choosing sets of uncorrelated features

Of course, not all of the above defined features are independent of each other. In order to find out which ones may be combined to sets of features the easiest way is to look at their cross correlation coefficients ρ. A typical example for $abs(\rho) \in [0\ 1]$ is shown in Fig. 4.7 based on a simulated ship with fixed orientation, but at many

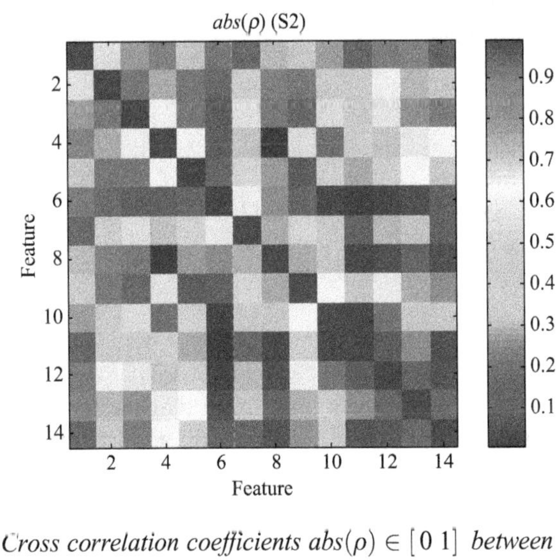

Figure 4.7 Cross correlation coefficients $abs(\rho) \in [0\ 1]$ *between all 14 features*

different range positions. From this one can conclude that good sets of uncorrelated features are 1, 4, 8, 14; or 2, 6, 14; or 7, 11, 14 among many other possibilities.

4.4 Ship ATR under the influence of multipath

4.4.1 What is multipath?

The task of detecting and classifying a ship very often has to be performed by sensors that are located close to the sea surface. This can either be onboard a ship or in a stationary location on the coast. Due to the shallow look angles and to the special conditions of the sea surface and of the atmosphere immediately above the sea surface (marine boundary layer), the influence of multipath and of atmospheric duct phenomena plays an important role when measuring a ship's radar signature. This is because scatterers are affected in a different way depending on their height above the sea surface. As a consequence, the amplitude relations between different scatterers at different heights on a ship depend strongly on the distance between the radar and the ship that has to be classified. This means that the HRR profiles and ISAR images that are the source of ATR features are not only a function of target aspect but also a function of the distance to the target under distinct atmospheric conditions and sea states [17]. In the ideal case of mirror-like reflection and without taking into account atmospheric effects, multipath is described by means of the path difference Δ between direct and reflected rays.

$$\Delta = \frac{2Hh}{s} \cdot \frac{1 + ((H^2 + h^2)/2R) \cdot (s^2/(2Hh(H+h)))}{1 - (2Hh/s^2)} \tag{4.1}$$

where:

H = antenna height
h = scatterer height
R = Earth's radius
s = slant range between antenna and scatterer

This takes into account only the monostatic contributions. There also exist two bistatic paths [17] antenna → sea surface → target → antenna and vice versa. These are of lesser intensity and are omitted here for simplicity.

The interference between the direct and the reflected rays leads to the complex voltage

$$V = A \cdot e^{i(kx-\omega t)}(1 + f \cdot e^{i(4\pi/\lambda)\Delta}) \tag{4.2}$$

where $f \in [0 \quad 1]$ takes into account the surface roughness via a reduced reflectivity. This is shown in Fig. 4.8 for $H = 19$ m and a radar frequency of 35 GHz ($\lambda = 8.6$ mm). If atmospheric effects and duct phenomena in the marine boundary layer are included, a closed analytical solution is no longer available. Rather, one has to depend on modelling software such as TERPEM [19], which yields results as

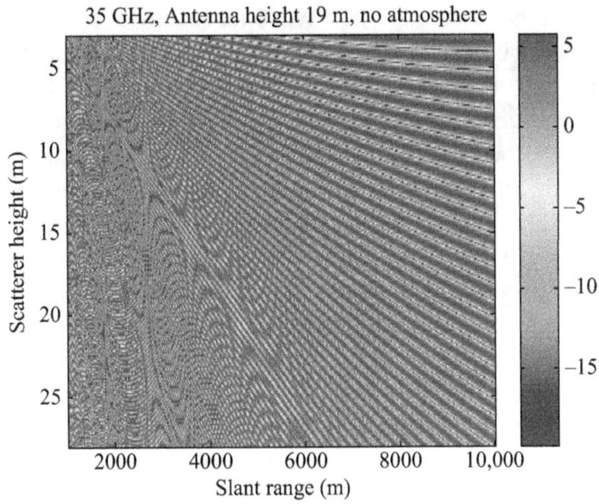

Figure 4.8 35 GHz propagation factors (dB) as given by (4.1) and (4.2), for pure multipath, antenna height 19 m and scatterer height versus distance

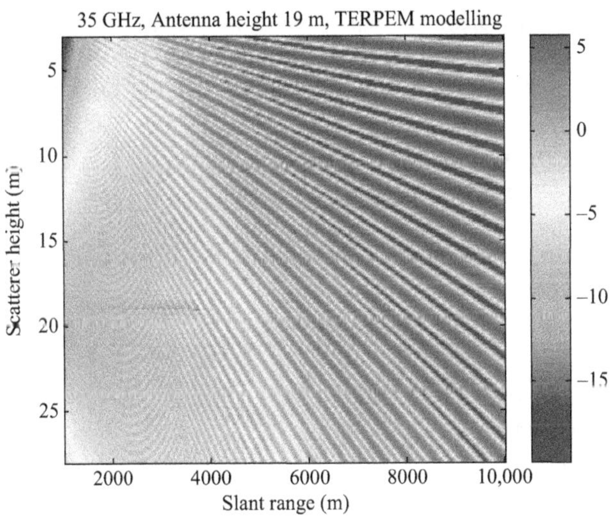

Figure 4.9 35 GHz propagation factors (dB) including atmospheric effects (TERPEM modelling), antenna height 19 m, scatterer height versus distance

shown in Fig. 4.9. The overall appearance, of course, is similar to the 'undisturbed' case, but the behaviour in detail may be completely different.

Reality as it manifests itself in measured HRR ship profiles is even more complicated. If we imagine the sea surface as a multitude of non-stationary reflecting facets with different inclinations then even the concept of one reflected ray interfering with the direct ray fails. Wave facets closer to the ship but inclined towards the radar as well as facets closer to the radar but inclined towards the ship may as well act as the origin of additional reflected paths. Also, facets with varying inclination may cause scatterers to show up that were not visible before: parts of the ship that have the same angle with respect to vertical as the wave facet has with respect to horizontal may lead to double bounce backscatter. These scatterers hence only are visible for certain sea states and wave conditions. One more effect should be mentioned. Depending on the sea state the ship will perform pitch and roll movements as well as being lifted or lowered as a whole. In that case the concept of a scatterer having a fixed height above the sea surface is only an approximation.

All these effects are much too complicated to be treated in this chapter. Here, we will limit ourselves to the simple case of a single reflecting point per scatterer and to scatterers at fixed heights. Some of the analysis will be done using a set of simulated ships in order to show the influence of multipath under controlled conditions. Sets of real measurements will also be used to demonstrate the additional difficulties that occur under real-life conditions. Figs. 4.10 and 4.11 show examples from real measurements and demonstrate how the RCS of scatterers at different heights and a complete HRR profile, respectively, vary as a function of range.

The multipath problem has not gained much attention because most ATR approaches are designed for airborne or spaceborne sensors that operate at steeper look-down angles and therefore do not suffer from multipath effects.

4.4.2 The problem of defining testing and training vectors

For classical land targets that have a stable signature it is straightforward to define references that are established by means of some training procedure and subsequently stored in a library. These references may be HRR profiles or ISAR images

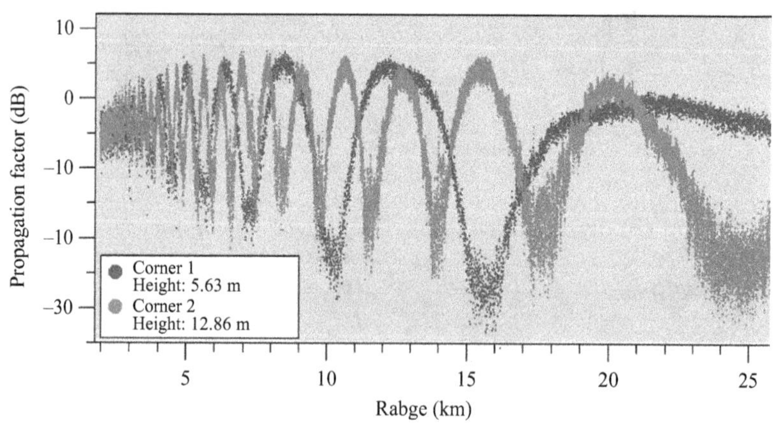

Figure 4.10 Multipath effect on two trihedrals at different heights above the sea surface

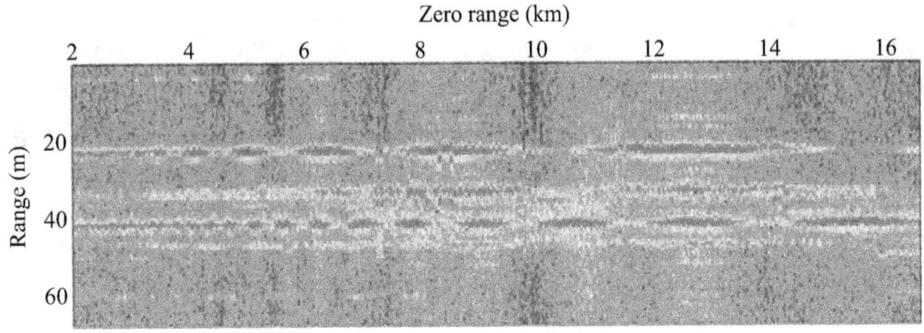

Figure 4.11 HRR profile of a ship as a function of the slant range between the antenna and the ship, from 2 km to 16 km, HH polarisation

for pattern matching, or some vector in feature space. Usually these references depend on the aspect angle and have to be defined for all azimuth angles from 0° to 360°, and desirably also for elevation angles out of an interval that is prescribed by the mission parameters. In the case of ships that are subject to multipath this approach will fail because even for a constant aspect angle the signatures will be a function of slant range. One might think to retain the azimuth dependence and replace the elevation dependence (because the depression angles are always close to 0°) by a range dependence thus avoiding an additional dimension to the 'data dome'. The basis for the range dependence could be the geometrical multipath equation as shown in (4.1) and (4.2). However, this equation is applicable only for the ideal case of a mirror-like sea surface without any atmospheric influence. In reality, there exists no closed expression by which to include all effects of the sea surface and the conditions of the marine boundary layer. This is only feasible by means of specialised software such as TERPEM [19] that requires exact knowledge of important input meteorological parameters like temperature and humidity profiles along the propagation path. These will vary from one mission to another. Moreover, they will not be accessible without the use of meteorological buoys. From this one sees that a reference library that includes multipath and atmospheric effects is not feasible and not realistic.

The only way out is to use 'true profile' (TP) references that pertain to the undisturbed ship signatures that are not influenced by multipath effects. But then immediately the problem arises that any test vectors that are determined locally will have no resemblance to these TP references and cannot be used together with them to calculate, e.g. Euclidian distances in feature space. The task, therefore, will be to construct test vectors that somehow are an approximation of TP test vectors in order to be comparable to TP reference vectors. In section 4.5.3 several approaches will be presented that try to achieve this. It will be found, however, that local approaches (i.e. combining groups of consecutive HRR profiles) are not sufficient to provide stable TP estimates. In Chapter 6 some possibilities will be discussed as to how to overcome this systematic shortage by means of rather unusual measures like multi-frequency or multi antenna systems.

4.5 Results

4.5.1 Length estimate

4.5.1.1 Results for L_a and L_b based on measurements of ship HRR profiles

Based on fully polarimetric data that were measured with the MEMPHIS radar at 35 GHz [20], several combinations of n_{clut} and N_{Thr} were analysed [13], the one using $n_{clut} = 140$ with $N_{Thr} = 10$ gave the best results. Fig. 4.12 shows the time behaviour of the length estimate for the first 1000 HRR profiles (distance to the ship 1.19–1.44 km). The mean value is 50.7 range cells corresponding to 76 m, with a standard deviation of 12.6 range cells (18.9 m). This large value reflects the extreme variability of the length estimate (LE). A similar behaviour is found for L_b (Fig. 4.13). Here again, looking at a large number of profiles yields estimates that show a very strong variability with a mean value of 42.1 cells corresponding to 63.2 m (standard deviation = 25 m) as can be seen from Fig. 4.4 for the first 10,000 profiles (distance to the ship from 1.19 km to 4.17 km).

It is interesting to examine the differences between the results. From Table 4.1 one can see that for all polarimetric combinations (except the cross-pol case) L_b yields systematically smaller values than L_a. As for X-pol the *T/C* is smaller than for the 'strong' parallel channels, it seems that L_b suffers more from the clutter contribution in the HRR cells. Also, the standard deviations of the LEs are smaller for L_a than for L_b in all cases. From this one can infer that L_a is the more robust approach with values that are closer to the true ones than L_b.

The cross correlation coefficient between L_a and L_b is close to 0 no matter which polarisation channel or combination of channels is used. Of course, the LE values are

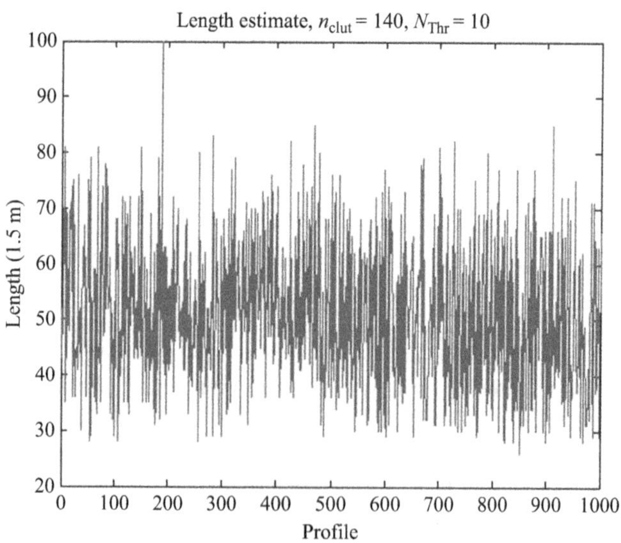

Figure 4.12 Length estimate L_a with distance to ship from 1.2 km to 1.4 km

Table 4.1 Length estimates (metres)

	L_a Mean	L_b Std. dev.	Mean	Std. dev.
HH	79.5	16.4	62.9	25.1
VV	76.5	17.0	60.5	23.1
X	64.3	33.1	83.8	36.5
HH + VV	67.7	13.2	57.0	19.1
HH + X	73.2	16.1	59.4	21.2
VV + X	72.4	15.6	58.2	20.2
HH + VV + X	66.4	**12.7**	56.4	**18.4**

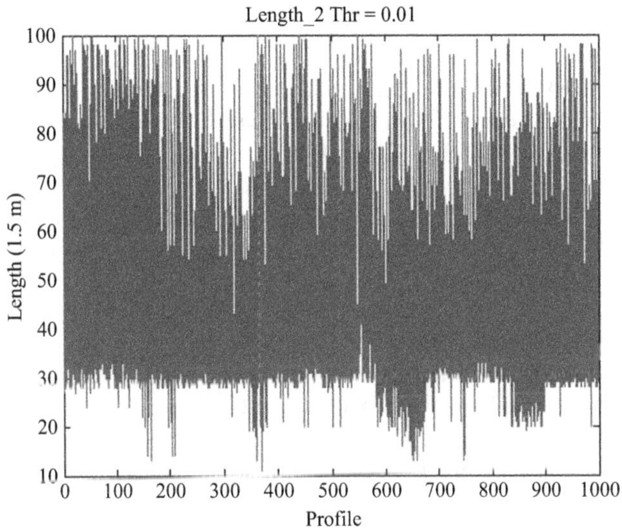

Figure 4.13 Temporal behaviour of L_b

of the same order of magnitude, but their variations are obviously not correlated, which may be due to the special character of the sea clutter, especially its spikiness.

The role of polarisation

After applying a careful polarimetric calibration based on so-called 'distortion matrices', three independent polarimetric channels are available, namely HH, VV, and the cross-pol channels X = HV = VH. These can be used independently or combined incoherently by summing up the power values. The length estimates (in metres) and especially their standard deviations (cf. Table 4.1) depend strongly on which channel or combination of channels is chosen. As one sees from the standard deviations, the combination of all three polarisation channels provides the smallest values, i.e. the most stable length estimates (marked by bold type). If only two channels are available then the combination of the two 'strong' channels HH + VV is recommended. As to the correctness of the LE, HH + X yields the best results, whereas the combination of three channels, although being the most robust one, tends to underestimate the ship's length by about 7 m.

Integration of HRR profiles

It has been demonstrated that the two different approaches to estimate a ship's length show a high variability, i.e. rather low robustness. Multipath is only one reason for this, another one being the spiky nature of sea clutter. In addition to polarimetry a second possibility to improve the LE robustness consists in integrating several consecutive HRR profiles by means of incoherently adding their powers. Based on the radar PRF and the ship's outbound velocity an integration of three consecutive profiles is possible when the shift is limited to less than about half a range bin [13]. Table 4.2 gives the results of mean LE and pertinent standard deviations for L_a (parameters $n_{clut} = 140$, $N_{Thr} = 10$ as before) and for L_b with three different thresholds (2%, 1% and 0.5%).

In the case of L_a the LE is slightly less robust than combining two or three polarimetric channels, but it is clearly more robust than the single channel estimate without integration. L_b on the other hand is obviously too small for threshold values of 1% or 2%. It seems as if the three corner reflectors with their separation of 40 m that had been mounted on the ship dominate the estimate. For a threshold of 0.5% L_b reaches values around 57 m as before (but still too small). However, in that case also the standard deviation is around 20 m as before, so the integration cannot be considered successful.

4.5.1.2 Simulation of ship HRR profiles

In addition to the measurements, some simulations were made using a 'generic' ship that is represented by a dozen main scatterers located at different heights and positions along the length profile of the ship. The cross-sections of the scatterers were determined at random between 100 m^2 and 1000 m^2, which are common values for a ship. This layout of scatterers remained fixed for the complete run during which the ship moved away from the radar from a starting distance of 1 km to a final distance of 35 km with a step size of 10 m from one profile to the next based on the TERPEM grid. The effective RCS values of all 12 scatterers varied due to multipath as a function of their height above the water and their distance to the radar (Fig. 4.14) as described by the propagation factors that are provided by TERPEM.

Fig. 4.14(a) shows all profiles between 1 km and 35 km distance. They were rectified with respect to the first profile in order to show the range-dependent RCS

Table 4.2 Mean LE and standard deviation for L_a and for L_b with three different thresholds (2%, 1% and 0.5%)

	Mean (LE) (m)	Std. dev. (LE) (m)
L_a	65.1	14.2
L_b (2%)	40.0	11.2
L_b (1%)	46.2	13.5
L_b (0.5%)	57.1	19.7

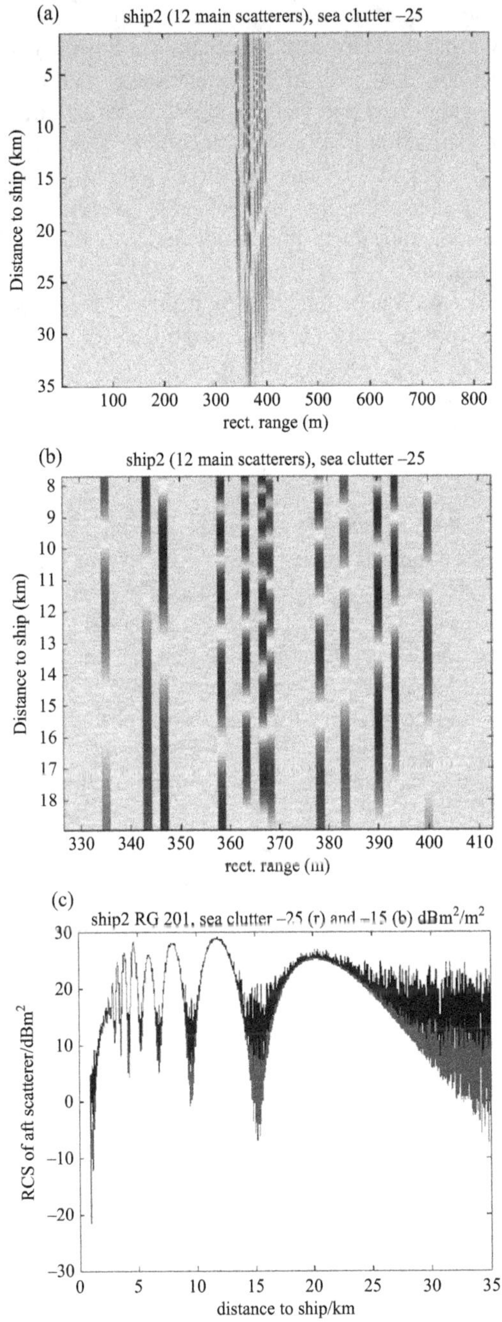

Figure 4.14 Simulated ship HRR profiles modified by multipath

variation due to multipath for each individual scatterer. The expanded version (Fig. 4.14(b)) shows that the stern scatterer is almost invisible between 14.7 km and 16 km distance; the bow scatterer suffers extinction as well between 16.3 km and 17.3 km distance. As the front and rear scatterers, in particular, make an important contribution to the estimation of the ship's length, it is expected that the LE will shrink between 14.7 km and 17.3 km. Fig. 4.14(c) shows as an example how the effective RCS of the stern scatterer behaves as a function of distance. Near the minima the scatterer disappears in the clutter and will not be detectable by any thresholding operation.

The profiles so created were analysed by means of the same algorithms as the measured data. The length estimate L_a (with parameters $n_{clut} = 400$ and $N_{Thr} = 20$ in profiles of length 500 cells) is shown in Fig. 4.15 for a sea clutter with $\sigma^0 = -25$ dBm^2/m^2. Due to the benign conditions of the simulation (e.g. no masking effects due to wave height) the results look quite good. As expected the LE is too small in cases where the rear or front scatterers are not discernible like at distances 14–18 km or beyond 25 km.

Fig. 4.15(b) shows an example of L_b for $N_{st} = 4$ and a threshold of 0.02. The large number of outliers with a tendency to overestimate the LE may be due to sea clutter effects. However, applying a moderate median filter of length 7 (Fig. 4.15(c)) leads to an efficient decrease of outliers. This shows that one individual estimate may not be very reliable using this approach. An appropriate filter, however, leads to stable values. The median of all 3000 LE values is 39 cells where the correct value would be 40. One also sees very well the expected LE reduction around 15 km and beyond 25 km, both due to multipath.

Obviously, the effect of LE fluctuations can be mitigated either by temporal integration or by making use of polarimetry. The latter yields more realistic estimates and provides more samples during each time-on-target. All approaches have in common that the length estimate usually is smaller than the physical/geometric range extent of the target due to self-masking effects at low depression angles.

4.5.2 PSM results

The PSM scheme described in section 4.3.2 was applied to the data of the approaching ship consisting of 400 consecutive HRR profiles of length $N_p = 208$. As we have only data from a single type of ship, i.e. one target class, the PSMs cannot be used for classification at the time being. The goal is rather to study the influence of multipath on the expected classification performance. This is done by testing each of the 400 HRR profiles against some reference PSMs of that same ship. As mentioned earlier, 20 PSMs are constructed at 20 different positions during the ship's approach by subdividing the series of 400 profiles into 20 packages of 20 profiles each. In addition, an average PSM is created as a mean over all 20 individual PSMs in order to judge how much the variability of HRR profiles due to multipath affects the FoM used for classification.

Fig. 4.16 shows four examples, namely reference PSMs 1, 11 and 20, and the averaged PSM. In Fig. 4.16(a), the FoM is plotted for all 400 HRR profiles tested

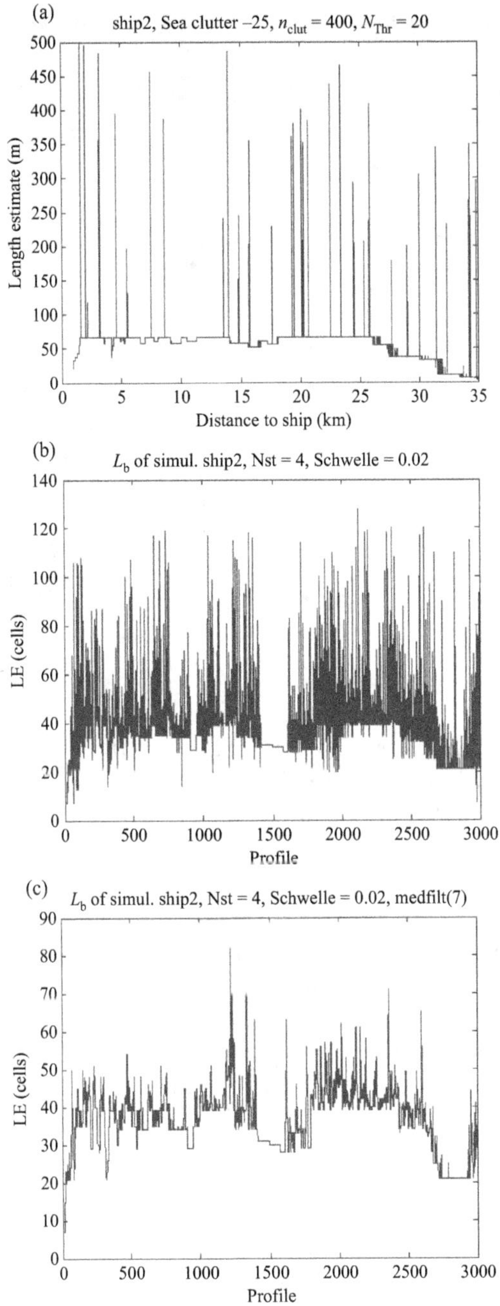

Figure 4.15 Length estimates L_a and L_b based on simulated data with sea clutter $\sigma^0 = -25$ dBm^2/m^2

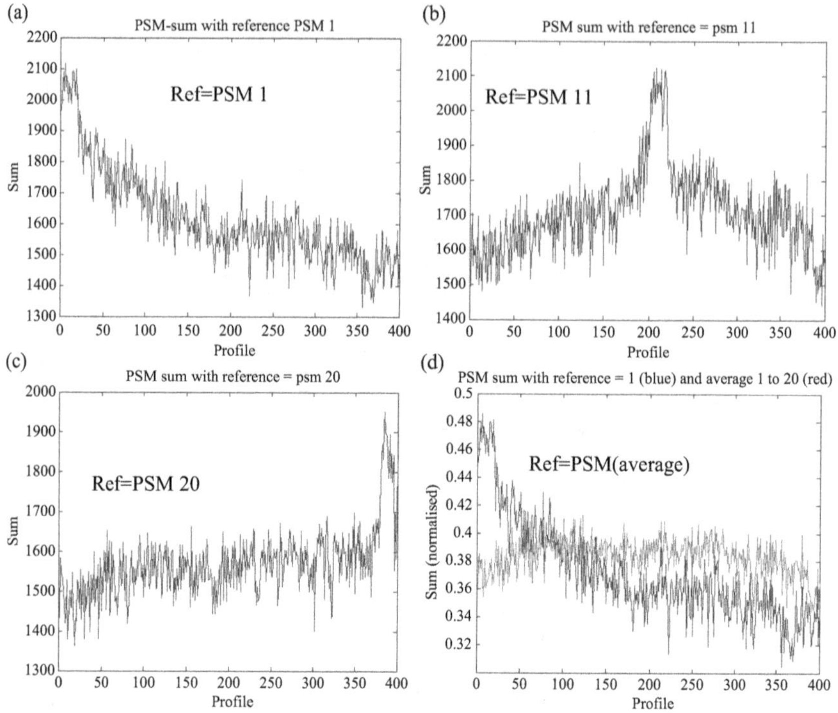

Figure 4.16 Three examples of the FoM ((a)–(c)). (d) The normalised FoM is shown in a comparison between PSM (lower) and the average PSM (upper)

with PSM 1 as reference. The highest FoMs occur at the beginning, which is not a surprise as these are the profiles from which reference PSM1 was constructed. However, only a little further away from these first 20 profiles, the FoM decays continuously to about 2/3 of the original value. This is a clear indication that the variability of the HRR profiles due to multipath is so high that a strong decrease in classification performance has to be expected.

Obviously, a reference PSM that is constructed for a certain target is not characteristic for that target at all ranges but only near the position where the reference was established.

The same result is obtained for the other PSMs 11 and 20 (Fig. 4.16(b) and (c)). The FoM is always highest around the reference range, and decays strongly farther away. If we look at the average PSM (Fig. 4.16(d), upper trace) then there is no preferred range any more, i.e. all HRR profiles under test independent of range show more or less the same (dis)similarity to the reference PSM. However, the FoMs reach only 70–80% of the former maximum values and thus likewise demonstrate a loss in classification performance potential.

4.5.3 Results based on geometrical, statistical and structural features

4.5.3.1 Measurements

For this analysis only one ship was available that was measured under controlled conditions at 17 GHz. The MARSIG radar [21] was installed, on the coast at a height of 19 m above the sea surface. In the beginning the trajectory of the target ship was directed away from the radar. After making a turn it approached the radar on a straight course for more than 1 km at distances between 7.5 km and 6.5 km. A full circle with a diameter of about 460 m concluded the measurement [16]. Here, we analyse the straight course towards the radar under constant front aspect. We select a series of 3600 HRR profiles that cover a total of 900 m in distance. In order to analyse how the HRR signature of the ship varies, six groups of 100 profiles each are extracted with a distance separation of about 160 m between each group. The six groups are B2, B8, B13, B16, B21 and B27, extracted from a total of 27 groups. Fig. 4.17 shows three examples that are taken from near the beginning, near the centre part and near the end of the straight trajectory. It is striking that there is hardly any similarity between the three groups. Although the front aspect of the ship was constant during the approach it seems as if its length did change. Obviously some of the scatterers near the bow and the stern that had disappeared due to multipath at far distance reappeared during the approach thus giving the impression of increased range extent. Also, looking at B13 as an example of one individual group (Fig. 4.17(b)) one can distinguish strong variability of individual scatterers although each group represents a distance interval of only 25 m. The scatterer near the bow shows four well-defined maxima, the scatterer near range gate 380 shows a strong maximum between profiles 1480 and 1500 and a pronounced oscillation in earlier profiles. Obviously it cannot be expected that a training feature set that was calculated at a certain distance will work well at a different distance.

In the following we will look how ATR test features vary within each group and from one group to the other. Next, reference features will be determined as the outcome of a training procedure. Two different approaches will be compared to each other. Finally, the feature references and the local test features will be used to classify the six groups against each other. Each group will be treated as a different target for this analysis. If the confusion matrices are 'flat', this would mean that the groups cannot be distinguished from each other, i.e. that the influence of multipath (and propagation effects) is not strong enough to change the ship's HRR signature significantly. On the other hand, if the confusion matrices show a pronounced 'ridge' in the main diagonal, this means that multipath will have a strong influence on ATR performance.

For the same three groups as in Fig. 4.17, Fig. 4.18 shows the behaviour of features ft1, ft2 and ft3 across the 100 profiles of each group. In order to calculate the feature values, the original HRR profiles were processed to generate well-defined peaks as described in Chapter 2. The most stable and most distinctive one is the geometrical feature ft1 (range extent), which essentially describes what can be

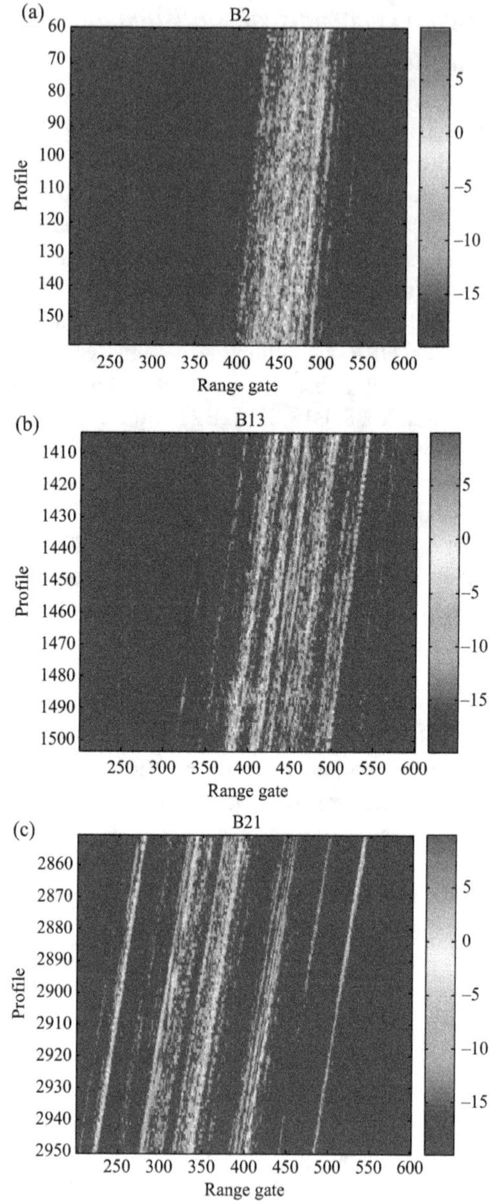

Figure 4.17 Measured HRR profiles (dB) for three distance intervals B2, B13 and B21 (100 profiles corresponding to 25 m) lying about 350 m apart

seen in Fig. 4.17. All other features show much stronger fluctuations with larger overlaps of their histograms, ft2 and ft3 being typical examples (Fig. 4.18). The strong fluctuations of the local feature values reflect the similarly strong variations of the HRR profiles from which they are derived.

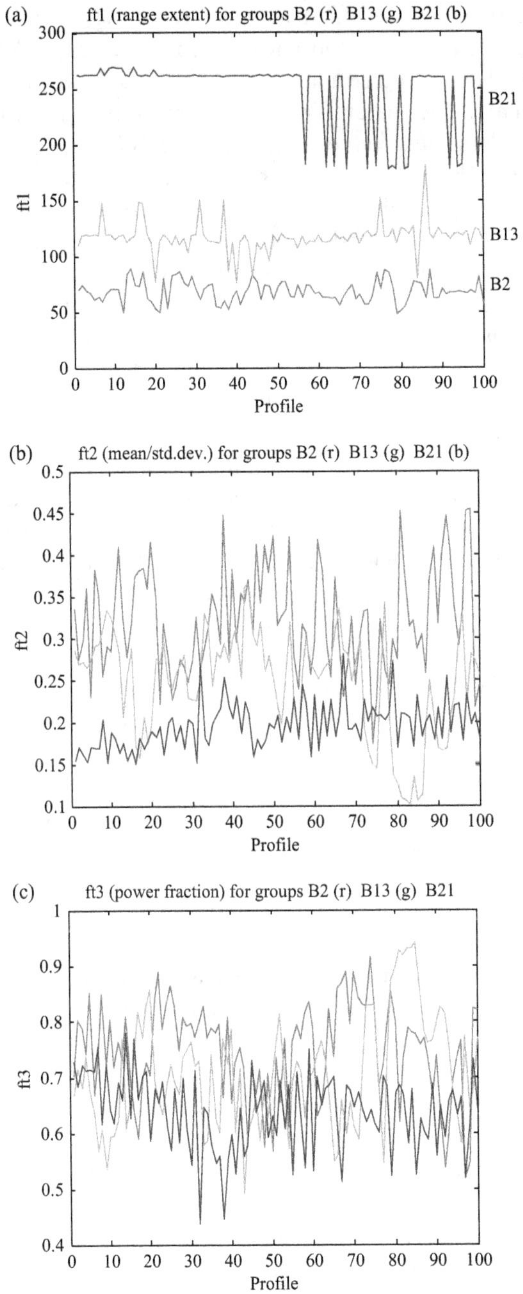

Figure 4.18 Comparison of feature history of features ft1 (a), ft2 (b) and ft3 (c) for three distance intervals B2, B13 and B21

For classification purposes one needs feature references that are the result of some training procedure. As we do not have any independent measurements or 'true profiles' we have to determine these references from each group of HRR profiles individually. The feature values that are derived from individual HRR profiles are considered test values, i.e. there are 100 test values per group as shown in Fig. 4.15. The training values have to be somehow representative of all of them; therefore, they should be based on one 'reference profile' that represents its group. There are two possibilities to determine such a reference profile as described in Chapter 2. One is to average all RCS values within each range cell, and the second one is to retain the maximum value within each range cell ('maxing'). Of course, all 100 profiles have to be carefully aligned for both approaches. The averaging process will lead to lower RCS values as compared to the 'true' profile (i.e. undisturbed by multipath effects), the 'maxing' process may lead to higher RCS values (up to 6 dB depending on the reflectance of the sea surface). This does not matter because the calculation of the features only depends on relative RCS values. The question, therefore, is how the two approaches influence the RCS relations between the scatterers. There is no simple answer to this because it depends on the distance to the ship and on the height of each scatterer as can be seen from the propagation factor diagrams (Figs. 4.8 and 4.9). The faster the fluctuation, the more likely the 'maxing' will lead to a stable reference profile. For a rough surface with a higher sea state leading to stronger motion of the ship, the single reflection multipath mechanism will no longer be realistic (cf. section 4.4.1). In that case the averaging approach may lead to a better reference profile.

These training references are now used together with the 100 test values from each group to construct confusion matrices by trying to classify the six groups against each other. Four sets of features are analysed here: set 1 (features 1 4 8 14), set 2 (features 26 14), set 3 (features 7 11 14) and set 4 (features 1 2 3 4 5 6 7). Sets 1 through 3 are composed of mainly uncorrelated features, while set 4 disregards any possible correlation. Fig. 4.19 shows a comparison between the (grey-scale coded) confusion matrices of all four sets using the AV references (upper row) and the MX references.

Table 4.3 summarises the P_{cc} percentages from the main diagonal of all eight confusion matrices.

One sees that the different feature sets show quite different classification performance. One should emphasise, however, that the task is not to find a set of best features for ship classification. Rather, the task is to demonstrate that there exist (sets of) features that are able to distinguish between the HRR signatures of the same ship at different distances. This means that the classification of this ship (and consequently any ship) against other ships will most likely fail. Both 'successful' feature sets (1 and 4) contain the 'range extent' (ft1). This may be a hint that it is mainly the geometry that suffers from multipath while statistical and structural features show less distinction. Comparing both rows of Fig. 4.19, i.e. AV and MX references, one may conclude that averaging seems to be more advantageous than 'maxing' in the case of real ship measurements, i.e. for real-life applications.

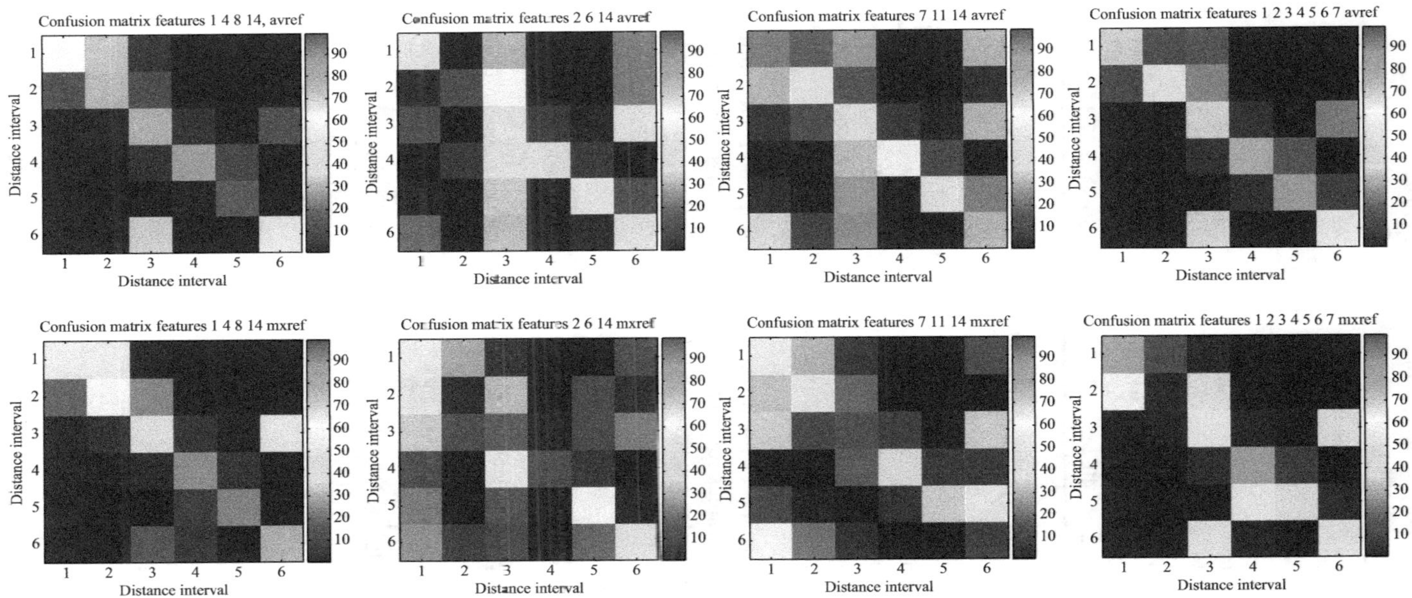

Figure 4.19 Confusion matrices (%) between six distance intervals for four sets of features with feature references obtained by averaging (top row) and by maxing (bottom row)

Table 4.3 $P_{cc}(\%)$ *values from main diagonal of confusion matrices*

Distance interval	Averaging reference				Maxing reference			
	Set 1	Set 2	Set 3	Set 4	Set 1	Set 2	Set 3	Set 4
1 (B2)	62	47	23	71	49	53	51	83
2 (B8)	76	13	47	66	58	4	48	4
3 (B13)	77	36	42	70	41	18	15	53
4 (B16)	85	42	57	77	90	14	68	87
5 (B21)	97	52	44	87	92	60	33	44
6 (B27)	66	45	31	65	77	36	9	51
Average	77.2	39.2	40.7	72.7	67.8	30.8	37.3	53.7

4.5.3.2 Classification based on simulated ships

A group of five ships (S1,..., S5) was simulated as described in Reference 15. It contains three long and two shorter ships, one rather high ship, two medium-high ships, and two with height up to 10 m. S1 is identical to the simulated ship used in section 4.5.1.2. In order to determine the classification performance, the five ships moved away from the radar starting at 1 km distance, and ending at 9 km distance. Every 10 m an HRR profile was determined on the basis of the scatterers' geometry and the propagation factors that were calculated using TERPEM with an antenna height of 19 m above the sea surface. Consequently, for each ship 800 HRR profiles were available from which 800 values for each of the 14 features were determined. Figs. 4.20 and 4.21 show examples for ship S3, which is 60 m long but rather flat with no scatterer higher than 10 m.

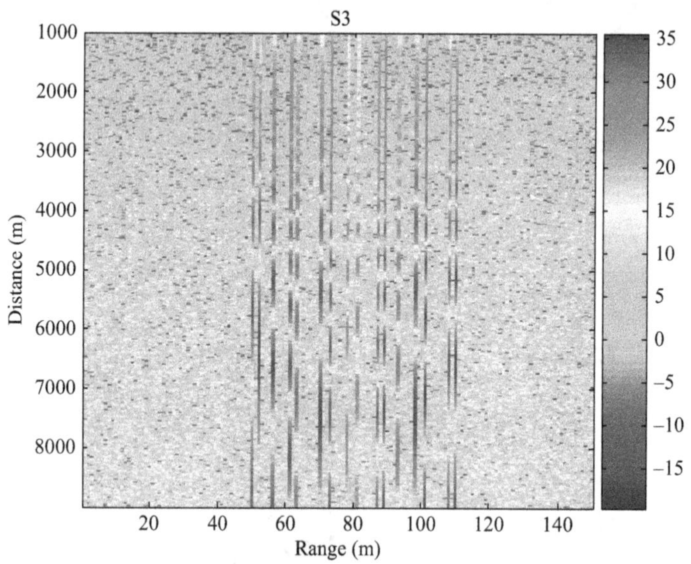

Figure 4.20 Aligned HRR profiles (dB) of ship S3 for distances from 1 km to 9 km

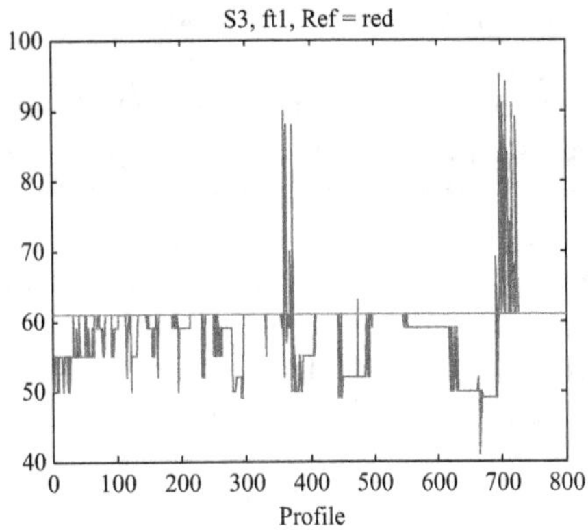

Figure 4.21 Values of features 1 and 3 (ship S3) as a function of distance (horizontal line, reference from Table 4.2)

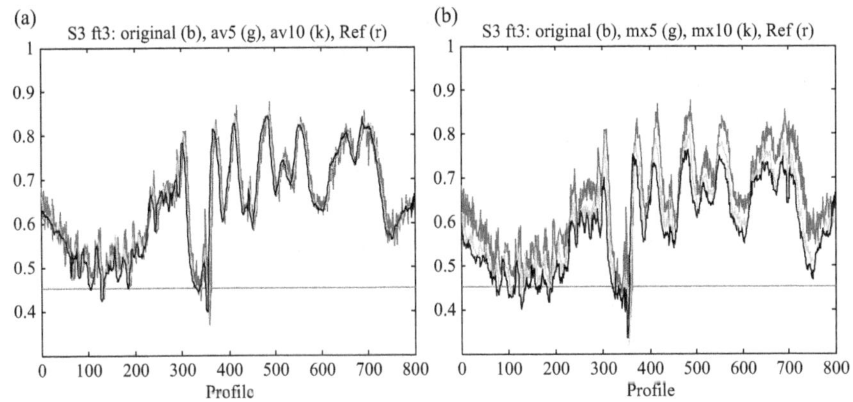

Figure 4.22 Comparison of different ways to compute the test features (S3, ft3): averaging (a) or maxing (b) for L = 1 (dark grey), L = 5 (light grey), L = 10 (black) (horizontal line = overall reference)

As can be seen from Fig. 4.21 the local feature values differ considerably from the 'true profile' references (horizontal line). This is found for all 14 features, the range extent (ft1) being one of the moderate examples. Consequently, it cannot be hoped that the single profile feature values will perform well when used for classification. Based on the cross correlation between features (Fig. 4.4), three sets of features were selected for classification, set 1 consisting of features 1, 4, 8 and 14; set 2 consisting of features 2, 6 and 14; and set 3 consisting of features 7, 11 and 14. For all three sets, confusion matrices between all five ships were constructed based on all 800 HRR profiles. In no case did the results show a pronounced main diagonal as expected. The question is how one can extract better test feature vectors from the available data. One possibility is the formerly mentioned method of grouping several consecutive HRR profiles either by averaging or by 'maxing'.

When we look at how the local feature estimates (i.e. test features) change depending on averaging or maxing we find as expected a tendency to come closer to the overall reference. Fig. 4.22 shows feature 3 of S1 as an example. The effect is more pronounced for maxing (Fig. 4.22(b)) than for averaging (Fig. 4.22(a)), and it increases with the number L of profiles that are grouped together. The examples show $L = 5$ (light grey) and $L = 10$ (black).

One may now go back to calculating confusion matrices based on the modified feature estimates that one gets from averaging or 'maxing' groups of L consecutive profiles. As it is not possible to show all the matrices here, and as we are mainly interested in the main diagonal, these values are shown in Table 4.4 for the feature set ft1, ft4, ft8 and ft14.

Listed are the percentages of correct classification of each of the five simulated ships. The first column gives the P_{cc} values based on test features that are derived from individual HRR profiles. Columns 2 and 3 show P_{cc} values based on test features that are derived from fivefold and tenfold averaged HRR profiles, respectively. Finally, columns 4 and 5 show P_{cc} values based on test features that are derived from fivefold and tenfold 'maxed' profiles. In all cases, the 'true profile' reference feature vector was used. As one sees, the 'good' classification of S1 has a tendency to become slightly worse while all the others show an improvement, which is more pronounced for 'maxing' than for averaging. Also, larger L is better than smaller L. Still, the classification results are not satisfactory.

We have already discussed that in the case of real ships the 'true profile' (TP) is not necessarily known (only when the ship is known from a signature library), and that it may be required to obtain an estimate from all available observations. It may even be advantageous to use such an estimate instead of the TP reference because the true profile actually is never seen during a real measurement. This explains the large deviation between local feature estimates and the TP reference as shown in Figs. 4.21 and 4.22. We, therefore, determine two alternative feature references that are based on averaging and 'maxing' all 800 HRR profiles, i.e. using an 8 km observation distance interval. This is a rather optimistic assumption, as in a real-life scenario the available distance intervals will tend to be

Table 4.4 Main diagonal values (%) of confusion matrices for five different approaches to calculate the test features (feature set fts 1, 4, 8, 14), 'true profile' reference

	Original	**av5**	**av10**	**mx5**	**mx10**
S1	68.6	67.5	67.5	67.0	64.0
S2	9.0	8.1	9.1	12.0	13.1
S3	15.1	15.4	18.3	18.0	23.3
S4	27.3	27.1	29.1	32.1	38.5
S5	64.5	71.1	72.6	66.6	67.6
Average	36.9	37.8	39.3	39.1	**41.3**

Table 4.5 Main diagonal values (%) of confusion matrices for three different ways to determine the test features (HRR profiles original/averaged L = 10/ 'maxed' L = 10) combined with two ways (averaging/maxing all 800 HRR profiles) to determine the feature reference

Ship	Original		Averaging $L = 10$		Maxing $L = 10$	
	avref	mxref	avref	mxref	avref	mxref
S1	77.3	62.8	75.5	56.4	71.3	53.4
S2	8.5	21.9	8.5	22.4	13.4	25.0
S3	9.8	22.8	11.3	22.1	14.9	30.1
S4	18.6	18.5	17.4	20.6	28.1	28.9
S5	73.5	78.5	83.1	88.6	79.9	84.3
Average	37.5	40.9	39.2	42.0	41.5	**44.3**

shorter mainly due to time constraints before making a decision. Table 4.5 shows the respective P_{cc} values (%), again only for the main diagonal of the confusion matrices using the same feature set as before.

The terms in row 1 refer to the way the test feature values are computed, i.e. from original individual HRR profiles or from tenfold averaged or tenfold 'maxed' HRR profiles. Each of these three cases is combined with the two possibilities of choosing the overall reference, either by averaging ('avref') or by 'maxing' ('mxref'). There is no clear tendency to be inferred from Tables 4.4 and 4.5. For ships S1 and S4, using the TP reference seems best, whereas for S2, S3 and S5 using the 'mxref' yields the best results. Therefore, it might be best to look at the average P_{cc} values in the last line, which summarise the results for all five ships. The highest value (44.3%) is reached for test feature vectors derived from maxing with $L - 10$, together with a feature reference that also was obtained by means of maxing. This is in contradiction to the findings from section 4.5.3.1 based on measurements. To what extent this depends on the properties of the sea surface remains an open question. It may well be that the smoother the sea surface, the more realistic the 'maxing' reference profile may become.

4.6 The mitigation of multipath effects on ship ATR

The results from Section 4.5 based on measurements as well as simulations show that TP references and local test vectors do not fit together. Trying to approximate true profiles by local operations like averaging or maxing of consecutive HRR profiles shows a certain improvement but is not sufficient. On the other hand (sections 4.5.3.1 and 4.5.3.2), using references that are obtained from a limited observation interval instead of TP references showed considerable improvement in classification performance in connection with local test feature vectors. However, these references cannot be established ahead and stored in a reference library. This leads to the conclusion that there is no way around TP references for ship classification. The main difficulty is then to obtain a realistic test feature vector that can be compared to the stored 'true' reference feature vector for the classification process. This can only be achieved by filling the 'gaps' in the propagation factor diagram in order to eliminate excessive oscillations of the scatterer amplitudes.

The multipath formula (4.2) takes into account the surface roughness via a reduced reflectivity by introducing a factor $f \in [0 \quad 1]$. For 100% reflection ($f = 1$), the resulting voltage may assume values between 0 (extinction) and 2, which corresponds to 6 dB increase in received power. For $f < 1$, the extinction is less pronounced, and the gaps that have to be filled are less deep. A typical example of the propagation factor (V^2|dB) is shown in Fig. 4.23, namely a cut of the 2D diagram for fixed scatterer height. From this one sees that the oscillation between minima and maxima is faster (i.e. the gaps are narrower) the closer the ship is to the antenna. Such a cut will be called a 'distance profile' (DP). It suggests a possibility of arriving at stable HRR profile estimates by filling in the gaps of extinction. Obviously this can only be achieved by means of properly integrating several distance profiles that have their minima and maxima at disjoint positions. From (4.2) one sees that this can be done by varying either Δ or λ. As the height of individual scatterers on a ship is not known, the applicable DPs are not known, either. Therefore, only the combination of HRR profiles obtained at several optimised values of λ or H offers a chance to approximate the unknown TPs. Using the complete propagation factor diagram seems to be the best way to perform that optimisation.

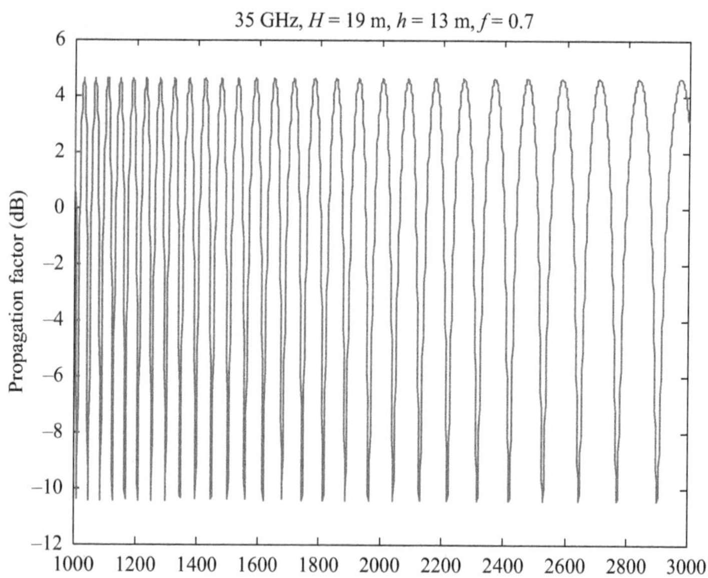

Figure 4.23 Geometric multipath propagation factor at 35 GHz for $H = 19$ m, $h = 13$ m and $f = 0.7$

Varying λ would mean using a multi-frequency radar that is able to switch its frequency from pulse to pulse or from sweep to sweep (for a rotating antenna). Varying Δ offers two more possibilities, namely to vary either the antenna height H or the slant range s for a fixed scatterer height h. Varying s is equivalent to temporal integration as long as the ship that has to be classified is approaching the antenna or receding from it. Of course, the antenna height H can normally not be changed but it is conceivable to install two or three antennas at different heights and to alternately use them to transmit and receive. It can be seen from Fig. 4.25 that height differences of less than 1 m are sufficient to achieve a considerable improvement in the HRR profile estimates.

A criterion is needed to judge the optimisation of combining several antenna heights or transmit frequencies. As the optimisation has to be valid for any scatterer height of the target ship, it is best to look at the oscillation of the pertinent distance profile, which may be described by its standard deviation. The latter can be used as an FoM in order to quantify the smoothing success.

4.6.1 Using several antennas

Let us first vary the antenna height H by using two or even more antennas. As an example Fig. 4.24 shows this FoM obtained as the result of combining up to ten distance profiles (i.e. as seen from up to ten antennas at different heights). This curve was obtained by starting at a reference antenna height of $H = 20$ m, then a second height was sought such that the FoM of the resulting combined (either by averaging or by maxing) DP became minimal. Then a third height was added in the same way. As one sees, the most significant reduction is already obtained by

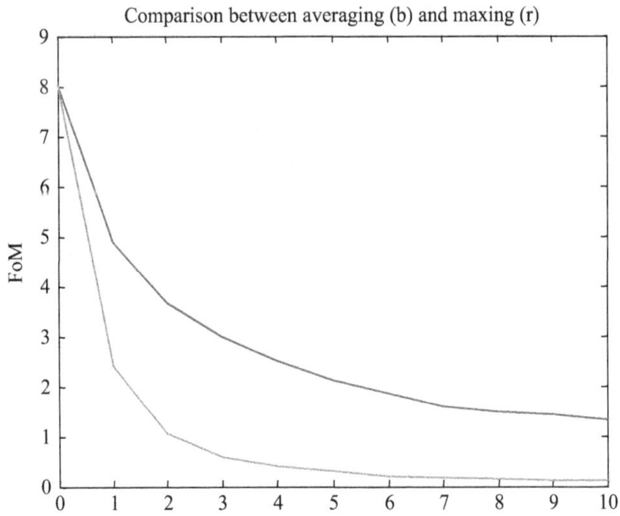

Figure 4.24 FoM obtained by combining n = 1, ..., 10 profiles with the original profile by means of averaging (dark grey) or 'maxing' (light grey)

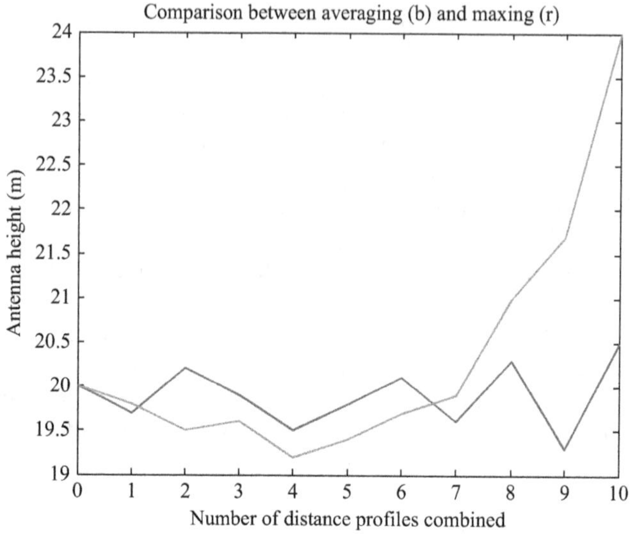

Figure 4.25 Heights of second antenna for optimised combination (dark grey: averaging, light grey: maxing) with first antenna at $H = 20$ m

combining only three profiles, which seems feasible in practice. The results shown in Fig. 4.25 suggest two antennas at heights $H = 19.7$ m and $H = 19.5$ m combined with the original antenna at $H = 20$ m to obtain minimal oscillation of the distance profile in Fig. 4.23. It is remarkable that the two or three antennas are grouped very close together.

Different ways to combine the distance profiles were analysed. Averaging in dB space (dark grey curve) was less successful in minimising the FoM than 'maxing' (light grey curve), i.e. retaining the maximum of any two values to be combined. It is obvious that this approach will also provide HRR profiles that are more similar to the 'true' (i.e. without multipath) profile of the ship and hence will result in more realistic estimates of the test feature vector and better classification results. As long as only the FoM is considered, averaging can be done in dB space to save computational effort. The FoM is only a vehicle in the optimisation process leading to certain optimal parameters. However, as soon as several profiles are combined for the purpose of calculating ATR features, the averaging has to be done in the linear domain. This is more efficient in filling the gaps. On the other hand, the 'maxing' can be done either in the linear or in the dB-domain as the 'max' and the 'log' operations are interchangeable.

4.6.2 Using several frequencies

A similar approach as for different antenna heights can be performed for different transmit frequencies. Fig. 4.26(a) shows the propagation factor diagram for all frequencies between 28 GHz and 36 GHz with a step size of 40 MHz. The parameters are: antenna height $H = 19$ m, scatterer height $h = 12$ m and roughness

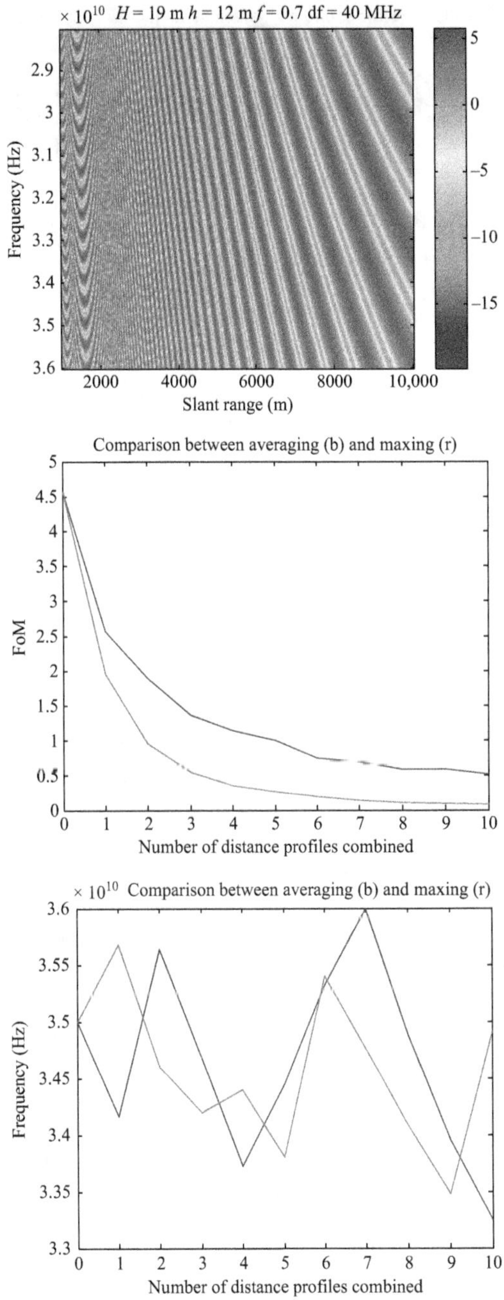

Figure 4.26 Propagation factor (a), FoM (b) and optimised transmit frequencies (c) for multi-frequency radar. Averaging (dark grey), maxing (light grey)

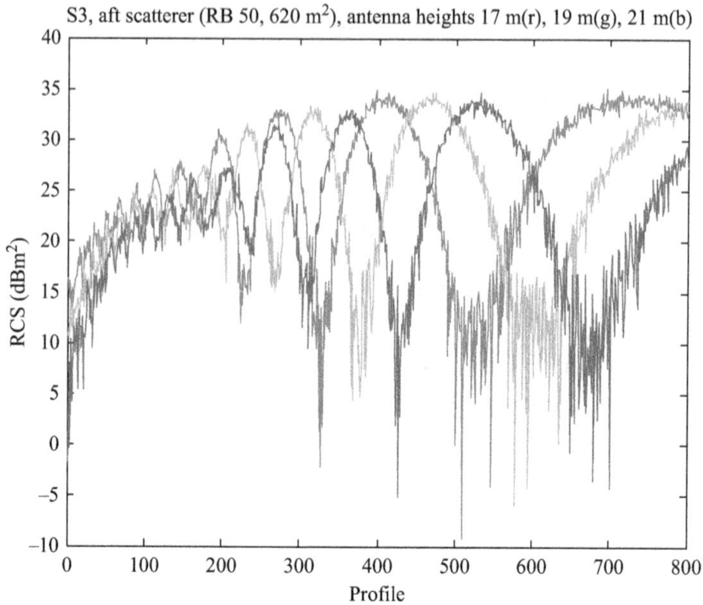

Figure 4.27 Distance profiles of aft scatterer of ship S3, seen with antenna heights 17 m, 19 m and 21 m

factor $f = 0.7$. Starting at a reference frequency $f_0 = 35$ GHz, a second frequency was sought such that the FoM of the resulting combined (either by averaging or by maxing) DP became minimal. Then a third frequency was added in the same way. Fig. 4.26(b) shows how the FoM decreases rapidly from its original value of 4.5 to a fraction of that, with maxing being more efficient than averaging. Finally, Fig. 4.26(c) shows which frequencies are the ones that are the basis of the optimisation. It is found that, although the whole range from 28 GHz to 36 GHz was searched, the optimal frequencies are astonishingly close to the reference frequency, not more than 400 MHz to 800 MHz away, which corresponds to less than 3%. Once again, as was found for multiple antenna heights, a combination of three frequencies is sufficient for a considerable improvement.

In order to analyse the averaging and maxing approach in more detail, the TERPEM modelling was performed for three different antenna heights, namely $H = 17$ m, $H = 19$ m (cf. Fig. 4.9) and $H = 21$ m. The five simulated ships S1, ..., S5 (cf. 5.3.2) were used to obtain estimates of the classification performance. As an example we look at the aft scatterer of S3 with an RCS of 620 m^2 located at $h = 4$ m above the sea surface. Fig. 4.27 shows the observed RCS as a function of distance for all three antenna heights. One clearly sees that the locations of the minima and maxima mostly are different although the three antenna heights were not optimised as described above. In Fig. 4.28 we see the results of averaging and maxing in all three cases: the RCS variability has been reduced considerably, averaging seems even smoother than maxing for certain distances. As positive interference can

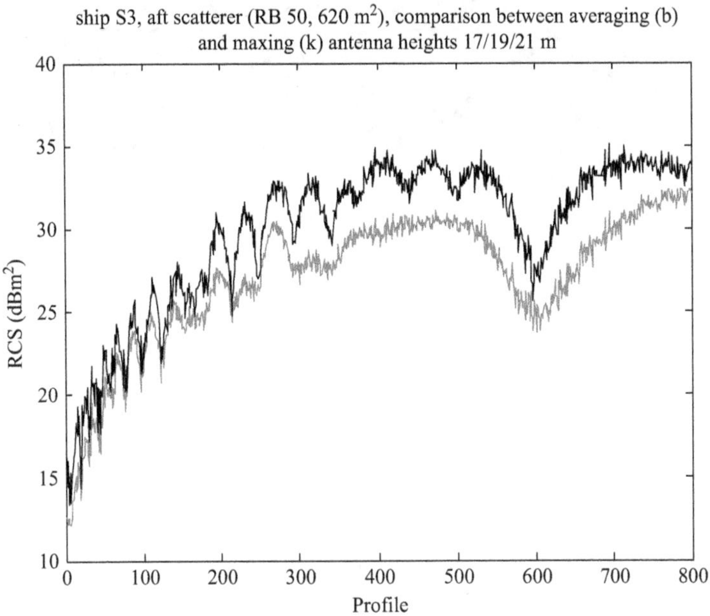

Figure 4.28 Results of averaging (grey) and maxing (black) applied to profiles from Figure 4.26

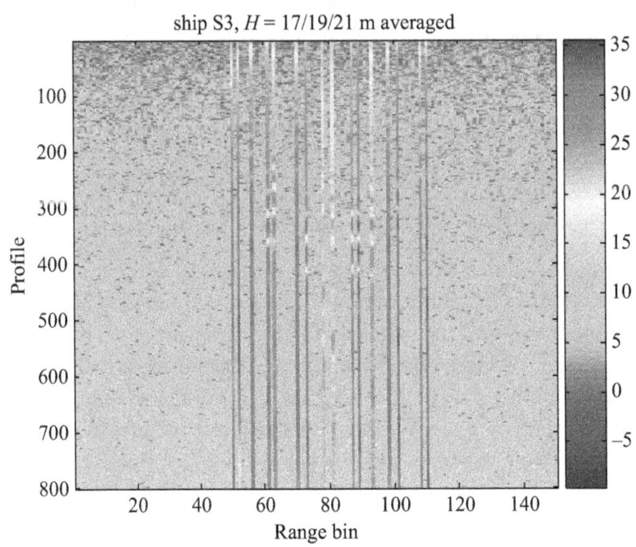

Figure 4.29 HRR profiles of ship S3 after averaging antenna heights 17/19/21 m

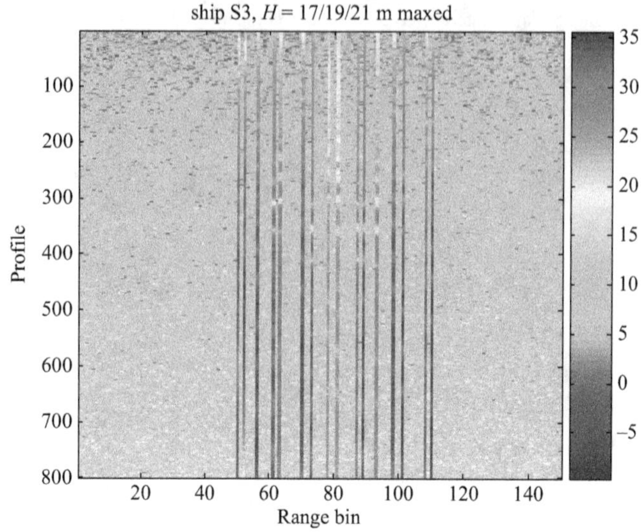

Figure 4.30 HRR profiles of ship S3 after maxing antenna heights 17/19/21 m

increase the RCS by a factor of 4. Therefore, there will be a systematic difference of up to 3 dB between the two approaches. The exact value depends on the unknown reflectivity f of the sea surface in (4.2). This difference does not play a role for feature calculation, as long as only relative RCS values are involved. Figs. 4.29 and 4.30 demonstrate the effect of averaging and maxing when all three antenna heights are combined and applied to a time series of HRR profiles like in Fig. 4.20 for ship S3. It is obvious that over extended distance intervals the HRR profile so obtained has a stronger resemblance with the 'true profile' than before. Consequently, the test features calculated from these profiles should be closer to their TP references in feature space.

4.6.3 Combining two antennas and two frequencies

After describing the use of several antennas or several frequencies, it is an interesting question whether the combination of the two approaches provides additional ATR performance. We will analyse this for the simplest case, namely the combination of two antennas and two frequencies. This means that we start with a reference antenna at height H_0 and frequency F_0 and add one more antenna at height H and a second frequency F. In order to limit the effort one has to choose one fixed scatterer height h and a fixed value for the reflectivity factor f. In the following, all examples are presented for $H_0 = 19$ m, $F_0 = 35$ GHz, a representative scatterer height $h = 12$ m and $f = 0.7$. Moreover, only examples based on 'maxing' are shown as this is the more promising approach to combine distance profiles, as was demonstrated previously. The distance range for the DPs on which the FoM is computed is limited to the interval [2000 m, 8000 m] in steps of 10 m.

H is now varied between $H_0 \pm 5$ m, i.e. 14 m and 24 m in steps of $dH = 0.1$ m, and F between $F_0 \pm 5$ GHz, i.e. 30 GHz and 40 GHz in steps of 50 MHz. For each pair (H, F) four DPs are determined, namely at (H_0, F_0), (H_0, F), (H, F_0) and (H, F). These are combined via maxing and the pertinent FoM (i.e. standard deviation of the combined DP) is determined. Fig. 4.31 shows the FoM grey-coded over the H–F plane. The maximum value, of course, occurs in the centre at $H = H_0$ and $F = F_0$ because here no smoothing takes place. The diagram looks rather, but not quite, symmetric. There is an interesting lobing structure indicating a certain correlation between optimal H and F values. It is striking that the smallest FoMs occur relatively close to the centre thus confirming the findings in sections 4.6.1 and 4.6.2. This means that even when combining two antennas and two frequencies, neither the H excursion nor the frequency excursion has to be large. The absolute minimum in Fig. 4.31 is located at $F = 34.55$ GHz and $H = 19.5$ m corresponding to $\Delta F = -450$ MHz and $\Delta H = +0.5$ m. The smallest FoMs are not found along the lines $H_0 = \text{const.}$ (i.e. vary only F) or $F_0 = \text{const.}$ (i.e. vary only H) but rather along with one of the lobes that are oriented diagonally. This means that working with two antennas *and* two frequencies is superior to either one possibility separately. This can easily be seen in Fig. 4.32 that shows a comparison between the original DP, the DP combined at two optimised frequencies, the DP combined at two optimised antenna heights and the DP with H and F optimised.

The optimised values $\Delta F = -450$ MHz and $\Delta H = +0.5$ m were found, as mentioned above for a fixed scatterer height $h = 12$ m and a fixed height of the first antenna $H_0 = 19$ m. Now, ships under investigation have dominant scatterers at many different heights, which contribute to the HRR profile. Also, the ATR radar

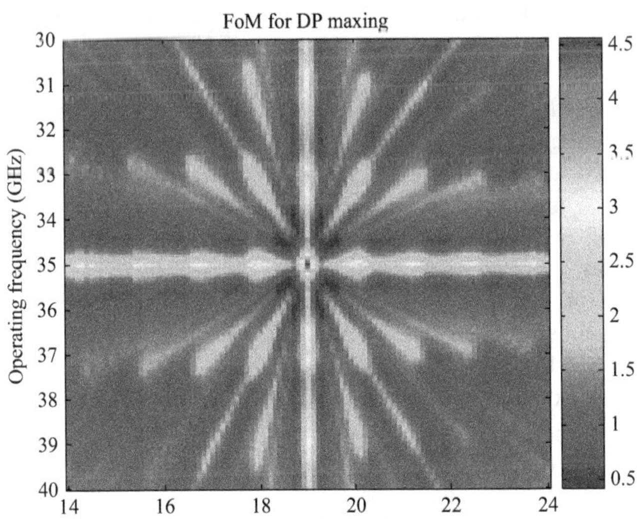

Figure 4.31 FoM for maxing DP's with two antennas at two frequencies

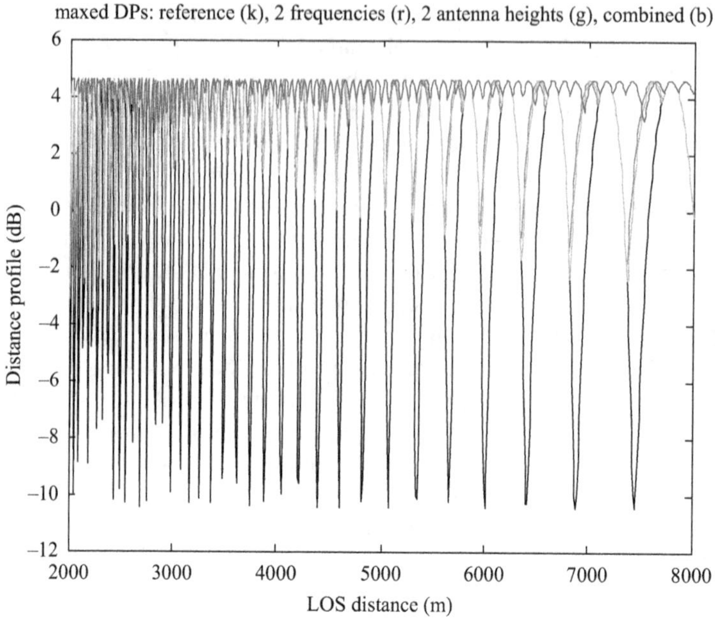

Figure 4.32 Original DP compared to different ways of smoothing

might be positioned at different heights in different scenarios (harbour, ship). It is therefore an interesting question how the minimum obtainable FoM is degraded when F_0, ΔF and ΔH are kept constant while H_0 and h are varied. An increase in the FoM would on the other hand mean an expected loss in ATR performance because the combined DPs used for feature construction are farther away from the true profile used for the feature references. The result is shown in Fig. 4.33. The behaviour of the minimum obtainable FoM is rather irregular but shows a very broad 'valley' around the original values of $H_0 = 19$ m and $h = 12$ m. This means that with a fixed set of parameters H_0, F_0, ΔF and ΔH one can expect to handle a wide range of realistic scenarios without loss of performance. However, as one sees towards the left part of the diagram, the FoM is increased considerably for low scatterer heights, especially below 5 m. From this one may conclude that for scenarios when small maritime targets are the main issue one should look for a different set of parameters that are adapted to $h \leq 5$ m. Also, when the expected target is likely to have a low bow or stern with strong scatterers on them that determine the ship's range extent, then also a loss in ATR performance has to be taken into account.

Fig. 4.29 showed how the HRR profile of the simulated ship S3 varies as a function of distance in the case of three antennas at heights 17 m, 19 m and 21 m. The same ship is presented in Fig. 4.34 using only two antennas, but in addition to two frequencies, both optimised. One sees that the combined HRR profile is very stable, even more stable than in Fig. 4.29, and should lend itself quite nicely for

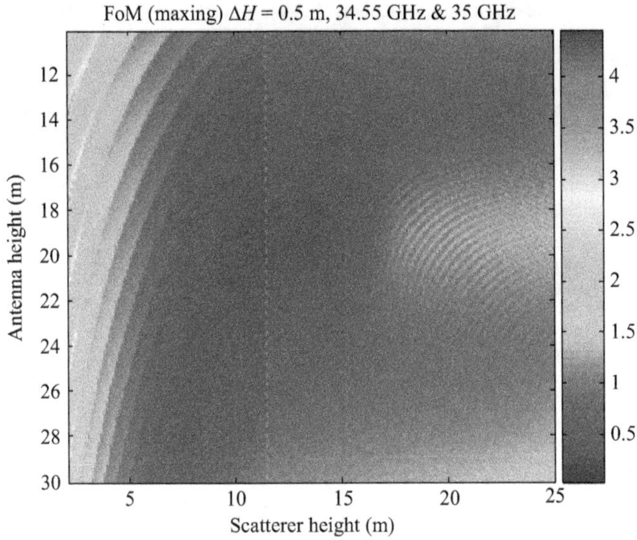

Figure 4.33 Minimum obtainable FoM as a function of antenna height H_0 and scatterer height h with f_0, ΔF and ΔH kept fixed

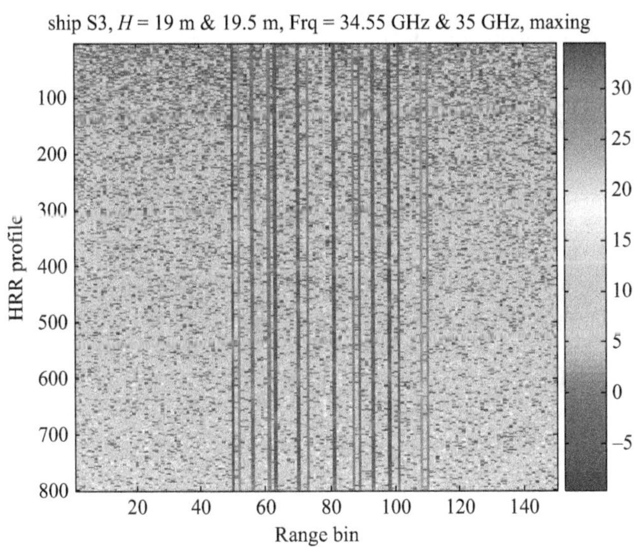

Figure 4.34 Combined (maxing) HRR profile of simulated ship S3 as a function of distance

feature construction. However, if we look into the behaviour of individual scatterers (Fig. 4.35) there still are distance intervals where the RCS is decreased considerably by up to 10 dB. This is more pronounced for scatterers closer to the sea surface as was expected from Fig. 4.33.

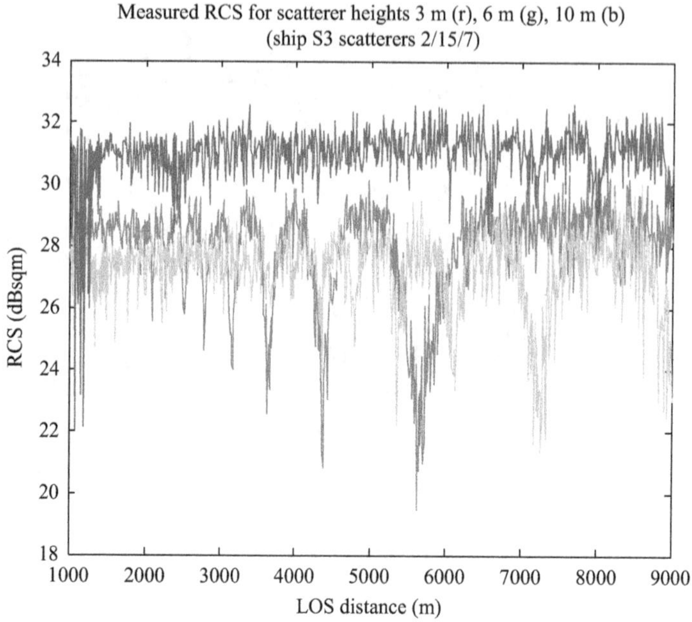

Figure 4.35 Observed RCS of scatterers 2 (h = 3 m), 15 (h = 6 m) and 7 (h = 10 m) as a function of distance

Another interesting question is if or how the optimised frequency and antenna shifts depend on the scatterer height for a given antenna height and transmit frequency. To answer this one has again to create diagrams like the one in Fig. 4.31 for every value of h and then determine the position of the minimum and the pertinent FoM value. One finds that the positions irregularly oscillate from left to right of H_0, F_0 or above and below this respective point. The reason is that the diagrams are almost but not exactly symmetric about the two axes $H = H_0$ and $F = F_0$. As we do not operate on a continuum of values but on a discretised grid with a certain spacing it may happen that the 'true' minimum falls between grid positions and an alternate minimum with almost identical value is found at a mirror position. Therefore, in Figs. 4.36 and 4.37 we plot only the magnitudes $|\Delta F|$ and $|\Delta H|$ versus the scatterer height h. For practical applications this means that the sign of the frequeny or antenna shifts is not crucial for the performance, the differences will be negligible.

The behaviour of $|\Delta H|$ versus h is shown in Fig. 4.36. The rather coarse grid spacing of $dH = 0.1$ m may be the reason that there is not a smooth curve. But the tendency becomes clear that optimised smoothing for scatterers closer to the sea surface requires a larger antenna shift. The same is true for the frequency shift as can be seen from Fig. 4.37. The reason in both cases is the same: for low scatterers the path difference between direct and reflected path is rather small. Therefore, a large frequency or antenna shift is needed to create a displacement of the distance

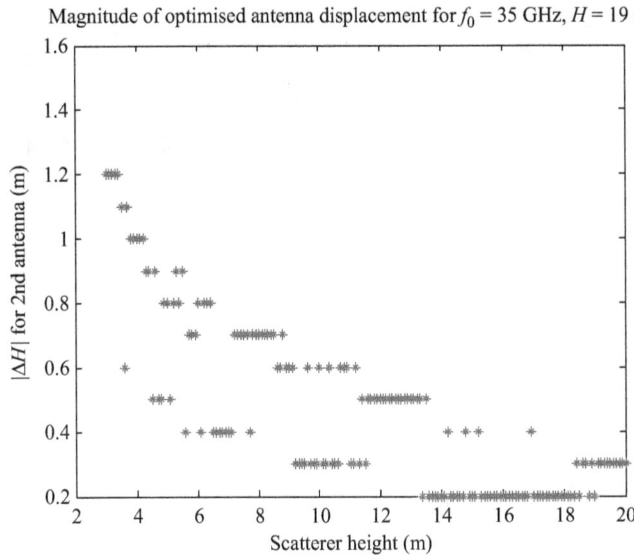

Figure 4.36 Magnitude of optimised antenna shift versus scatterer height

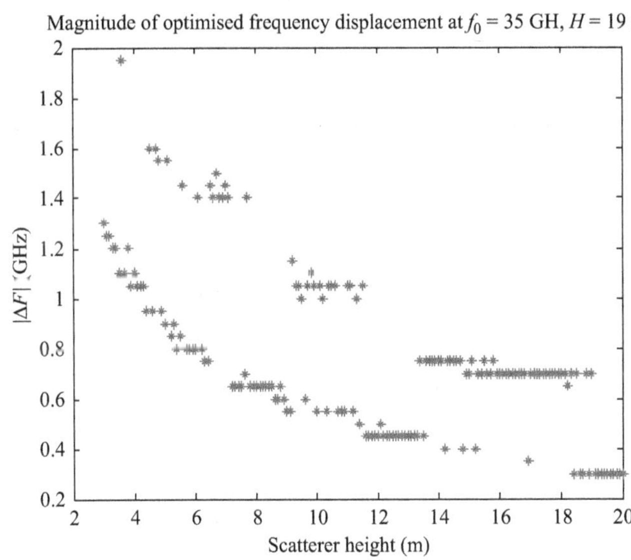

Figure 4.37 Magnitude of optimised frequency shift versus scatterer height

profile that is sufficient to fill in the gaps. Fig. 4.38 shows how the minimum obtainable FoM behaves as a function of scatterer height: there is a sharp increase from 0.25 to 0.38 for scatterers close to the sea surface (3–6 m) and then an almost flat plateau, which stays below 0.42 for all scatterers between 6 m and 20 m. In

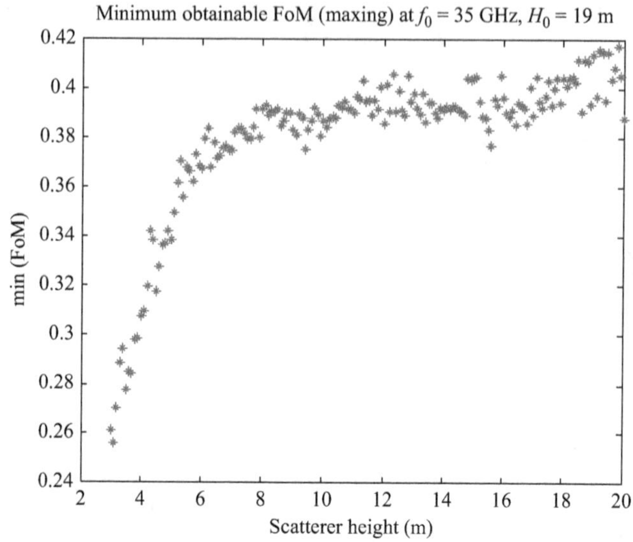

Figure 4.38 Minimum obtainable FoM (maxing) versus scatterer height

view of the maximum FoM of ≈4.6 these differences are negligible and indicate satisfactory DP smoothing. If a recommendation is needed for a fixed set of values $|\Delta H|$ and $|\Delta F|$ that provides good performance without adaptation to a specific scenario, one should look at the histograms of $|\Delta H|$ and $|\Delta F|$. One finds $|\Delta H| \approx 0.2$ m, . . . , 0.3 m and $|\Delta F| \approx 0.7$ GHz, which should cover successfully most cases in practice (except small boats).

4.6.4 Classification improvement via multi-frequency and/or multi-antenna approach

Let us first assess the advantage of the multi-antenna approach for test feature calculation. We consider 11 different cases: all three antenna heights 17/19/21 m individually (cases 1–3), averaging 17&19, 17&21, 19&21, 17&19&21 (cases 4-7), and maxing 17&19, 17&21, 19&21, 17&19&21 (cases 8–11). Fig. 4.39 shows the rms distance between all 800 values of feature 9 and its TP reference for all five simulated ships. Clearly the single-antenna cases have the largest deviations, followed by the two-antenna cases. Using three antennas leads to the smallest deviations between test and reference values (cases 7 and 11). There is no significant difference between averaging and maxing. The ranking of the five ships remains consistently independent of the approach chosen. The behaviour shown in Fig. 4.39 is typical for features 1–5 and 9–14. Only features 6–8 show no improvement by averaging or maxing. Their rms distances are more or less constant for all 11 cases. These features are more based on relative distances between scatterers while the others rather are related to amplitude ratios and therefore are more sensitive to multipath. The results from Fig. 4.39 suggest an improvement in classification performance based on the multi-antenna approach.

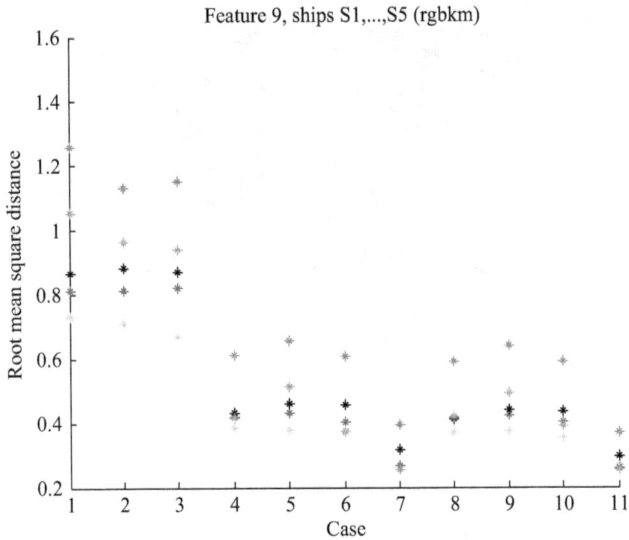

Figure 4.39 RMS distance between local feature estimates and TP reference for 11 cases, ships S1, …, S5 feature 9

The results obtained on the behaviour of test features will now be used to assess the classification performance based on confusion matrices $\boldsymbol{C}$. The targets are the five simulated ships S1, …, S5 giving rise to 5×5 confusion matrices. The feature references will be the TP references for reasons described above. The test features will be constructed from single-antenna data as well as from multi-antenna data either via averaging or via maxing such that 11 different cases will be compared to each other as defined above. For this comparison an optimised set of 4 out of 14 features was used [22] consisting of features 1, 2, 10 and 11. Fig. 4.40(a)–(c) demonstrates how the classification performance varies from the single-antenna to the three-antenna case.

There is an obvious improvement with increasing number of antennas. Still, the classification is not satisfactory. The main reason is that the three antenna heights are not optimised as described earlier in the chapter.

The values of 17 m, 19 m and 21 m were chosen arbitrarily and served only to demonstrate the principle. For realistic applications, they should be much closer to each other as demonstrated in section 4.6.1 (Fig. 4.25). Ships S4 and S5 have almost the same length (44 m and 40 m, respectively). Obviously, feature 1 (range extent), which is the most important feature, is not capable of distinguishing between the two ships. On the other hand, S2 is misclassified to a high percentage as S1, although they differ considerably in length (S1: 65 m, S2: 74 m). It seems that the scatterer geometry of S2 is such that too often either the stern or the bow scatterer 'disappears' and does not contribute to the range extent estimate. The results for all 11 cases and for both FoMs are summarised in Fig. 4.41.

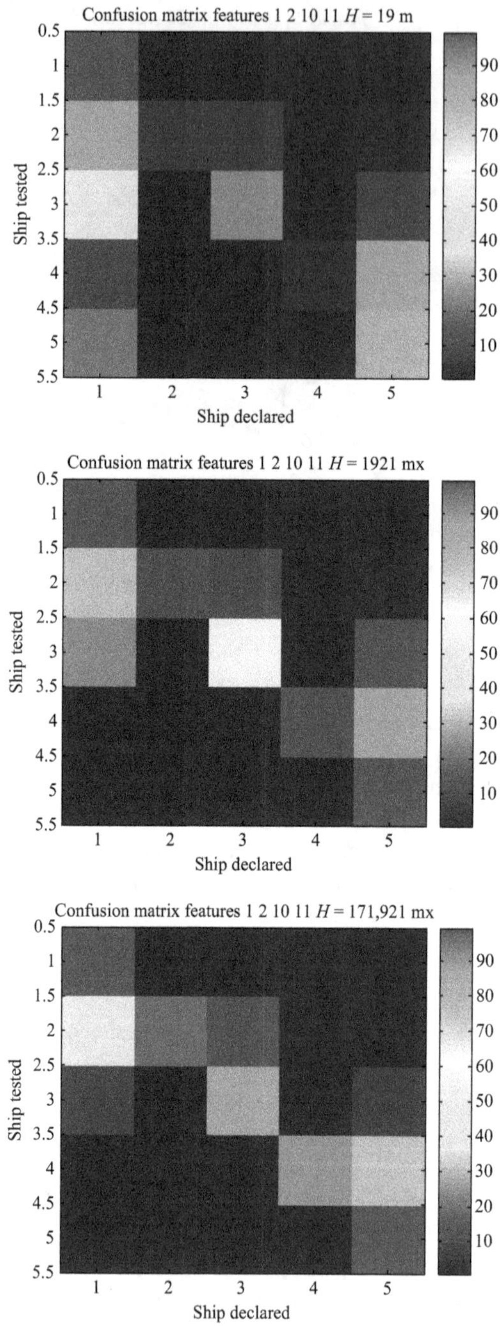

Figure 4.40 *Confusion matrices between ships S1, …, S5 with feature set* [1 2 10 11] *for single antenna (a), two antennas (b) and three antennas (c) combined via maxing*

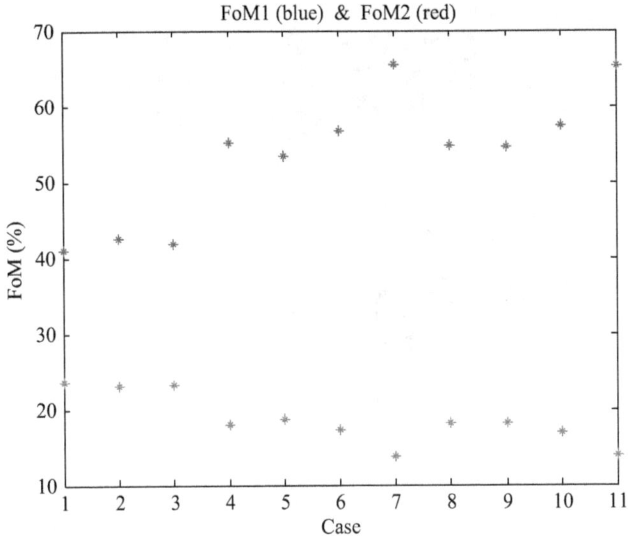

Figure 4.41 FoM1 (upper) and FoM2 (lower) for 11 cases, feature set [1 2 10 11]

The advantage of combining several antennas becomes evident; the performance increase from one to two antennas is about 15% in P_{cc} plus another 10% when a third antenna is added. These numbers can certainly be improved further when optimised antenna heights are chosen. The difference between averaging and maxing is negligible (as was already indicated in the results from Fig. 4.11) so that the approach that requires less processing (i.e. maxing) can be recommended.

For comparison, next the case of combining two antennas and two frequencies will be considered. Fig. 4.42 shows the confusion matrix between the simulated ships S1,..., S5 using the same set of features (1, 2, 10, 11) as in Fig. 4.40. The parameters used are $H_0 = 19$ m, $F_0 = 35$ GHz, $\Delta H = 0.5$ m and $\Delta F = 450$ MHz. Compared to Fig. 4.41 (three antennas) there is a certain improvement (only one instead of two misclassifications), the overall values FoM1 = 62.6% and FoM2 = 14.96% are more or less comparable to the three-antenna case in Fig. 4.41. The misclassification of S3 (declared as S5) may be due to the fact that the five scatterers closest to bow and stern are only between 3 m and 6 m above the sea surface. As we saw above, the values chosen for ΔH and ΔF are not well suited for the low scatterers and thus have a strong influence on the length estimate (feature 1). This makes S3 (60 m long) more similar to S5 (40 m long with 9 m high scatterers at bow and stern), because the closest and farthest high scatterers of S3 are located at $L = 11$ m ($h = 8$ m) and $L = 51$ m ($h = 8$ m), exactly 40 m apart!

4.7 Summary

Based on real measurements (at 17 GHz and 35 GHz) and simulations (at 35 GHz) of ship HRR profiles it has been shown how the influence of multipath and

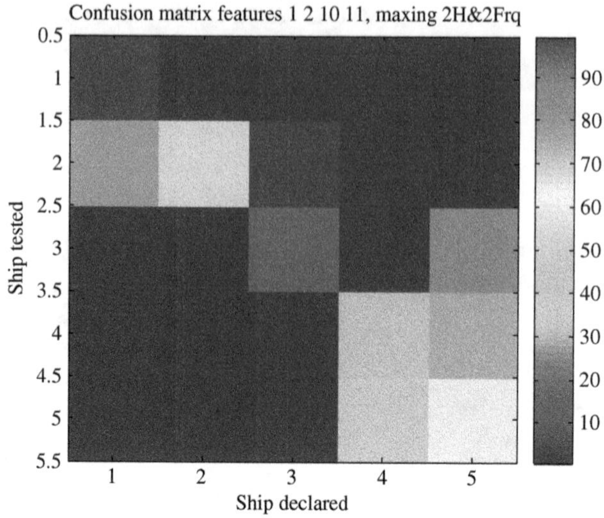

Figure 4.42 Confusion matrix between ships S1, ..., S5 with feature set [1 2 10 11]*, two antennas and two frequencies*

propagation effects at shallow depression angles will inevitably lead to a strong variability of the ship's signature as a function of distance. This variability will affect HRR profiles as well as ISAR images both of which are the basis for constructing ATR features. Two different approaches (averaging and 'maxing') to establish stable test HRR profiles were analysed. None of them is optimum and comparable to TP reference profiles. Also, estimates of reference profiles from longer observation intervals as an approximation to 'true profiles' were analysed. They show an increase in classification performance; however, they are not suitable to be stored in a reference library as a surrogate for the true profiles. When the latter are used as references, the main problem remains with the test features that are derived from single HRR profiles and naturally show the same strong fluctuation as these. A temporal integration of several consecutive HRR profiles leads to a certain stabilisation of the test features and results in improved ATR performance [13]. However, this is not sufficient, and moreover time may be a critical issue. New approaches [22] are mainly aimed at filling the 'gaps' of the propagation factor diagram either by operating at several radar frequencies or with several antennas at different heights. Technically, both approaches can easily be realised. Two frequencies and two antenna heights seem sufficient to achieve a substantial improvement in ATR performance without causing excessive additional cost thus offering a viable solution to the multipath problem. Antenna and frequency offsets may be adapted to the expected types of ships (especially to distinguish between big and high ships or small vessels). However, a 'typical' set like $|\Delta H| \approx 0.2$ m, ..., 0.3 m and $|\Delta F| \approx 0.7$ GHz will provide satisfactory performance in most cases.

References

[1] D. Pastina, P. Lombardo, A. Farina and P. Daddi, 'Super-Resolution of Polarimetric SAR Images of Ship Targets', *Signal Processing*, Vol. 83, pp. 1737–1748, 2003.

[2] M. Sciotti, D. Pastina and P. Lombardo, 'Exploiting the Polarimetric Information for the Detection of Ship Targets in Non-Homogeneous SAR Images', *Proceedings of IGARSS 02 Conference*, Vol. 3, pp. 1911–1913, Toronto, Canada, 24–28 June 2002.

[3] G. Panagopoulos, V. Tsagaris and V. Anastassopoulos, 'Using Synthetic Aperture Radar Data to Detect and Identify Ships', SPIE online doc. no. 10.1117/2.1200802.1062, 2008.

[4] R. Touzi, R. K. Raney and F. Charbonneau, 'On the Use of Permanent Symmetric Scatterers for Ship Characterization', *IEEE Trans. Geosci. Remote Sens.*, Vol. 42, No. 10, pp. 2039–2045, October 2004.

[5] G. Margarit, J. J. Mallorqí and X. Fábregas, 'Single-Pass Polarimetric SAR Interferometry for Vessel Classification', *IEEE Trans. Geosci. Remote Sens.*, Vol. 45, No. 11, pp. 3494–3502, November 2007.

[6] N. Bon, G. Hajduch, A. Khenchaf, R. Garello and J.-M. Quellec, 'Recent Developments in Detection, Imaging and Classification for Airborne Maritime Surveillance', *IET Signal Processing*, Vol. 2, No. 3, pp. 192–203, 2008.

[7] D. Pastina and C. Spina, 'Multi-Feature Based Automatic Recognition of Ship Targets in ISAR', *IET Radar Sonar Navigat.*, Vol. 3, pp. 406–423, 2009.

[8] D. Pastina and C. Spina, 'Multi-Feature Based Automatic Recognition of Ship Targets in ISAR Images', *Proceedings of 2008 IEEE Radar Conference*, Rome, Italy, May 2008.

[9] S. Musman, D. Kerr and C. Bachmann, 'Automatic Recognition of ISAR Ship Images', *IEEE Trans. Aerosp. Electron. Syst.*, Vol. 32, pp. 1392–1404, 1996.

[10] M. M. Menon, E. R. Boudreau and P. J. Kolodzy, 'An Automatic Ship Classification System for ISAR Imagery', *The Lincoln Laboratory Journal*, Vol. 6, pp. 289–308, 1993.

[11] N. Cotuk, S. Ture and M. Cetin, 'Application of Point Enhancement Technique for Ship Target Recognition by HRR', *SPIE Proceedings 'Algorithms for Synthetic Aperture Radar Imagery'*, Vol. 5095, pp. 185–193, September 2003.

[12] H. Schimpf and T. Dessauvagie, 'The Estimation of Target Length from Radar High Range Resolution Profiles', *FGAN-FHR Technical Report No. 54* (05-2003), Wachtberg, May 2003.

[13] H. Schimpf and H.-H. Fuchs, 'The Influence of Multipath on the Classification of Ships', *Proceedings of International Radar Symposium IRS 2009*, p. 597, Hamburg, Germany, 9–11 September 2009.

[14] H. Schimpf, H.-H. Fuchs and A. Maresch, 'The Classification of Ships in the Presence of Multipath', *NATO-Symposium SET-160 on 'NCI/ATR in Air*

Ground and Maritime Applications Based on Radar and Acoustics', Vol. 20, Athens, Greece, 11–12 October 2010.

[15] H. Schimpf, 'The Influence of Multipath on Ship ATR Performance', SPIE Proceedings Vol. 8049, paper no. 38, *Defense, Security, and Sensing 2011 Conference*, Orlando, FL, April 2011.

[16] H. Schimpf and H.-H. Fuchs, 'Analysis of ATR Features for Non-Cooperative Ground-Based Classification of Ships', *Proceedings of International Radar Symposium IRS 2010*, pp. 148–152, Vilnius, Lithuania, 16–18 June 2010.

[17] H. Essen, H.-H. Fuchs, G. Lindquist and A. Pagels, 'Influence of Atmospheric Propagation in the Maritime Boundary Layer on the Measured RCS of Ships', *Proceedings of SPIE Conference on Remote Sensing*, Vol. 7476-02, Berlin, Germany, 31 August–3 September 2009.

[18] E. A. Shtager, 'An Estimation of Sea Surface Influence on Radar Reflectivity of Ships', *IEEE Trans. Antennas Propagat.*, Vol. 47, pp. 1623–1627, October 1999.

[19] *TERPEM User Guide* (2nd edition), Signal Science Ltd., Abingdon, UK, December 2005.

[20] H. Schimpf, H. Essen, S. Boehmsdorff and T. Brehm, 'MEMPHIS – A Fully Polarimetric Experimental Radar', *Proceedings of IGARSS 02 Conference*, Toronto, Canada, 24–28 June 2002.

[21] H. Essen, H.-H. Fuchs, R. Sommer, A. Wahlen, W. Johannes and D. Janssen, 'MARSIG, Entwicklung eines Experimental Radars für Schiffs-Signaturmessungen', FGAN-FHR Jahresbericht, 2008, pp. 102–104.

[22] H. Schimpf, 'The Mitigation of the Influence of Multipath on the Ground-Based Classification of Ships', *Proceedings of International Radar Symposium IRS 2011*, pp. 797–802, Leipzig, Germany, September 2011.

Chapter 5
Effects of image quality on target recognition

Leslie M. Novak

5.1 Introduction

SAR image quality has a significant effect on the performance of SAR automatic target recognition systems. Template-based classifiers and model-based classifiers both require well-focused high resolution imagery in order to accurately match an observed target image to a database of stored templates or features such as peak-scatterer locations, etc. High resolution SAR requires accurate motion compensation in order to form well-focused target images, and errors in motion compensation can yield images with poor image quality such as excessive cross-range image smearing or blurring. Section 5.2 demonstrates the use of phase-gradient processing [1,2] to refocus SAR target imagery that has been degraded by cross-range smearing. ATR performance of the MSTAR model-based classifier [3–5] is investigated; the probability of correct classification (P_{cc}) is compared using target imagery having significant cross-range smearing versus target imagery that has been refocused using phase-gradient algorithm (PGA) processing. It is demonstrated that model-based ATR performance is improved considerably by applying PGA-processing to cross-range blurred SAR target imagery prior to passing the target imagery to the ATR.

Section 5.3 investigates the use of high resolution, full-polarisation SAR imagery for stationary target recognition. The data used in this study were gathered by the Lincoln Laboratory MMW SAR [6,7]. Optimum speckle reduction using PWF processing [8,9] of the full-polarisation data is shown to significantly enhance the image quality of the SAR imagery, and the detection performance of a multi-stage (detection, discrimination and classification) ATR system is demonstrated to be improved using PWF-processed imagery.

Section 5.4 investigates the ability of Lincoln Laboratory's high-definition imaging (HDI) algorithm [10,11] to improve the image quality and enhance the resolution of SAR imagery. The technique is shown to improve image quality by enhancing the resolution of well-focused complex SAR imagery [12]. A ten-target, template-based classifier is described and its classifier performance is presented using SAR imagery having 0.3 m × 0.3 m, 0.5 m × 0.5 m and 1.0 m × 1.0 m resolutions; classifier performance is quantified in terms of confusion matrices and probability of correct classification (P_{cc}). Next, enhanced resolution imagery is formed from the original 0.3 m × 0.3 m and 1.0 m × 1.0 m data using Lincoln Laboratory's HDI algorithm – this processing improves (approximately) the resolution of the data to 0.15 m × 0.15 m and 0.5 m × 0.5 m, respectively; and the image speckle noise is reduced. The improvement in the performance of the template-based classifier due to using HDI-processed data is demonstrated [13,14].

In section 5.5 we investigate the effects of data gaps (interruptions) in SAR phase-history data due to interruptions in the SAR data collection [15,16]. We show examples of SAR imagery formed using phase-history data containing various types of gapping patterns – and we demonstrate that SAR image artefacts induced by interruptions in SAR phase-history data are mitigated using compressed sensing image formation [17], resulting in excellent quality complex SAR imagery.

Section 5.6 summarises the results and conclusions of these studies, and finally the references used in these research studies are provided at the end of the chapter.

5.2 Improving ATR performance via PGA image quality enhancement

This section summarises the results of an ATR performance study using high resolution SAR imagery gathered by the Lynx SAR [18,19]. In this study, the imagery used was gathered at a nominal resolution of 0.15 m by 0.15 m in spotlight mode; a contiguous sequence of seven SAR images was used in the study. Fig. 5.1 shows these seven images. Visually, these SAR images appear to have very good image quality (IQ); however, the results of our target recognition studies indicate that SAR Image 1 has the best image quality and SAR Image 7 has the poorest image quality.

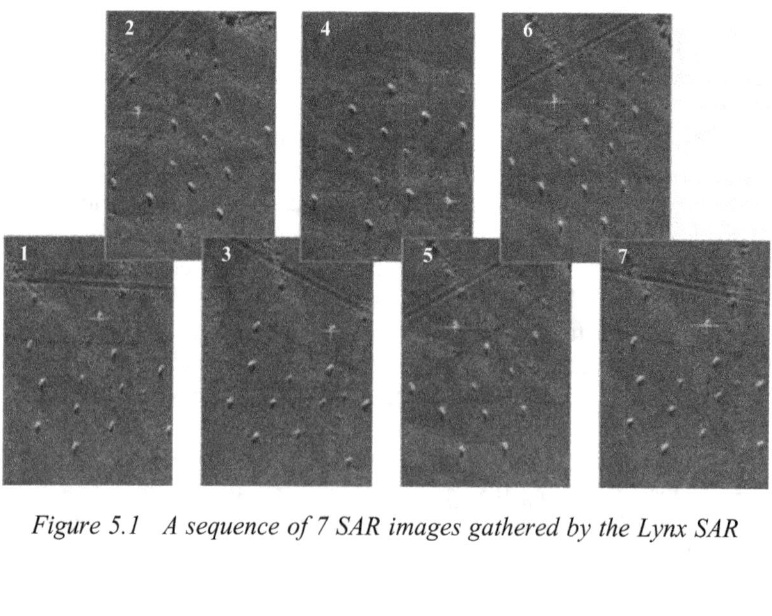

Figure 5.1 A sequence of 7 SAR images gathered by the Lynx SAR

A side-by-side comparison of SAR Image 1 with SAR Image 7 is shown in Fig. 5.2. Comparing the bright scatterer located on the uppermost target shows that Image 7 has significant cross-range blurring, most likely due to uncompensated platform motion. Our analyses will show that the average cross-range scatterer width = 17 pixels for the scatterers in Image 1, whereas the average cross-range scatterer width = 36 pixels for the scatterers in Image 7. These 'image quality feature' values were calculated from the image data during our target recognition studies. As we will show, the importance of this observation is directly related to the performance of an ATR system. In this study the model-based MSTAR target recognition system was used to classify the individual targets in each image. The classifier was designed to recognise 20+ military targets. The target array in these studies, as shown in Figs. 5.1 and 5.2, contained twelve military targets, and seven of these targets were contained in the classifier's set of 20+ targets.

A CFAR detector [20,21] was used to detect the targets located in these Lynx SAR images. Each of the seven targets contained in the classifier's 20+ target set was presented to the model-based classifier; thus, a total of 49 target images were

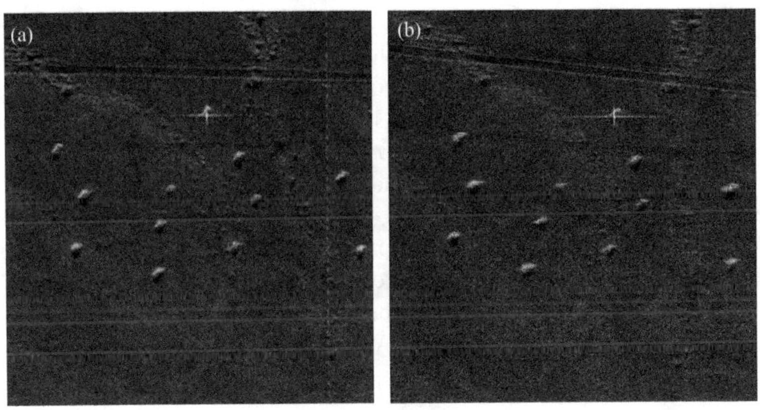

Figure 5.2 Lynx SAR images; Image 1(a) and Image 7(b)

Table 5.1 Classifier performance, Image 1 versus Image 7

Target	Image 1	Image 7	Image 7 (PGA3)
T72	'T72'	'BRDM2' (45)	'T72' (7)
2S1	'2S1'	'BMP2' (51)	'2S1' (19)
M60	'M60'	'BRDM2' (47)	'M60' (21)
M2	'M2'	'M2' (21)	'M2' (17)
BMP2	'BMP2'	'M113' (37)	'BMP2' (11)
BDRM2	'BDRM2'	'BRDM2' (21)	'BDRM2' (15)
M113	'M113'	'M113' (27)	'M113' (9)
	Width (Avg.) = 17	Width (Avg.) = 36	Width (Avg.) = 14
	$P_{cc} = 7/7$	$P_{cc} = 3/7$	$P_{cc} = 7/7$

input to the classifier. Table 5.1 presents the target recognition results obtained for Image 1 (column 2) versus the results obtained for Image 7 (column 3). Column 1 gives the true target identities. As the table shows, each of the targets contained in Image 1 was correctly classified ($P_{cc} = 7/7$); however, four of the targets from Image 7 were incorrectly classified (compare column 3 versus column 1); thus, for this image, $P_{cc} = 3/7$.

Next, the targets from Image 7 were refocused using phase-gradient processing. Each target's brightest scatterers were CFAR detected and aligned as required by the PGA algorithm and averaged – an average cross-range scatterer width was calculated from the average of the brightest scatterers. Table 5.1 summarises the cross-range scatterer widths obtained for each target, and also gives the average widths obtained for the seven targets in Image 1 and Image 7. As stated previously, Image 1 has an average cross-range width = 17 pixels and Image 7 has an average cross-range width = 36 pixels. As shown in Table 5.1, column 4 tabulates the 'classifier calls' and the average cross-range widths obtained after applying three iterations of phase gradient focusing to Image 7. After PGA3 focusing, each of the targets in Image 7 was correctly classified ($P_{cc} = 7/7$) – and the average cross-range width for the 7 targets was reduced to 14 pixels.

A side-by-side comparison of SAR target images from Image 7 is presented in Fig. 5.3. The original target image (a) shows significant cross-range image blurring; the well-focused target image (b) is the same target after reprocessing the complex image data using three iterations of the PGA algorithm. As shown in the figure, the target scatterers in the PGA-processed image are very nicely focused. Note also that the contextual background features are more visible in the PGA-processed image.

Fig. 5.4 shows a side-by-side comparison of another target from Image 7. Similar image quality improvements are observed for this target.

Table 5.2 summarises the model-based classifier's performance for each of the seven target images processed. As the data indicates, the image quality

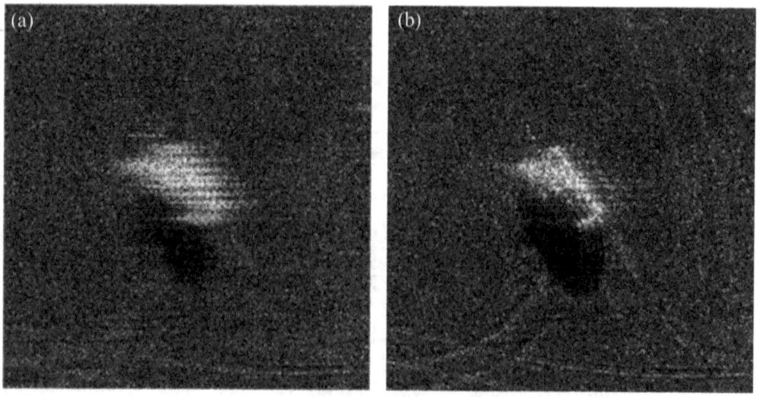

Figure 5.3 Target extracted from Image 7; original (a) and after PGA3 (b)

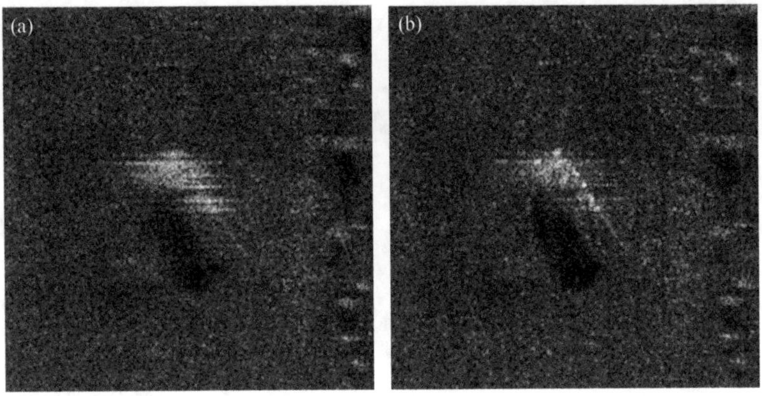

Figure 5.4 Another target extracted from Image 7; original (a) and after PGA3 (b)

Table 5.2 Average target scatterer widths in 'number of pixels' and correct classification probabilities for Lynx images

Image number	Avg. width (original)	Avg. width (after PGA3)	P_{cc} (original)	P_{cc} (after PGA3)
1	17.0	11.5	7/7	7/7
2	21.8	13.8	7/7	7/7
3	23.2	16.1	6/7	7/7
4	24.1	11.8	6/7	7/7
5	26.4	16.0	5/7	5/7
6	31.5	12.7	6/7	6/7
7	35.5	14.1	3/7	7/7
Averages	25.6	13.7	40/49	46/49

improvement achieved via PGA focusing of the SAR target images has increased the MSTAR model-based classifier P_{cc} from 81.6% to 93.9%.

An alternative image focusing algorithm based upon minimising the image entropy is described in References 22 and 23; however, results presented in Fig. 5.5 indicate that minimum entropy focusing requires using at least ten iterations of the entropy minimisation algorithm in Reference 23. However, the entropy minimum is achieved using only three iterations of the PGA algorithm; thus, PGA processing seems to be the preferred SAR image focusing technique.

5.3 Improving ATR performance using high resolution, PWF-processed full-polarisation SAR data

This section investigates the use of high resolution, full-polarisation SAR imagery for stationary target recognition. The data used in this study were gathered by the Lincoln Laboratory MMW SAR. Fig. 5.6 shows an example of SAR image; this

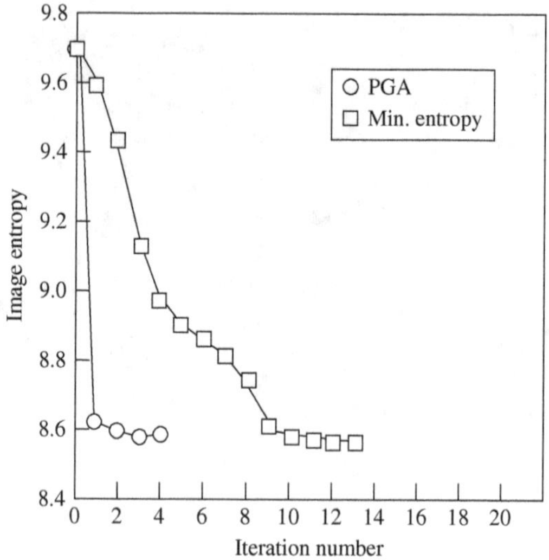

Figure 5.5 Image entropy; PGA versus minimum entropy algorithm

Figure 5.6 Ka-band SAR image of an office building located in Lincoln, Massachusetts; image resolution is 0.3 m × 0.3 m, SAR data is full polarisation (HH, HV, VV), PWF-processed

figure shows a 0.3 m × 0.3 m resolution, full-polarisation image of an office building located in Lincoln, Massachusetts. The high resolution data and full-polarisation processing allow for near-optical image quality. The SAR image in Fig. 5.6 was formed by using PWF processing, which optimally combines the HH, HV and VV polarisation components into a minimum-speckle intensity image [24].

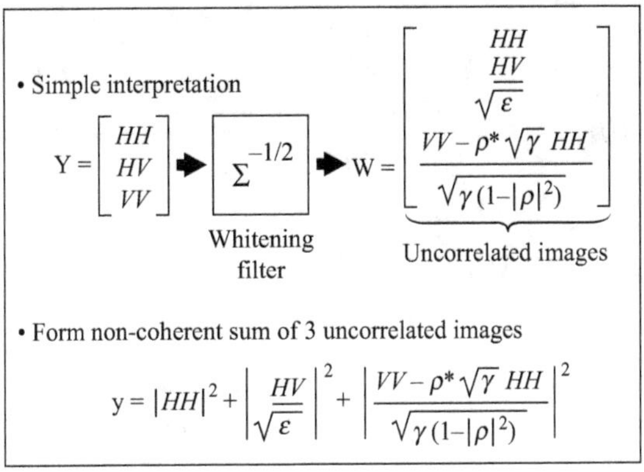

Figure 5.7 PWF (minimum-speckle) processing

The SAR image was obtained under clear weather conditions (and during off-hours, which explains the lack of vehicles in the parking lot). However, the quality and resolution of the SAR image would not have been degraded by dense fog or thick cloud cover. Thus a SAR sensor has a significant advantage over optical sensors: SAR image quality is essentially unaffected by rain or cloud cover weather conditions and the SAR sensor can be used during the day or at night. In addition, SAR sensors can image large areas of the ground from a long distance.

A simplified description of PWF processing is presented via the block diagram shown in Fig. 5.7. PWF processing enhances the quality of the imagery in two ways: (1) the amount of random speckle in the imagery is minimised and (2) the edges of objects such as the building are more sharply defined. As a result, PWF-processed imagery is visually much clearer than single-polarisation-channel imagery. The PWF transformation shown in Fig. 5.7 is derived in Reference 8, and is briefly developed as follows. We assume that the radar measurement vector, $\boldsymbol{Y}$, consists of three complex elements: HH, HV and VV. Therefore,

$$\boldsymbol{Y} = \begin{pmatrix} HH \\ HV \\ VV \end{pmatrix} = \begin{pmatrix} HH_{\mathrm{I}} + jHH_{\mathrm{Q}} \\ HV_{\mathrm{I}} + jHV_{\mathrm{Q}} \\ VV_{\mathrm{I}} + jVV_{\mathrm{Q}} \end{pmatrix}$$

where HH_{I} and HH_{Q}, e.g. are the in-phase and quadrature components, respectively, of the complex HH polarisation measurement. The joint probability density function (PDF) of the complex vector $\boldsymbol{Y}$ is given by the expression

$$f(\boldsymbol{Y}) = (1/\pi^3|\Sigma|)\exp\left(-\boldsymbol{Y}^H\Sigma^{-1}\boldsymbol{Y}\right)$$

where $\Sigma = E(\boldsymbol{Y}\boldsymbol{Y}^H)$ is the polarisation covariance of the vector $\boldsymbol{Y}$. The vector, $\boldsymbol{Y}$ is zero mean, $E(\boldsymbol{Y}) = 0$. The covariance matrices we use for clutter data take the following form in a linear-polarisation basis:

$$\Sigma = \sigma_{HH}\begin{pmatrix} 1 & 0 & \rho\sqrt{\gamma} \\ 0 & \varepsilon & 0 \\ \rho^*\sqrt{\gamma} & 0 & \gamma \end{pmatrix}$$

where:

$$\sigma_{HH} = E\left(|HH|^2\right), \quad \varepsilon = \frac{E\left(|HV|^2\right)}{E\left(|HH|^2\right)}, \quad \gamma = \frac{E\left(|VV|^2\right)}{E\left(|HH|^2\right)} \quad \text{and}$$

$$\rho = \frac{E(HH \times VV^*)}{\sqrt{E(|HH|)^2 E(|VV|)^2}}$$

We consider how to process the three complex measurements – HH, HV and VV, i.e. the vector $\boldsymbol{Y}$ – into pixel intensity in a way that minimises speckle. The measure of speckle we use is the ratio of the standard deviation of the image pixel intensities to the mean of the intensities$(s/m) = \text{std dev}(y)/\text{mean}(y)$, where the image intensity is $y = \boldsymbol{Y}^H \boldsymbol{A} \boldsymbol{Y}$ and we seek the matrix $\mathbf{A}$ that minimises the speckle measure. We have the following useful identities: (1) $E(\boldsymbol{Y}^H \boldsymbol{A} \boldsymbol{Y}) = \text{trace}\ (\Sigma\, \boldsymbol{A}) = \sum_{k=1}^{3} \lambda_k$, and (2) $\text{var}(\boldsymbol{Y}^H \boldsymbol{A} \boldsymbol{Y}) = \text{trace}\ (\Sigma \boldsymbol{A})^2 = \sum_{k=1}^{3} \lambda_k^2$, where $\lambda_k,\ k = 1, 2, 3$ are the eigenvalues of matrix $(\Sigma\, \boldsymbol{A})$. By using the approach of Cadzow [25], we can show that matrix $\boldsymbol{A}_{\text{opt}}$ that minimises the speckle measure must make the eigenvalues of $(\Sigma\, \boldsymbol{A})$ all equal, i.e. $\lambda_1 = \lambda_2 = \lambda_3$. Therefore, we select as the optimum solution, $\boldsymbol{A}_{\text{opt}} = \Sigma^{-1}$. Note that any constant multiple of $\boldsymbol{A}_{\text{opt}}$ is also optimum since the s/m is invariant to scale change. Note that this solution is equivalent to applying a whitening filter to the polarisation vector $\boldsymbol{Y}$ prior to forming the image, and Fig. 5.7 presents a simplified block diagram representation of the solution. In the figure, $\boldsymbol{Y}$ is the complex vector containing the three linear-polarisation measurements. The polarisation whitening filter, Σ^{-1}, transforms the (correlated) linear basis into an orthonormal polarisation basis, W. In this new basis, the three complex polarisations are uncorrelated and have equal average power. The PWF image is y, the non-coherent sum of the three uncorrelated intensity images.

Fig. 5.8 shows a single-polarisation-channel image (HH) of the same scene shown previously in the PWF image of Fig. 5.6. The speckle inherent in single-polarisation SAR imagery obscures much of the detail visible in the PWF SAR image. As a result, PWF-processed imagery improves the detection and classification performance of an ATR system.

The following paragraphs summarise the performance of the detection, discrimination and classifier stages of the SAR ATR system using full-polarisation PWF imagery versus single-polarisation HH imagery. In these studies, a total of 56 km^2 of high resolution clutter data were processed by the ATR algorithms, along with 136 test target images. The targets were M48 tanks and M55 howitzers, which have similar dimensions. The target data and clutter data were collected at Stockbridge, New York; the clutter data contained a significant number of man-made discretes such as buildings, roads, cars, electrical power lines, etc. The target training data were an independent set of target images and the target test data were collected under realistic deployment conditions. Fig. 5.9 shows performance

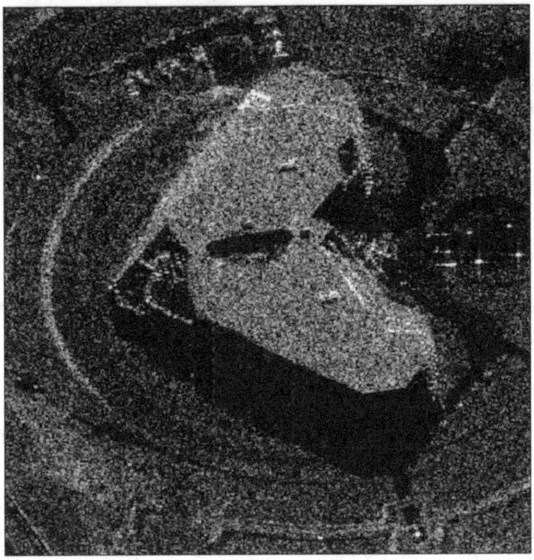

Figure 5.8 Ka-band HH-polarisation SAR image of the office building; image resolution is 0.3 m × 0.3 m

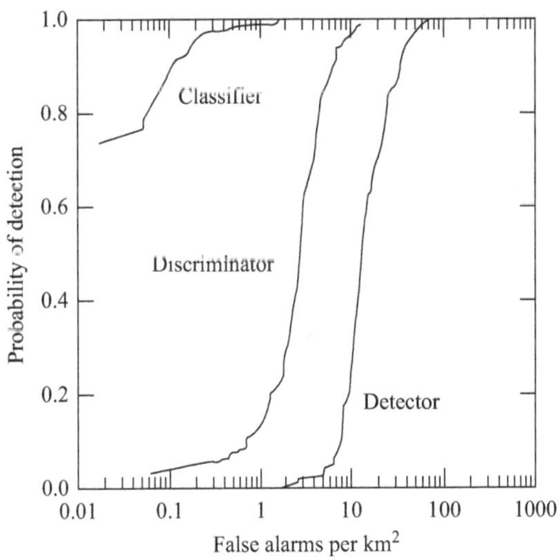

Figure 5.9 Performance ROC curves for the three stages of the ATR system using 0.3 m × 0.3 m PWF data

Receiver Operating Characteristic (ROC) curves at all three ATR stages (detection, discrimination and classification). The detector we used was a two-parameter CFAR algorithm [20]. First the detection performance curve was generated; then an operating point on that curve was selected and the detected objects (targets and false alarms) obtained at that operating point were passed to the discrimination stage to obtain the discrimination performance curve. Next, an operating point on the discrimination curve was selected and the detected objects obtained at that operating point were passed to the classifier stage to obtain the classification performance curve. Finally, to compare results in a consistent manner, we set the operating points for the detector at detection probability $P_d = 1.0$, for the discriminator at $P_d = 1.0$ and for the classifier at $P_d = 0.9$.

In the discrimination stage, the location and orientation of each detected object were determined. Then, discrimination features (textural, size, contrast and polarisation-based features) were calculated [26,27] and the best subset was used by the discriminator [28] to reject both natural clutter false alarms and man-made clutter 'discretes'. In the classification stage, a 2D template matcher was applied to those detected objects that had passed the discrimination stage. In the classification stage, the mean-squared error threshold was set to produce a probability of detection $P_d = 0.9$. Therefore, of the 136 test targets input to the classifier, a total of 122 test targets were detected (i.e. classified as either a tank or a howitzer). The pattern matching classifier was also applied to the clutter false alarms that passed the discrimination stage; these clutter false alarms were either rejected by the classifier (i.e. declared 'clutter') or classified (erroneously) into tank or howitzer categories. Classification performance results were tabulated summarising the ability of the classifier to reject clutter false alarms and assign the remaining detected objects into tank and howitzer categories. Fig. 5.9 shows the performance Receiver Operating Characteristics (ROC curves) at all three stages of the ATR system using PWF-processed data. Note that the discrimination stage reduced the number of false alarms by approximately a factor of 5 compared to the detection stage; the set of discrimination features used were the fractal dimension, percent pure even, standard deviation, diameter, mean CFAR and normalised rotational inertia (see Reference 29 for details). Finally, the classifier further reduced the number of false alarms by more than an order of magnitude.

The false alarms at each stage were categorised as either man-made 'discretes' or natural clutter false alarms. Table 5.3 gives the breakdown with respect to the type of false alarms detected for each stage using PWF data. The results presented in Table 5.3 also indicate the clutter rejection capabilities of the classifier. In the classification stage, with the probability of detection $P_d = 0.9$, nearly 99% of the

Table 5.3 Clutter false-alarm statistics/categories; PWF data, 0.3 m × 0.3 m resolution

Stage	Detection probability	Number of false alarms	Man-made discretes	Natural clutter false alarms
Detection	1.0	3457	1760	1697
Discrimination	1.0	709	444	269
Classification	0.9	6	4	2

Table 5.4 Clutter false-alarm statistics/categories; HH data, 0.3 m × 0.3 m resolution

Stage	Detection probability	Number of false alarms	Man-made discretes	Natural-clutter false alarms
Detection	1.0	8739	2266	6473
Discrimination	1.0	3758	1928	1830
Classification	0.9	229	83	146

clutter false alarms presented to the classifier were rejected as being non-targets. Finally, the false-alarm density achieved at $P_d = 0.9$ is calculated from the data in the table to be approximately 0.1 false alarms per km^2 (6 false alarms were detected in 56 km^2 total clutter area).

Table 5.4 presents the performance of all three stages of the ATR system for single-polarisation 1 ft × 1 ft resolution HH data. Note that the discrimination stage does not significantly reduce the false-alarm density. This result occurs partially because of the large amount of speckle contained in single-polarisation data, which prevents accurate estimates of the discrimination features. Additionally, the use of polarisation-based discrimination features such as percent pure even is not possible with single-polarisation data. The false alarms at each stage are categorised as either man-made discretes or natural clutter false alarms. Table 5.4 gives the breakdown of each stage into types of false alarms; as the data shows, the number of false alarms coming out of the detection stage was significantly greater than that achieved by using PWF data. Most notably, the number of natural clutter false alarms increased by nearly a factor of 4 (6473 natural clutter false alarms) resulting in a total of 8739 false alarms for HH data. By using the best subset of discrimination features for HH data (standard deviation, weighted fill, mean CFAR and mass) the discriminator rejected slightly more than half of the false alarms. Nearly all the false alarms rejected were from the natural clutter category – only a small fraction of the false alarms from man-made discretes were rejected. Table 5.4 also indicates that the template-matching classifier is able to reject most of the clutter discretes even though a single-polarisation channel is being used. At the final detection probability of $P_d = 0.9$, nearly 94% of the false alarms presented to the classifier were rejected as being non-targets. However, the performance is significantly worse than that achieved by using PWF imagery. The false-alarm density achieved at $P_d = 0.9$ is calculated from the data in the table to be approximately four false alarms per km^2 (229 false alarms in 56 km^2 total clutter area).

Fig. 5.10 presents a comparison of the end-to-end detection performance of the ATR system using 0.3 m × 0.3 m resolution PWF data versus 0.3 m × 0.3 m resolution HH data from the Stockbridge dataset. Overall, the end-to-end performance of the ATR system is degraded when using HH data compared to using PWF data. Note that the false-alarm density with PWF data is over an order of magnitude lower than with HH data.

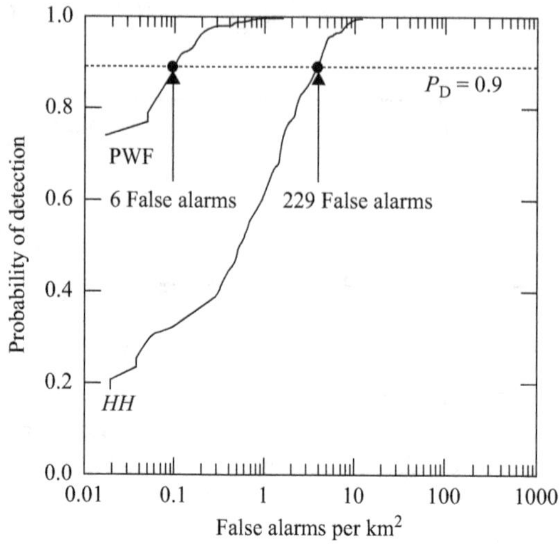

Figure 5.10. Input-to-output detection curves for the ATR system, 0.3 m × 0.3 m PWF data versus HH data

5.4 Improving ATR performance via high-definition image processing

In this section we present an approach that has been shown to improve the ATR performance of the template-based classifier using complex SAR imagery that has been resolution-enhanced using Lincoln Laboratory's high-definition image (HDI) processing [10]. The SAR imagery used in these studies was gathered in the fall of 1995 at the Redstone Arsenal in Huntsville, Alabama, by the Sandia X-band (9.6 GHz) HH-polarisation SAR. Fig. 5.11 shows a typical SAR image of the target array imaged by the sensor. The radar illumination is from the top; therefore, radar shadows appear below tall objects. A road can be seen in the lower right corner of the image. The plus (+) signs represent corner reflectors that were placed in the scene and used for image registration. The data comprise a large set of military targets imaged over 360° of aspect. In these studies the recognition performance of a template-based mean-square-error classifier was evaluated using imagery of the 18 distinct targets contained in the dataset.

Photos of the individual targets in the target array are shown in Fig. 5.12; included are three versions each of the BMP2 armored personnel carrier, the M2 armored personnel carrier and the T72 main battle tank. The T72 tanks contain significant differences from tank to tank; T72#2 has barrels mounted on the rear of the target; T72#3 does not have skirts along the side of the target. The BMP2 and M2 armoured personnel carriers have minor differences in target-to-target configuration. We trained a ten-target classifier and then evaluated the ability of the classifier to recognise and classify all 18 targets shown in the figure. Of the eight

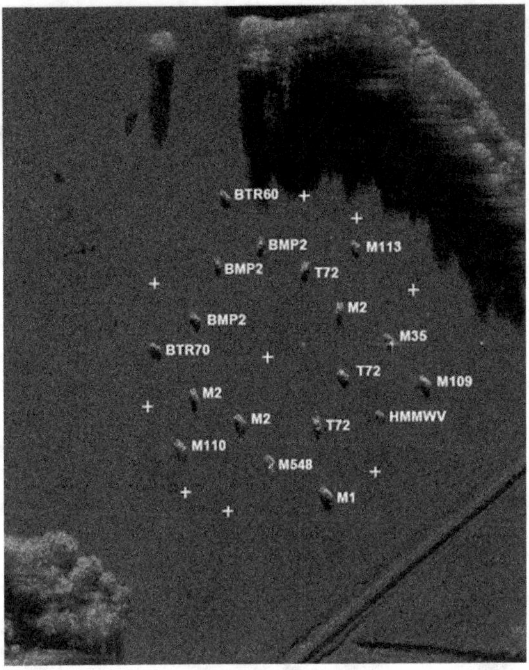

Figure 5.11 SAR image of the MSTAR target array; ground truth superimposed, '+' signs denote point scatterers

test vehicles, the HMMWV and M35 were used as confuser vehicles; the classifier was not trained to recognise these vehicles. In our presentation of results, the confuser vehicles should be classified as unknown. The SAR imagery used in these studies was gathered by the Sandia SAR sensor in support of the DARPA-sponsored Moving and Stationary Target Acquisition and Recognition (MSTAR) program [3]. The data comprise military targets imaged over 360° of aspect angle in spotlight mode and approximately 30 km^2 of natural and cultural clutter gathered in stripmap mode. Fig. 5.11 shows a typical 0.3 m × 0.3 m resolution SAR image of the target array and the identity of each target.

Fig. 5.12 shows photos of the eighteen target set, which includes three versions each of the BMP2 armoured personnel carrier, the M2 infantry fighting vehicle and the T72 main battle tank. The T72 tanks differ noticeably from each other – T72 2 has fuel drums mounted on the rear of the target, while T72 3 lacks skirts along the side of the target. The BMP2 and M2 infantry fighting vehicle have minor differences in target-to-target configuration. We constructed 72 classifier templates per target, covering approximately 360° of aspect per target; the total number of classifier templates was 720. The classifier was initially tested using the training data images as test inputs, providing a sanity check on the algorithm code.

Our initial classifier evaluations used non-HDI-processed data to establish a baseline with which the performance using HDI-processed data could be compared;

Figure 5.12 Photos of the 18 target vehicles used in these classifier studies

the improvement in classifier performance using HDI-processed data was then evaluated. Performance results are presented in terms of classifier confusion matrices, which show the number of correct and incorrect classifications achieved; the confusion matrices are summarised in terms of a probability of correct classification metric.

Table 5.5 is the classifier confusion matrix for the 0.3 m $\times$ 0.3 m resolution data. When the classifier was tested using the training data, perfect classifier performance was achieved. When the classifier was tested using the independent test

Table 5.5 Classifier performance (0.3 m × 0.3 m); $P_{cc} = 93.9\%$

	NUMBER OF TARGETS CLASSIFIED AS										
	BMP2	**BTR60**	**BTR70**	**M109**	**M110**	**M113**	**M1**	**M2**	**M548**	**T72**	**UNKNOWN**
BMP2#1	256										
BMP2#2	252										3
BMP2#3	255							1			
BTR60		256									
BTR70			256								
M109				256							
M110					256						
M113						256					
M1							255				
M2#1								256			
M2#2								251	2		2
M2#3								248	3		3
M548									256		
T72#1										256	
T72#2										232	23
T72#3										245	6
HMMWV	4	1				13		3			223
M35								1	6		248

#2, #3 = INDEPENDENT TEST DATA **M35, HMMWV = CONFUSERS**

data, nearly perfect classifier performance was achieved ($P_{cc} = 93.9\%$). Note, however, that the performance for T72#2, which contained extra barrels on the rear of the tank, resulted in 39 images out of the 255 total declared unknown. The performance for T72#3 (which did not have skirts along the sides of the target) was nearly perfect; only 4 images out of the 251 total were declared unknown. At this resolution, the classifier rejected a large number of confuser vehicles (438 images out of the total of 499).

Table 5.6 shows the classifier confusion matrix for 0.5 m × 0.5 m resolution data. The probability of correct classification for data at this resolution (calculated using only the independent test vehicles and the confuser vehicles) is 84.1%. At this resolution, the M35 truck was misclassified only 13 times out of the 255 total M35 test images. The HMMWV, however, was misclassified most of the time (only 61 HMMWV images were declared unknown).

Table 5.7 shows the classifier confusion matrix for the 1.0 m × 1.0 m resolution data. For these specific targets at this resolution, we observe a very large degradation in classifier performance. The probability of correct classification degraded to 45.4%. Note, however, that nearly perfect classifier performance was achieved when the classifier was tested on the training data; this result shows the importance of testing classifiers using *independent target test data*.

Fig. 5.13 presents a side-by-side comparison of M35 truck images: the left image (a) was formed using conventional 2D FFT SAR processing of 1.0 m × 1.0 m resolution data and the right image (b) is the corresponding HDI-processed

Table 5.6 Classifier performance (0.5 m × 0.5 m); $P_{cc} = 84.1\%$

	NUMBER OF TARGETS CLASSIFIED AS										
	BMP2	**BTR60**	**BTR70**	**M109**	**M110**	**M113**	**M1**	**M2**	**M548**	**T72**	**UNKNOWN**
BMP2#1	256										
BMP2#2	235	1						8		7	3
BMP2#3	237	1				1		11		4	3
BTR60		255									1
BTR70			256								
M109				256							
M110					256						
M113						255					1
M1							256				
M2#1								256			
M2#2				4	5			233		4	9
M2#3				2	5	1		239		4	5
M548									254		2
T72#1										256	
T72#2					3		8	3		217	24
T72#3				1			2	1		241	6
HMMWV	29	8	1	9	3	115		14	3	1	61
M35	1			3	3	1			5		242

#2, #3 = INDEPENDENT TEST DATA

M35, HMMWV = CONFUSERS

Table 5.7 Classifier performance (1.0 m × 1.0 m); $P_{cc} = 45.4\%$

	NUMBER OF TARGETS CLASSIFIED AS										
	BMP2	**BTR60**	**BTR70**	**M109**	**M110**	**M113**	**M1**	**M2**	**M548**	**T72**	**UNKNOWN**
BMP2#1	255							1			
BMP2#2	129	15	21	8	1	39	3	15		18	5
BMP2#3	135	7	10	5	2	43		25		18	11
BTR60		255				1					
BTR70			256								
M109				256							
M110					256						
M113	1					253					2
M1				1		1	253				
M2#1						1		255			
M2#2	14	3	2	37	14	36	5	121		16	7
M2#3	21	6		18	6	47	9	118		27	4
M548									256		
T72#1						1				255	
T72#2	7	12	3	19	9	23	17	32	1	94	38
T72#3	19	8	6	23	5	12	14	19		127	18
HMMWV	48	7	4	4		145		13	1		22
M35	8			12	3	29	4	4	13	1	175

#2, #3 = INDEPENDENT TEST DATA

M35, HMMWV = CONFUSERS

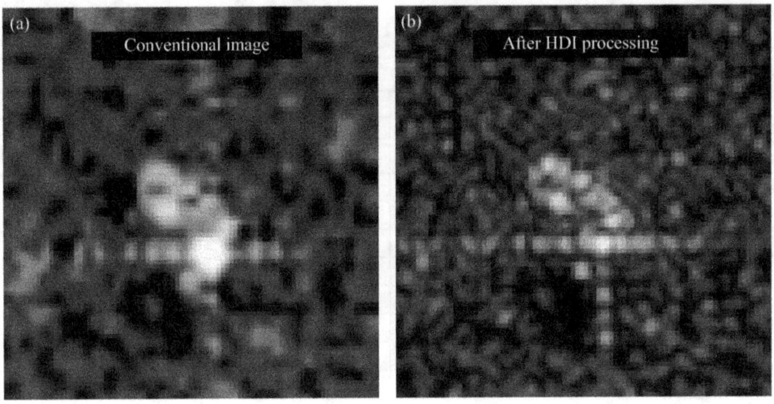

Figure 5.13 SAR images of an M35 truck: (a) 1.0 m × 1.0 m 2D FFT processed and (b) 1.0 m × 1.0 m HDI-processed

1.0 m × 1.0 m image. This image comparison validates that HDI-processing does significantly improve the image quality of SAR imagery – and therefore should result in improved ATR performance. Visually, the right image in Fig. 5.13 shows more clearly focused target scatterers, which should result in improved recognition of the target.

Table 5.8 shows the classifier confusion matrix for HDI-processed 1.0 m × 1.0 m resolution data (after HDI processing, the resolution of the data is approximately 0.5 m × 0.5 m). Comparing the results of Table 5.8 with the results of Table 5.7 shows a dramatic improvement in classifier performance. The probability of correct classification using HDI-processed data has increased by approximately 30% over that achieved with conventionally processed 1.0 m × 1.0 m resolution data; the probability of correct classification has increased from 45.4% to 73.4%. With HDI-processed 1.0 m × 1.0 m data, the number of rejected confuser vehicles increased from 197 images to 321 out of a total of 499 images. Although HDI processing of 1.0 m × 1.0 m data has resulted in a significant increase in the probability of correct classification (P_{cc} = 73.4%), performance using conventionally processed 0.5 m × 0.5 m resolution data gave somewhat better probability of correct classification (P_{cc} = 84.1%).

Fig. 5.14 presents a visual comparison of M35 truck images. The left image (a) was formed using conventional 2D FFT SAR processing of 0.3 m × 0.3 m resolution data; the right image (b) is the corresponding HDI-processed 0.3 m × 0.3 m image. This image comparison demonstrates that HDI-processing also improves the image quality of the high resolution SAR imagery – and the improved image quality should result in improved ATR performance. This conjecture is validated by the results presented in the following classifier confusion matrix (Table 5.9).

Table 5.9 shows the classifier confusion matrix for HDI-processed 0.3 m × 0.3 m resolution data (after HDI processing, the resolution of the data is approximately 0.15 m × 0.15 m). Comparing the results of Table 5.9 with the results of Table 5.5 shows somewhat-improved classifier performance; the probability of

Table 5.8 Classifier performance (1.0 m × 1.0 m + HDI); P_{cc} = 73.4%

	NUMBER OF TARGETS CLASSIFIED AS										
	BMP2	**BTR60**	**BTR70**	**M109**	**M110**	**M113**	**M1**	**M2**	**M548**	**T72**	**UNKNOWN**
BMP2#1	256										
BMP2#2	202	4	2			1		17		10	19
BMP2#3	195	2	4	2		4		22		17	10
BTR60		254									2
BTR70			256								
M109				256							
M110					256						
M113						255					1
M1							255				
M2#1								256			
M2#2	4		2	13	11	2	1	188		24	10
M2#3	8	1	2	16	5	3	1	180		27	13
M548									256		
T72#1										256	
T72#2	3			5		1	7	10		190	39
T72#3	5	1		3	4		3	10		211	14
HMMWV	25	8	8	1		101		8			93
M35	1			3	1	6			14		230

#2, #3 = INDEPENDENT TEST DATA

M35, HMMWV = CONFUSERS

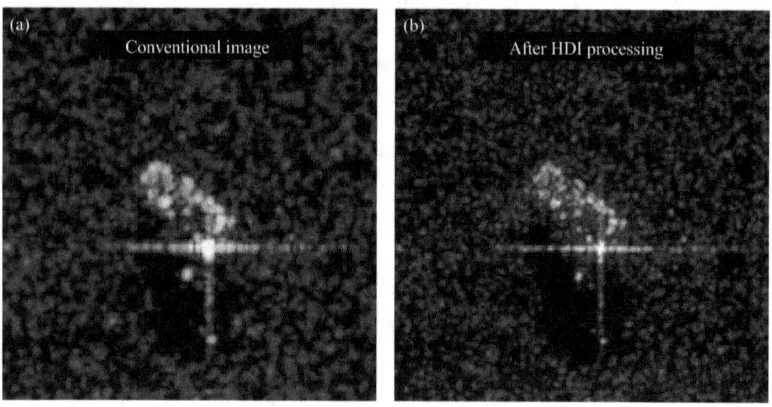

Figure 5.14 SAR images of an M35 truck: (a) 0.3 m × 0.3 m 2D FFT processed and (b) 0.3 m × 0.3 m HDI-processed

correct classification using HDI-processed data has increased to 96.4%, an improvement of 2.5% over the conventionally processed data – and with HDI-processed 0.3 m × 0.3 m data, the classifier rejected a larger number of confuser vehicles (471 images out of the total 499).

Table 5.9 Classifier performance (0.3 m × 0.3 m + HDI); $P_{cc} = 96.4\%$

	NUMBER OF TARGETS CLASSIFIED AS										
	BMP2	**BTR60**	**BTR70**	**M109**	**M110**	**M113**	**M1**	**M2**	**M548**	**T72**	**UNKNOWN**
BMP2#1	256										
BMP2#2	252										3
BMP2#3	255							1			
BTR60		256									
BTR70			256								
M109				256							
M110					256						
M113						256					
M1							255				
M2#1								256			
M2#2								251	2		2
M2#3								248	3		3
M548									256		
T72#1										256	
T72#2										232	23
T72#3										245	6
HMMWV	4	1				13		3			223
M35								1	6		248

#2, #3 = INDEPENDENT TEST DATA

M35, HMMWV = CONFUSERS

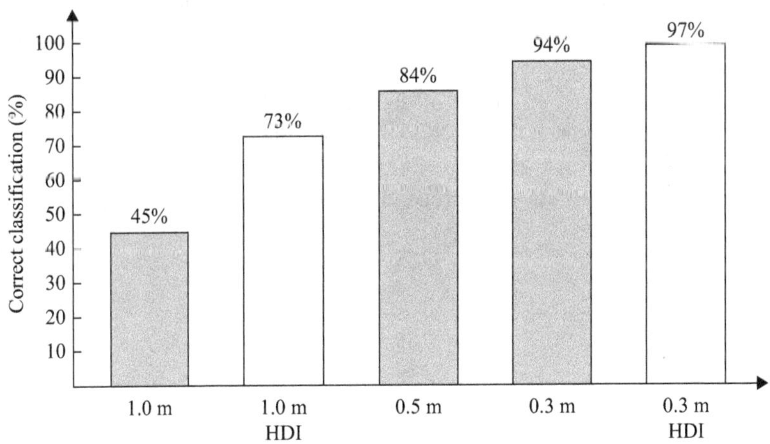

Figure 5.15 Summary of template-based ten-target classifier performance versus resolution

Fig. 5.15 presents a bar chart summarising the probability of correct classification (P_{cc}) versus SAR image resolution for the template-based ten-target classifier. The corresponding classifier confusion matrices for the results presented in Fig. 5.15 are given in Tables 5.5 through 5.9.

In section 5.3 we showed that significantly enhanced SAR image quality is obtained by processing full-polarisation complex SAR imagery using the polarisation whitening filter (PWF). Not only is the visual quality of PWF- formed SAR imagery clearer and nicely speckle-reduced compared with single-polarization imagery, the resulting ATR system performance is significantly improved. Each stage of the ATR system (detection, discrimination and classification) was demonstrated to have improved performance compared to the single-polarisation system [9]. In section 5.4 we showed that significantly enhanced SAR image resolution is also obtained by processing single-polarisation complex SAR imagery using Lincoln Laboratory's HDI algorithm. This image enhancement algorithm effectively improves the resolution of the original SAR imagery by (approximately) a factor of 2 – and the image speckle is also somewhat reduced. Thus, the probability of correct classification achieved using HDI-processed imagery in the ATR stage of the system is appropriately increased [13]. However, the ATR performance benefits achievable by combining both PWF and HDI processing of full-polarisation data have not yet been investigated and quantified. It seems such an approach should lead to additional improvements in SAR image quality and ATR performance.

A simplified block diagram of the combined PWF/HDI processing approach is shown in Fig. 5.16. The proposed approach is based on transformation of the full-polarisation (HH, HV and VV) complex data into an orthonormal polarisation basis via the PWF processing shown in Fig. 5.7; HDI processing is then applied to each of the complex orthonormal polarisations; the enhanced resolution HDI images are non-coherently averaged, resulting in the speckle-reduced, improved resolution SAR image.

Fig. 5.17 shows an example of the combined PWF and HDI processing obtained using the approach described in the block diagram (Fig. 5.16). Note that in forming the PWF/HDI image shown in Fig. 5.17, each pixel of an original complex image is super-resolved via HDI processing into a 3 × 3 array of amplitude-only

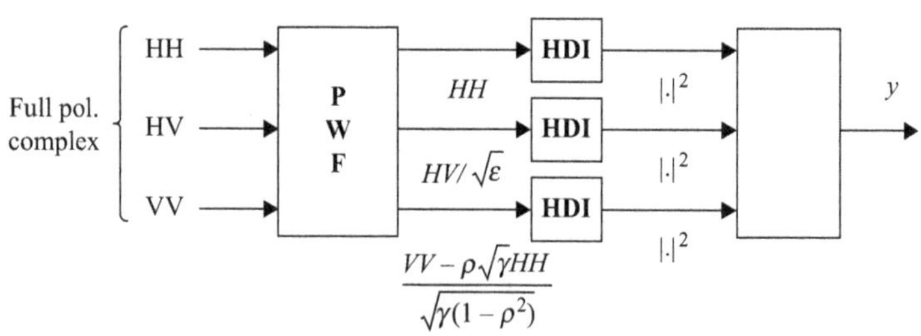

$$y = |HH|^2 + |HV/\sqrt{\varepsilon}|^2 + \left|\frac{(VV - \rho\sqrt{\gamma}HH)}{\sqrt{\gamma(1-\rho^2)}}\right|^2$$

Figure 5.16 Block diagram of an approach to perform combined HDI and PWF processing of full-polarisation SAR data

Figure 5.17 MMW SAR image, 0.3 m × 0.3 m resolution, full polarisation (HH, HV, VV), PWF/HDI-processed

pixels; the resulting image is visually clearer, has enhanced resolution and has additional speckle reduction.

5.5 Reconstruction of interrupted SAR imagery

In this section we investigate the effects of data gaps (interruptions) in SAR phase-history data due to interruptions in the SAR data collection [15]. We show examples of SAR imagery formed using phase-history data containing various types of gapping patterns, and we demonstrate that SAR image artefacts induced by interruptions in SAR phase-history data are mitigated using compressed sensing image formation, resulting in excellent quality complex SAR imagery.

Recent advances in radar technology allow modern day radar systems to employ active array antennas with transmit/receive modules at each antenna element. Active array technology equips the airborne radar with an agile beam, allowing the radar to accommodate multiple operational modes such as search, track, target recognition, synthetic aperture (SAR/ISAR) imaging, etc. These multiple modes may compete for radar resources, leading to an interrupted phase-history data collection for use in the formation of a high resolution SAR image. Typical coherent integration times may extend to several tens of seconds, thus potentially tying the radar resources to the single mode of operation for a long uninterrupted period of time. Depending on an aircraft's tactical situation, other high-priority modes may need to be serviced in a timely manner, and thus, it may be necessary to interrupt the SAR data collection randomly or periodically.

Such interruptions leave data gaps in the coherent SAR phase history, which can significantly degrade the resulting SAR image quality.

This section presents a summary of our most recent investigations into the effects on SAR image quality of various types of interruptions in the SAR phase-history data collection. We have focused on the use of compressive sensing approaches that might yield reconstructed SAR imagery having minimal loss in image quality. Specifically, we present the results of an image reconstruction study using a basis pursuit de-noising (BPDN) algorithm. Our studies demonstrate that excellent SAR image quality can be achieved using phase-history data that has been degraded by various types of interruptions. These studies used very good image quality data gathered by the Sandia X-band SAR for the MSTAR ATR program. We show that BPDN image formation of interrupted phase-history data yields reconstructed SAR imagery having *excellent* image quality.

The CS-based image reconstruction approach we use, referred to as 'basis pursuit de-noising', is described in References 30 and 31. Our study using BPDN processing for reconstruction of SAR imagery from gapped phase-history data is described as follows. We start with a complex SAR image having good image quality and we generate a 'pseudo' phase-history data array by taking the 2D Fourier transform of the SAR image. We then emulate the gapped phase-history data by appropriately removing range profiles from the (pseudo) phase-history, thereby mimicking unobserved aspects. We model the gapped phase-history data as $\underline{y} = RF\underline{x}$, where $\underline{y} \in C^m$ is the interrupted SAR phase-history and $\underline{x} \in C^{n^2}$ is an $n \times n$ spatial reflectivity image, both lexicographically ordered as vectors. The 2D Fourier transform operator is denoted by $F : C^{n^2} \rightarrow C^{n^2}$, and the aspect selection operator is denoted by $R : C^{n^2} \rightarrow C^m$.

For image reconstruction we use BPDN, defined as follows:

$$\hat{\underline{x}} = \arg_{\underline{x}} \min \|\underline{x}\|_1 \ s.t. \ \|\underline{y} - RF\underline{x}\|_2 \leq \sigma$$

where the sparsity of $\underline{x}$ is enforced on the magnitude of the spatial reflectivity, i.e.

$$\|\underline{x}\|_1 = \sum_i \sqrt{(R_e[x_i])^2 + (I_m[x_i])^2}$$

Fig. 5.18 shows examples of the SAR phase-history data after being interrupted by several types of gapping patterns; the figure shows examples of a single contiguous gap, a periodic gapping pattern and a random gapping pattern; each of these interruptions comprises 100 total unit-width interruptions.

Fig. 5.19 shows the SAR images obtained using Matched Filter (Taylor weighted, 2D FFT) image formation processing of the original uninterrupted phase-history data (a), the single contiguous interruption (b), the periodic interruptions (c) and the random interruptions (d); (interrupt patterns shown in Fig. 5.18). Note that the original uninterrupted SAR image contains a set of five military targets located in a homogeneous clutter background (a grassy field). Also visible in the image are a two good quality corner reflectors (metal spheres) – note the sharpness of these scatterers. The image quality of the images formed using matched filter processing

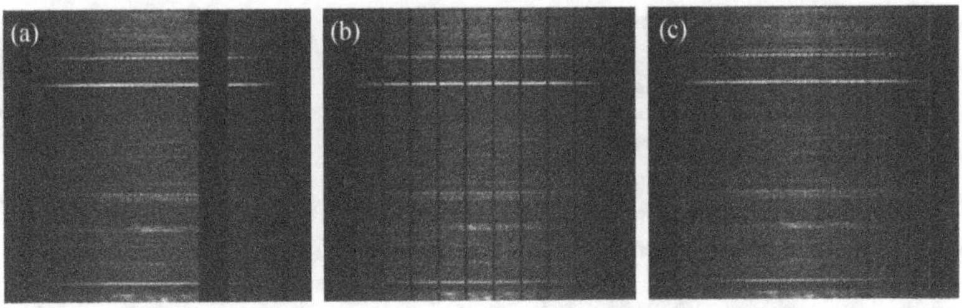

Figure 5.18 Range profiles of phase-history data filled with zeros during interrupts; total interrupt width is 100 in 1024 profiles: (a) single gap, (b) periodic interrupts (10 gaps, each of width 10) and (c) 100 random gaps

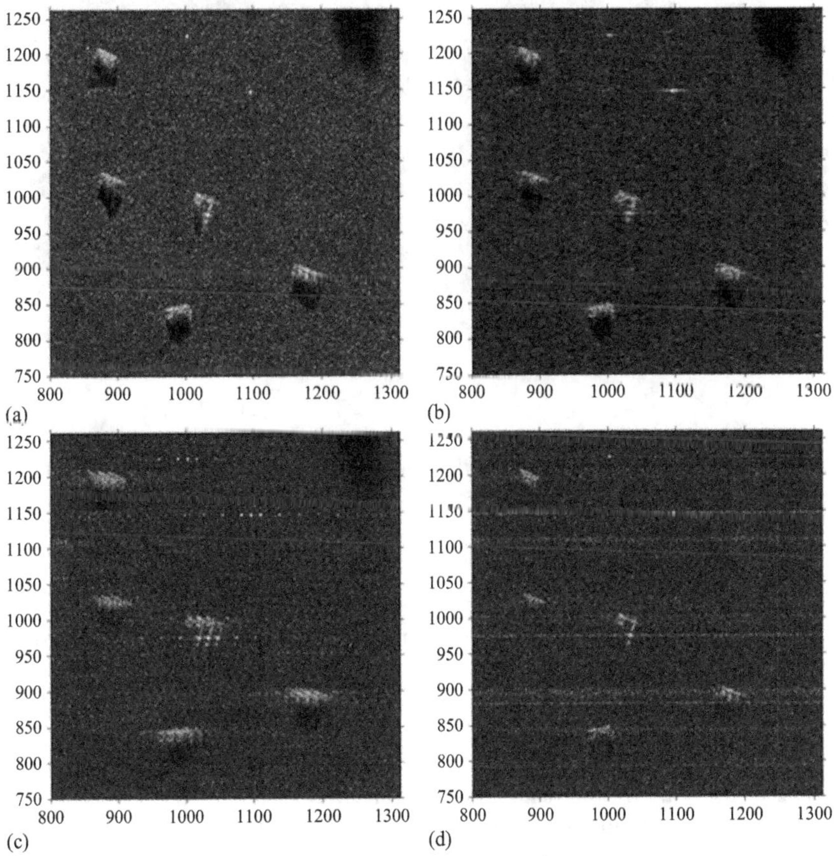

Figure 5.19 Matched filter image reconstruction: (a) uninterrupted phase-history,(b) single contiguous interruption, (c) periodic interruptions and (d) random interruptions

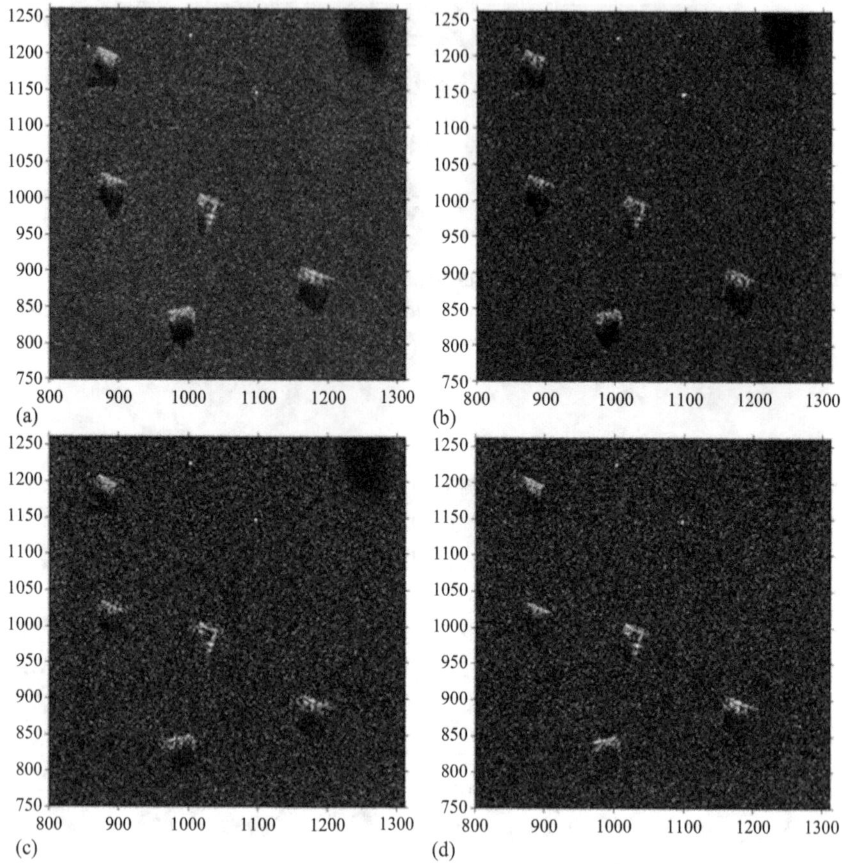

Figure 5.20 BPDN image reconstruction: (a) uninterrupted phase-history, (b) single contiguous interruption, (c) periodic interruptions and (d) random interruptions

of interrupted SAR phase-history data shown in Fig. 5.19(c) and (d) is significantly degraded compared to the original SAR image shown in Fig. 5.19(a). For example each of the targets in Fig. 5.19(c) appears significantly blurred – and reliable recognition of the targets in this image by either an automatic algorithm or an image analyst is most likely not possible.

Fig. 5.20 shows the SAR images obtained using BPDN image formation processing of the phase-history data arrays displayed in Fig. 5.18. We use the SPGL1 code [32] to numerically evaluate the BPDN solution. Note that the SAR image obtained from the BPDN-processed uninterrupted phase-history data (Fig. 5.20(a)) is essentially identical to the original uninterrupted SAR image shown in Fig. 5.19(a) – and the corresponding BPDN reconstructed images obtained from the interrupted phase-history data arrays have *excellent* image quality (see Fig. 5.20(b)–(d)). Additional details of these studies can be found in Reference 17.

Figs. 5.19 and 5.20 compared images obtained from two image formation approaches (BPDN and IFFT2) used to construct SAR imagery from various types of gapped phase-history data. First we evaluated the quality of the images formed using the baseline SAR image formation approach, i.e. standard 2D FFT (denoted IFFT2) processing of the phase-history data. These studies gave us an indication of the type and magnitude of the image distortions caused by various phase-history gapping patterns. Next we evaluated the quality of images formed using the BPDN approach. From visual observations of the SAR images formed using IFFT2 processing it was conjectured that reliable recognition of targets in these images by either an image analyst or an ATR algorithm is probably not possible. However, it appears that BPDN reconstructed imagery is clear enough for the image analyst to recognise the targets in the image – and an ATR algorithm should also be able to provide accurate recognition results. In order to quantify these conjectures, we performed a simple ATR study. We calculated a similarity metric between the distorted target images and the original, full aperture (non-gapped) target images. Fig. 5.21 shows these smaller sized chips of each of the five targets contained in the above sub-images. We performed a study of the similarity of each of these five reference target chips compared with these same targets in the IFFT2 and BPDN SAR images. The 'structural similarity index metric' (SSIM) calculated between the gapped IFFT2-formed images and the BPDN-formed images versus the original reference images provide the similarity measure [33,34]. Fig. 5.20 shows images of the target chips (reference chips) extracted from the non-gapped SAR image (Fig. 5.19(a)).

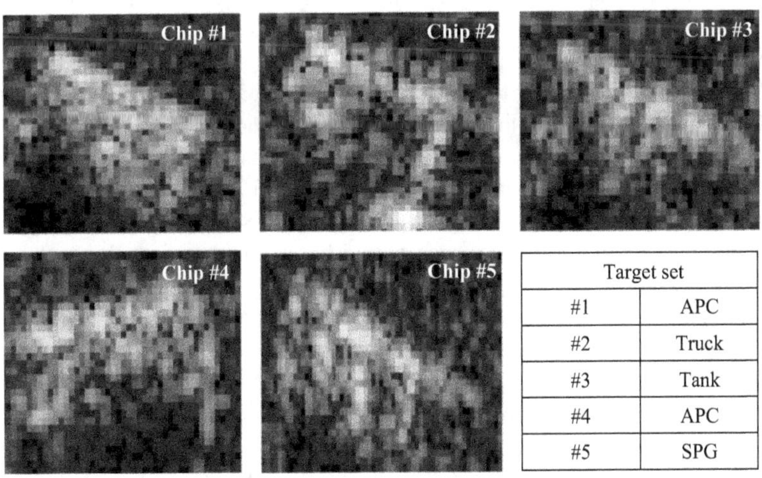

Target set	
#1	APC
#2	Truck
#3	Tank
#4	APC
#5	SPG

Figure 5.21 Target chips extracted from original, full aperture SAR image (Figure 5.19(a))

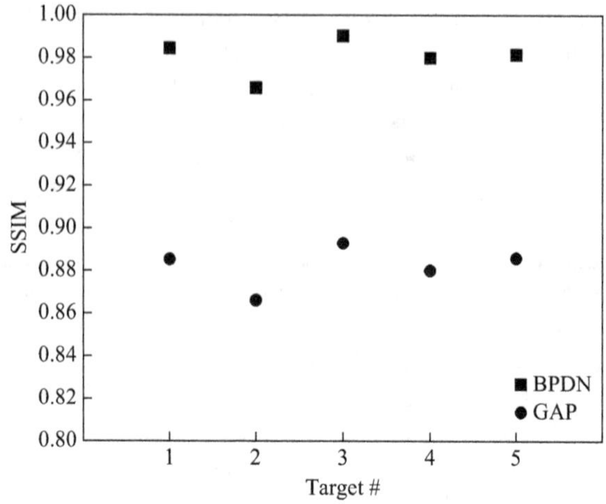

Figure 5.22 SSIM index values, BPDN versus IFFT2 images of targets 1–5; periodic gaps, total width = 350

Fig. 5.22 presents the results of our simple ATR studies. In these studies we employed the image similarity metric developed in References 33 and 34. This new similarity metric is referred to as the SSIM and has been shown to provide unambiguous image similarity results when mean-squared error measures are totally ambiguous. Each of the target reference chips shown in Fig. 5.21 was processed as follows: the SSIM index was calculated between these full aperture reference target chips and the corresponding *periodically gapped* images; the target chips were scanned over the periodically gapped images (Figs. 5.19(c) and 5.20(c)) and the maximum SSIM value was selected as the most similar object in the scene. Fig. 5.22 presents the similarity measures calculated for each of the five targets. As the data indicates, for periodic gaps with total gap width = 350, an average similarity metric ~0.97 was obtained for the BPDN-processed targets, implying that an ATR algorithm should obtain good recognition performance against the BPDN-processed SAR imagery. An average similarity index (SSIM) of ~0.91implies much less accurate recognition performance would be obtained against the periodically gapped IFFT2-processed imagery and numerous false target declarations would be obtained. For the more difficult case of periodic gaps with total gap width = 700, an average similarity ~0.90 was obtained for the BPDN-processed targets, whereas the average similarity measure for IFFT2-processed imagery was ~0.60. These similarity measures imply that BPDN-processed imagery would provide fairly good target recognition performance, whereas IFFT2-processed imagery would be significantly degraded, implying that very poor target recognition performance would be obtained. The results of this simple ATR study are preliminary and should be validated using a larger dataset of targets and an automatic target recognition system such as the model-based MSTAR system.

5.6 Summary and conclusions

Phase gradient SAR image focusing was demonstrated to provide well-focused imagery; cross-range smearing of the imagery was significantly reduced, resulting in higher probability of correct classification as demonstrated by a 20+ target model-based classifier. HDI processing was demonstrated to improve the image quality of complex SAR imagery; the effective resolution of SAR imagery was shown to be increased as demonstrated by the improved P_{cc} achieved by a ten-target template-based classifier. 2D FFT image formation processing of interrupted SAR phase-history data was shown to yield SAR imagery containing significant artefacts and degraded image quality; CS-based image formation processing (BPDN) was shown to mitigate these image artefacts and produced complex SAR imagery having excellent image quality.

References

[1] C. Jakowatz, D. Wahl, P. Eichel, D. Ghiglia and P. Thompson, *Spotlight-mode Synthetic Aperture Radar: A Signal Processing Approach*, Kluwer Publishers Norwell, MA, 1996.

[2] D.E. Wahl P.H. Eichel D.C. Ghiglia and C.V. Jakowatz Jr., 'Phase Gradient Autofocus – A Robust Tool for High-Resolution SAR Phase Correction', *IEEE Trans. Aerosp. Electron. Syst.*, Vol. 30, No. 3, July 1994.

[3] Moving and Stationary Target Acquisition Recognition (MSTAR), Program Review, Denver, CO, November 1996.

[4] E. R. Keydel, S. W. Lee and J. T. Moore, 'MSTAR Extended Operating Conditions: A Tutorial', *Proceedings of SPIE, Algorithms for Synthetic Aperture Radar Imagery III*, Vol. 2757, pp. 228–242, 1996.

[5] T. D. Ross and J. Mossing, 'MSTAR Evaluation Methodology', *Proceedings of SPIE, Algorithms for SAR Imagery VI*, Vol. 3721, p. 705, Orlando, FL, 1999.

[6] J. Henry, 'The Lincoln Laboratory 33 GHz Airborne Polarimetric SAR Imaging System', *IEEE National Telesystems Conference*, p. 353, Atlanta, GA, 26–27 March 1991.

[7] L. Novak, 'Target Detection Studies Using Fully Polarimetric Data Collected by the Lincoln Laboratory MMW SAR', *IEE International Radar Conference*, Brighton, UK, 12–13 October 1992.

[8] L. Novak, M. Burl, R. Chaney and G. Owirka, 'Optimal Processing of Polarimetric Synthetic Aperture Radar Imagery', *Lincoln Lab. J.*, Vol. 3, p. 273, 1990.

[9] L. Novak, S. Halversen, G. Owirka and M. Hiett, 'Effects of Polarization and Resolution on the Performance of a SAR ATR System', *Lincoln Lab. J.*, Vol. 8, No. 1, pp. 49–68, 1995.

[10] G. Benitz, 'High Definition Vector Imaging', *Lincoln Lab. J.*, Vol. 10, No. 2, 1997.

[11] G. Benitz, 'High-Definition Vector Imaging for Synthetic Aperture Radar', *31st Asilomar Conference on Systems, Signals & Computers*, Vol. 2, pp. 1204-1209, Pacific Grove, CA, 2–5 November 1997.

[12] L. Novak, G. Benitz, G. Owirka and L. Bessette L. 'ATR Performance Using Enhanced Resolution SAR', *SPIE*, Vol. 2757, pp. 332–337, 1996.

[13] L. Novak, G.R. Benitz, G.J. Owirka and J.D. Popietarz, 'Classifier Performance Using Enhanced Resolution SAR Data', *IEE International Radar Conference*, Edinburgh, UK, 1997.

[14] L. Novak, G. Owirka and A. Weaver, 'Automatic Target Recognition Using Enhanced Resolution SAR Data', *IEEE Trans. Aerosp. Electron. Syst.*, Vol. 35, No. 1, January 1999.

[15] J. Salzman, D. Akamine, R. Lefevre and J. C. Kirk, 'Interrupted Synthetic Aperture Radar (SAR)', *IEEE Aerosp. Electron. Syst. Mag.*, Vol. 17, No. 5, pp. 33–39, May 2002.

[16] R. Tough, K. D. Ward and P. Shepherd, 'Modelling and Enhancement of Interrupted Synthetic Aperture Radar Imagery', *IET Radar, Sonar Navigat.*, Vol. 1, No. 6, pp. 407–413, December 2007.

[17] I. Stojanovic, W. C. Karl and L. Novak, 'Reconstruction of Interrupted SAR Imagery for Persistent Surveillance Change Detection', *SPIE Conference on Algorithms for Synthetic Aperture Radar Imagery*, Baltimore, MD, April 2012.

[18] S. Tsunoda, F. Pace, J. Stence, M. Woodring, W.H. Hensley, A.W. Doerry and B.C. Walker, 'Lynx: A High Resolution Synthetic Aperture Radar,' *SPIE*, Orlando, FL, 1999.

[19] W. H. Hensley, A. W. Doerry, B. C. Walker, S.I. Tsunoda, F. Pace, J. Stence and M. Woodring 'A High Resolution, Light-Weight Synthetic Aperture Radar for UAV application', *45th Tri-Service Radar Symposium*, Monterey, CA, 22–24 June 1999.

[20] G. Goldstein, 'False-Alarm Regulation in Log-Normal and Weibull Clutter', *IEEE Trans. Aerosp. Electron. Syst.*, Vol. 19, p. 84, 1973.

[21] E. Kelly, 'An Adaptive Detection Algorithm', *IEEE Trans. Aerosp. Electron. Syst.*, Vol. AES-22, pp. 115–127, 1986.

[22] T. Kragh, 'Minimum-Entropy Autofocus for Three-Dimensional SAR Imaging', *Proceedings of SPIE 7337, Algorithms for Synthetic Aperture Radar Imagery XVI, 73370B*, 28 April 2009.

[23] T. Kragh, 'Monotonic Iterative Algorithm for Minimum-Entropy Autofocus', *ASAP Conference*, Lincoln Laboratory, 2006.

[24] L. Novak, M. Burl and W. Irving, 'Optimal Speckle Reduction for Enhanced Target Detection', *IEEE Trans. Aerosp. Electron. Syst.*, Vol. 26, No. 2, pp. 293–305, 1990.

[25] J. Cadzow, 'Generalized Digital Matched Filtering', '*Proceedings of 12th Southeastern Symp. on System Theory*, p. 307, Virginia Beach, VA, 19–20 May 1980.

[26] M. Burl, G. Owirka and L. Novak, 'Texture Discrimination in Synthetic Aperture Radar Imagery', *23rd Asilomar Conference on Signals, Systems, and Computers*, p. 399, Pacific Grove, CA, 30 October–1 November 1989.

[27] D. Kreithen, S. Halversen and G. Owirka, 'Discriminating Targets from Clutter', *Lincoln Lab. J.*, Vol. 6, p. 25, 1993.

[28] K. Fukunaga, R. Hayes and L. Novak, 'The Acquisition Probability for a Minimum Distance One-Class Classifier', *IEEE Trans. Aerosp. Electron. Syst.*, Vol. 23, p. 493, 1987.

[29] L. Novak, G. Owirka, W. Brower and A. Weaver, 'The Automatic Target Recognition System in SAIP', *Lincoln Lab. J.*, Vol. 10, No. 2, 1997.

[30] S. Chen, 'Basis Pursuit', PhD Dissertation, Stanford University, November 1995.

[31] S. Chen, D. Donoho and M. Saunders, 'Atomic Decomposition by Basis Pursuit', *SIAM J. Sci. Comput.*, Vol. 20, pp. 33–61, 1998.

[32] E. van den Berg and M. Friedlander, 'SPGL1: A Solver for Large-Scale Sparse Reconstruction', June 2007, http://www.cs.ubc.ca/labs/scl/spgl1.

[33] Z. Wang and A. Bovic, 'Mean Squared Error: Love It or Leave It? A New Look at Signal Fidelity Measures', *IEEE Signal Process. Mag.*, Vol. 26, No. 1, pp. 98–117, January 2009.

[34] Z. Wang and A. Bovic, 'A Universal Image Quality Index', *IEEE Signal Process. Lett.*, Vol. 9, No. 3, pp. 81–84, March 2002.

Chapter 6
Comparing classifier effectiveness

Joachim Schiller and Karlhans Rosenbach

6.1 Introduction

The classification and even, if possible, identification of any object or target under observation unambiguously at a distance and in such short times that a decision and an 'adapted' reaction are possible was, is and will be a most important task in the military and also security domains. There is an ongoing effort in many countries all over the world, and of course also in NATO, to improve the existing solutions. One approach, the obvious one, is the idea to allow any object entering observed areas to identify itself actively on request. Such systems have been installed for many years in the military domain and are called 'Q&A' systems or 'identification friend or foe' (IFF) systems. For air targets, and if radar is used as the sensor system, aircraft have been equipped for many years with systems like Mark X or Mark XII in different modes which were able to answer an interrogating pulse from the observer with a predefined code revealing, in addition to other information, its identity. Systems like that are also used in the civilian air traffic control (ATC) and because they support the interrogator they are called 'cooperative identification systems'. However, they show some important drawbacks.

First of all they require active support by the observed target, which is by no means self-evident. There are lots of reasons why a target under consideration does not answer on request:

- The equipment might be malfunctioning or not even be available (civil targets).
- The activated code might be wrong.
- The object may not have received the request or the interrogating system may not receive the answer (think of shadow zones).
- The target is not willing to reveal its identity.

In particular, the last point is an important concern in military and, increasingly important in the last decade, in security applications. So if a target is not willing to cooperate, the classification or identification needs to be achieved without its cooperation, i.e. non-cooperatively. Non-cooperative target identification (NCTI) is thus the only method to solve the above-mentioned problems, at least in principle. It is the only method with a potential for positive identification of non-answering objects, which could be neutral ones or foes. It is a most important step towards 'fratricide avoidance' and a clear step to increasing situational awareness.

6.2 NCTI studies

Since the late 1980s, possible methods for NCTI against air targets using radar have been investigated. The methods discussed have been (among others):

- determination of RCS,
- looking for resonances,
- jet-engine modulation,
- radar imagery of the targets.

Especially for aircraft powered by turbines, jet-engine modulation (JEM) was identified as a very interesting approach. But JEM shows a lot of drawbacks, so the next step was looking for the potential of radar imagery. In this chapter we will concentrate on the latter application. The data we will present are based on trials that were performed in the time span from 1992 to 2006 with many aircraft having been observed by the German FHR TIRA system which will be described roughly below.

The aim of this chapter is to deduce, based on real data, what are the main important steps in the process of NCTI and what are the drawbacks and limitations and where it is essential to put main emphasis. There are many efforts, and at the same time many hopes, concentrated on the idea of finding, like a *deus ex machina*, the one classifier solving all the problems, but is that really the important step?

6.3 Measurements

6.3.1 TIRA system

Starting in 1992 and in close cooperation with the NATO Research Study Group (RSG) 12 on Non-Cooperative Air Target Identification (in which Germany participated and in whose successor groups Germany continues to participate) real, in-flight target data were collected during extensive measurement campaigns using the FGAN (today Fraunhofer FHR) TIRA (tracking and imaging radar) system located at Wachtberg, Germany. Wachtberg is located just a few kilometres south of Bonn in the middle of Germany. Following the investigation of the JEM technique we stepped forward to study the use of radar imagery for the task of NCTI of aircraft. The group was, and still is, convinced that, based on theoretical studies, there is no way around test and validation of the developed ideas with *real* data to find out about the usefulness, robustness and confidence in real applications. It is also the only way to validate the assumptions made and it normally brings you 'back to earth' concerning some expectations.

In Fig. 6.1 an aerial photograph of the Fraunhofer compound at Wachtberg is shown where you can easily identify the radome of our TIRA system. The TIRA system works with a tracking unit in the L-band (1.3 GHz) and a co-located

Figure 6.1 Aerial photograph of the Fraunhofer compound at Wachtberg

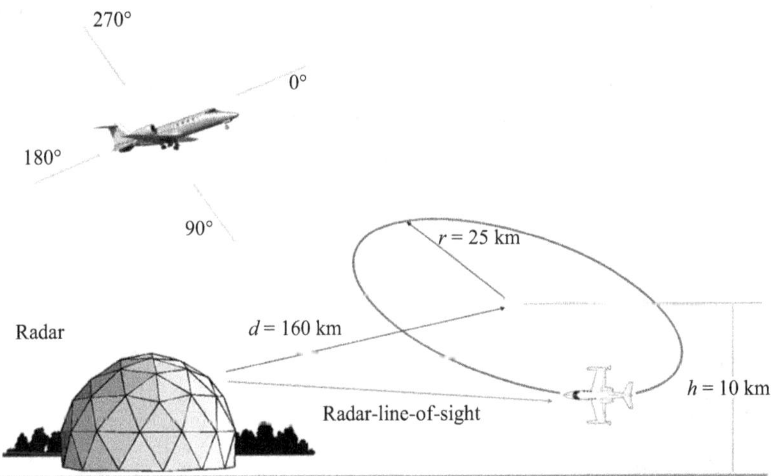

Figure 6.2 Trials geometry

imaging system in the Ku-band (16.8 GHz). The antenna has a diameter of 34 m and the bandwidth we used for our measurements was 800 MHz leading to a resolution of 18 cm.

As area for the trials we selected a temporary reserved area (TRA) over Volkel in the Netherlands (thanks to our Dutch colleagues and the Dutch military authorities), at a distance of about 160 km from TIRA, that fitted well to the blind zone of TIRA of about 100 km. Since this TRA was, of course, limited in diameter and because we looked for target data at all aspect angles we decided to use circular flights of the targets with a diameter of the circle of 50 km. In Fig. 6.2 a schematic presentation of the trials geometry is shown.

6.3.2 Targets

We started with the most challenging situation to find out about the capabilities of radar for air-target NCTI: we took data of different aircraft which are virtually members of the same target class as far as length, shape, velocity, manoeuvrability and other factors are concerned. Beginning in 1992 and followed by campaigns in 1993, 1997/1998, 2001, 2002, 2006 we established a comprehensive database for NCTI investigations. For this discussion, only a subset of six targets out of this database was selected for data security classification reasons among others.

6.4 Idea of classification

The goal of target classification for classes such as:

- in air applications: airliner, business aircraft, transporter, bomber, fighter aircraft, helicopter and
- in ground applications: tank, truck, missile launcher, jeep, civilian car, motorcycle, etc.

is firstly not to *identify* the targets directly but to *separate them* into the different classes. This is a very important message, i.e. that the main goal of a classifier is to find discriminating parameters, which may not be the prominent ones.

The investigations will be based on the so-called 'features' which are, in our radar case, properties of the objects that can be measured using a radar, e.g.:

- radar cross section (RCS),
- signal modulations,
- resonances,
- 1D and 2D target images,
- other radar properties.

6.4.1 Appropriate features

Quite a lot of investigations have been undertaken to find out what might be the information in the radar returns which will be distinctive enough for target description and separation and how to extract this information from the received radar signal. We kept in mind the approaches from the sonar community. In some ways they are in a more favourable position as long as they rely on a passive mode, i.e. just listen to the sound or noise which is emitted by the target under consideration. In the area of electromagnetic signals this situation is more comparable to the electronic warfare (EW) case with electromagnetic support measures (ESM) systems 'listening' to the electromagnetic emissions of a target. In addition to restrictions concerning detection distances and visibility, optics also has a more 'natural' potential for target identification than radar, so what should be used in the radar case?

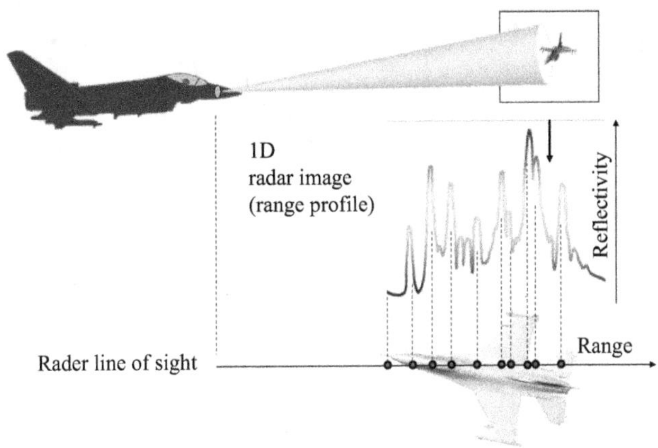

Figure 6.3 Schematic representation of the principles of HRR and 2D ISAR radar imagery [Courtesy of Dr Tanja Bieker, Fraunhofer FHR]

As shown above there are some parameters which might be used for classification purposes, such as:

- radar cross section,
- resonances,
- modulations,
- radar images of the target.

In this discussion, we will concentrate on the most advanced parameter, i.e. radar images.

6.4.2 *HRR and 2D ISAR*

High range resolution (HRR) profiles are 1D representations of the target's scatterer distribution projected on the radar 'line-of-sight' (see Fig. 6.3).

They form 1D images and can, in principle, be established by just one radar pulse of appropriate bandwidth. The bandwidth of the radar signal determines the resolution via

$$\Delta r = c/2B \tag{6.1}$$

where:

Δr = resolution in range (often also called 'slant range')
B = bandwidth of the signal
c = speed of light

This means that the distance between the so-called 'range bins', i.e. the sample-values in range, is separated by this distance in space.

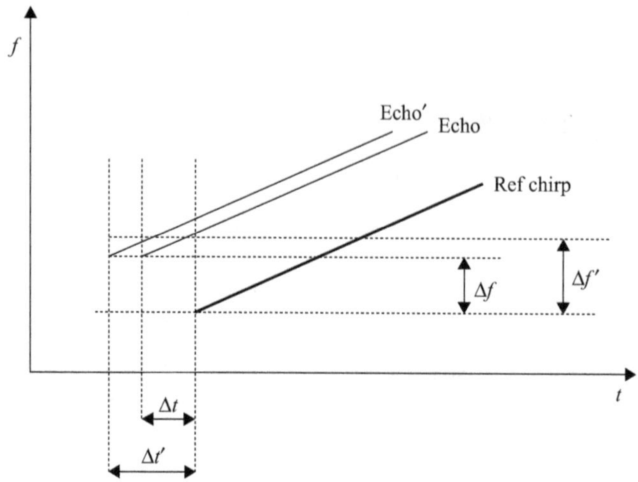

Figure 6.4 De-ramping technique [2]

There are different methods to gain such a 'range profile' from a target using radar, one of the most popular ones showing some advantages concerning sample frequency and processing time is the 'de-ramping technique' [1] which can be applied to the received signal in the analogue domain. In this case the emitted waveform, a chirp signal of appropriate bandwidth, is multiplied with the received signal and as a result frequency lines show up where the frequencies are proportional to the time delay between the emitted and the received signal (see Fig. 6.4). Subsequent use of an FFT as a second step then generates the HRR profile.

If the target is observed over some time using a number of radar pulses, a sequence of such HRR profiles can be measured. This is shown for a real measurement of an aircraft in Fig. 6.5(a).

From such a sequence of HRR profiles a 2D radar image can be formed using the inverse synthetic aperture radar (ISAR) technique. The technique is based on the same ideas as the synthetic aperture radar (SAR) technique with the difference that in SAR the radar is moving to establish the aperture while in ISAR the radar may rest but the target builds up the aperture. More details concerning the ISAR technique can be found in Reference 3 but the principal idea behind the ISAR technique will be discussed. By applying an appropriate transformation the target is transferred into a 'system-at-rest', i.e. its translational movement is zero. However, if the line-of-sight from the radar and the velocity vector of the target are not co-linear, i.e. if the closest point of approach (CPA) is not zero, a rotational motion of the target survives and a 'turntable' situation results. Imagine a target on a turntable, rotating over some degrees. If you illuminate the target with a radar, you will measure returns which show a Doppler-frequency offset proportional to:

- the rotational speed,
- the illuminating frequency,
- the distance of the scatterer from the rotation-axis.

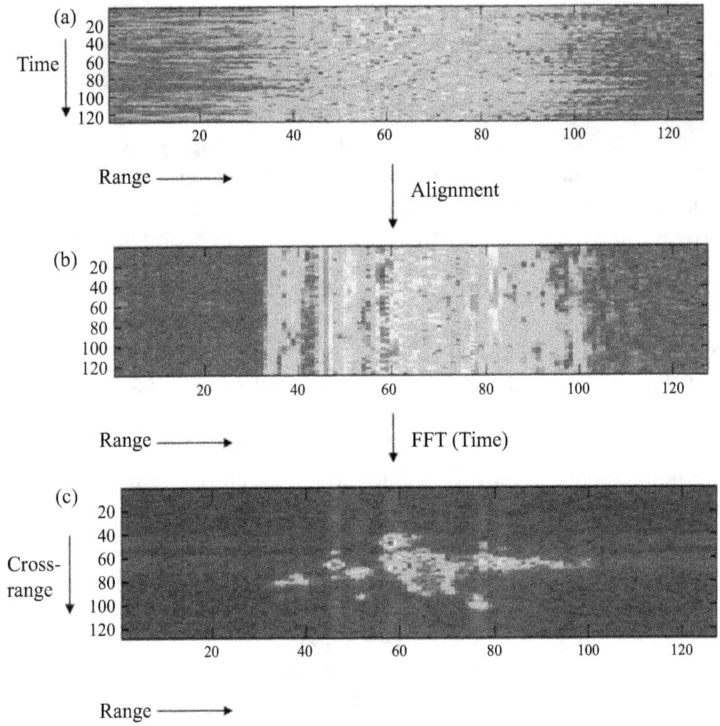

Figure 6.5 Formation of a 2D radar image

Thus, if in a range cell of the HRR profile a scatterer of the target exists over the time of observation, its phase will change which is equivalent to a frequency shift of the received signal. So performing an FFT over each range cell in the time direction, for all range cells containing a scatterer, a peak in the frequency dimension is formed. This frequency dimension can now be interpreted as the 'cross-range' dimension perpendicular to the 'slant-range' dimension visible in the HRR profiles.

In Fig. 6.5(a) it can easily be seen that the 'raw' sequence of the HRR profiles is not correctly aligned, i.e. there exist shifts in range resulting from the target motion and perhaps some imperfections of the radar system. So the first step on the way to a 2D ISAR image is the step of alignment of the HRR profiles. This has been performed in Fig. 6.5(b).

In practice there is another step necessary, before the 2D image can be formed, which is the step of phase compensation between the sequence of profiles. There are different techniques available and validated, one of which is the so-called 'dominant scatterer' technique in which one cell of the HRR profile is selected and its phase is corrected to zero over the time. (NB: As a result, this special cell becomes the rotation centre.) The phases of all the other range cells are corrected appropriately. If after that step the FFT is formed over time, a 2D ISAR image results, as depicted in Fig. 6.5(c).

A serious problem for practical applications is inherent in the ISAR data: as mentioned above the effect depends on a rotation of the target. In normal applications the observer has no direct knowledge of the rotational speed, ω. However, imagine the turntable situation: the phase derivative with respect to time, i.e. the Doppler frequency of a scatterer, depends on its velocity in the direction of the observer. On a turntable this velocity is increasing linearly with the distance of the scatterer to the centre of rotation, r. So we achieve the same Doppler frequency for a scatterer if the product of ω and r stays constant, following the simple equation

$$f_{\text{Doppler}} = f_{\text{radar}} 2\omega r / c \tag{6.2}$$

In ISAR, we have no control and also basically no knowledge regarding ω (this is different from the SAR case in which the motion of the observing platform is known). As a result, the correct so-called 'cross-range scaling' cannot be determined in real applications without additional information. Normally it is done by some reasoning about the most probable target motion (e.g. a car on a street may keep its velocity when it is going in a curve) or some rough tracking information on the target, but a sudden change in velocity may not be simple to detect. This effect can be seen in the sequence of ISAR images of an aeroplane observed continuously over just some seconds. Here the change in ω leads to the change in cross-range extension and the image is 'filled in' (see Fig. 6.6).

There have been some techniques proposed to mitigate the problem or even to solve it using higher order derivatives of the observed phase variations [4], but for practical measurements its application has been shown to be problematic.

The problem of uncertain cross-range scaling has a remarkable influence on the classification/identification capabilities of the classification techniques based on 2D-ISAR imagery and we spent considerable effort attempting to find features and/or techniques invariant to cross-range scaling. A feature-based approach will be demonstrated later (see section 6.6), while an alternative technique will be presented now.

6.4.3 2D ISAR template correlation classifier

As discussed above the problem of uncertain cross-range scaling is a crucial one for the capabilities of classification/identification based on 2D ISAR imagery. Throughout our investigations we realised that this might be the most important problem of all in using 2D ISAR images for classification. The reason that the cross-range scaling is uncertain of course directly influences the classification

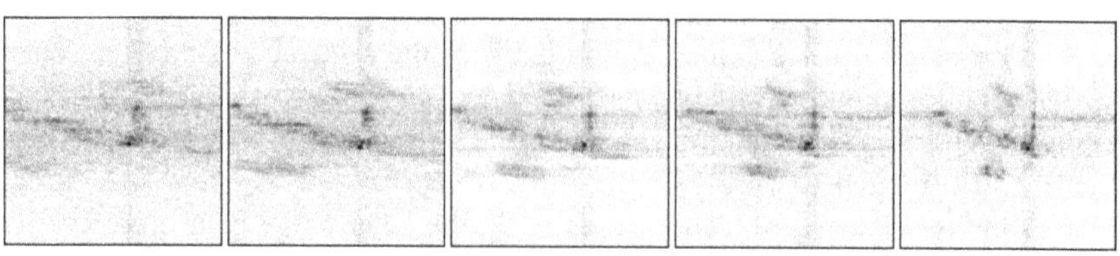

Figure 6.6 Effect of cross-range scaling uncertainty (time intervals of 1 s)

results, which can easily be seen by looking at the most common classification technique based on a correlation, where the measured 'test' image is compared with images in a database. It is quite obvious that differences in cross-range scaling lead to images with different stretch factors (see Fig. 6.6), which will not match perfectly, even if they show identical targets.

To mitigate the effect of uncertainty in the rotational speed, ω, which, from our experimental experience, can be of the order of up to a factor of 2, we developed a novel classification approach which will be presented here. The key idea is to cut the test image into small quadratic templates in the range and cross-range directions and to shift each of these templates in the cross-range direction over the reference image selected for classification. At one position it will fit best and that is exactly the position where the basic shift between the test and the reference image is a minimum.

Fig. 6.7 shows the principal idea in a 1D example symbolising the cross-range information at a pre-determined range distance. In this case the test image is simply a vector. If this test image is compared with a stretched version in the database we are faced with the situation depicted in Fig. 6.7. Here we cut the test image into nine templates which find their shifted and stretched counterparts in the nine templates of the stretched database image (marked by 1′, 2′, etc.).

At the best fitting position, i.e. a position where the shift has been balanced as well as possible, there still remains a difference in length between the template and the reference. However, for the small template, the remaining stretch effect (the cross-range template stretch) between itself and the reference is reduced by a factor of rs as compared with a correlation of the whole test and reference profiles (total cross-range stretch). Thus rs is given by

$$rs = n/x \tag{6.3}$$

where:

x = Length of the template (number of pixels) in the cross-range direction of the test image

n = Number of target pixels in the cross-range direction of the reference image

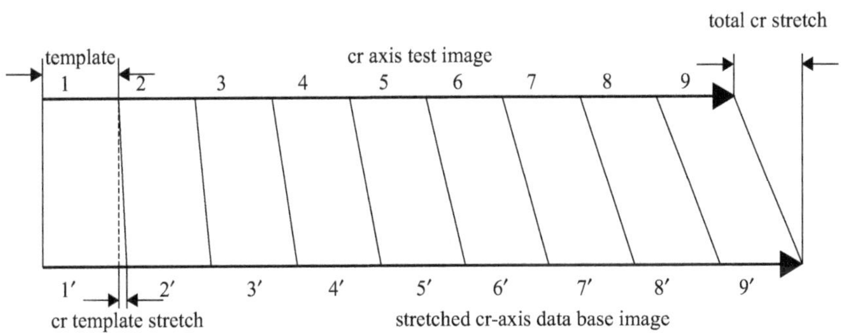

Figure 6.7 Schematic sketch of a template classifier to minimise the influence of unknown stretch

The classification procedure now is as follows:

1. start with a (quadratic in range and cross-range) template of x samples (in practice a factor of a quarter of the number of pixels in cross-range direction proved to be optimum) at, e.g. the upper left corner of the test image at maximum range (range cell m),
2. shift this template over the 2D reference image in the cross-range direction with an increment of one pixel,
3. perform at each position the cross-correlation between the template and the reference,
4. store the correlation maximum of this first range position of all cross-range positions,
5. shift the quadratic window one range resolution cell down to range cell $m - 1$, selecting by that a new quadratic template, and start with the same procedure at the left side of your image,
6. store the correlation maximum from all shifts in cross-range position,
7. continue until you reach lower left corner of the image, i.e. shortest range,
8. sum up the stored correlation maxima of each range line.

This sum is of course a measure for the similarity between the test and reference image and it proved to be widely independent of a stretch between the test and reference image. The capability of this approach will be shown under sections 6.5.4 and 6.6.

6.4.4 Selection of radar parameters

It was not known from the beginning what should be the optimal selection of waveform, bandwidth, carrier frequency and polarisation of the probing radar; this would be (hopefully) results of our investigations, but we made life simple, we just asked for the maximum. Getting data from our TIRA radar we decided to use a bandwidth of 800 MHz (comparing to 18 cm of resolution). The radar signal is circularly polarised, i.e. the axis of the electric and magnetic field is rotating. We had no choice concerning the waveform; TIRA uses a chirp or sweep, i.e. a frequency-modulated waveform. This waveform has some nice properties, so it is in principle possible to measure the range and the Doppler of a target at the same time to a satisfying accuracy (see discussion on ambiguity functions [5]). The Pulse Repetition Frequency (PRF) could be chosen to be 56 or 112 Hz, as necessary for our purposes.

6.5 Classification scheme

In Fig. 6.8 the generic processing/classification scheme is depicted. It consists of the different steps of the signal processing starting from raw data and ending with the application of the classifier.

In the beginning of the process some raw data are delivered by a sensor system. In our case these data are HRR profiles or 2D ISAR images, i.e. already pre-processed data. The next step is another step of pre-processing, where artefacts like

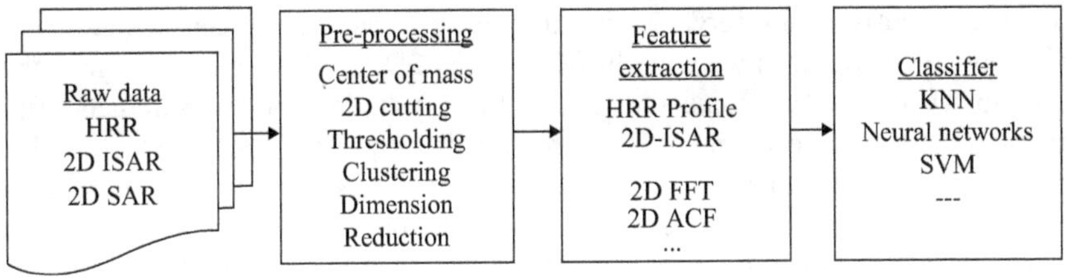

Figure 6.8 Processing scheme

outliers can be removed (while being careful not to destroy valuable information!). Other pre-processing steps for further signal conditioning may follow before extraction of (hopefully) discriminant features will take place. After that (including further pre-processing) the classification/identification step follows.

6.5.1 Pre-processing unit

The purpose of the pre-processing unit is to prepare the data for the following processing steps. This means that, as far as possible, noisy parts and artefacts should be reduced and the images should be prepared for the later comparison with data in a database.

As in the following steps care has to be taken not to remove valuable information. In the case of an aircraft powered by turbines, the JEM lines for example (see Fig. 6.9(a), the broad band near the rear of the target in the cross-range direction) could be looked upon as a noisy signal part that should be removed *or* as a typical signal part carrying valuable information. In principle this image

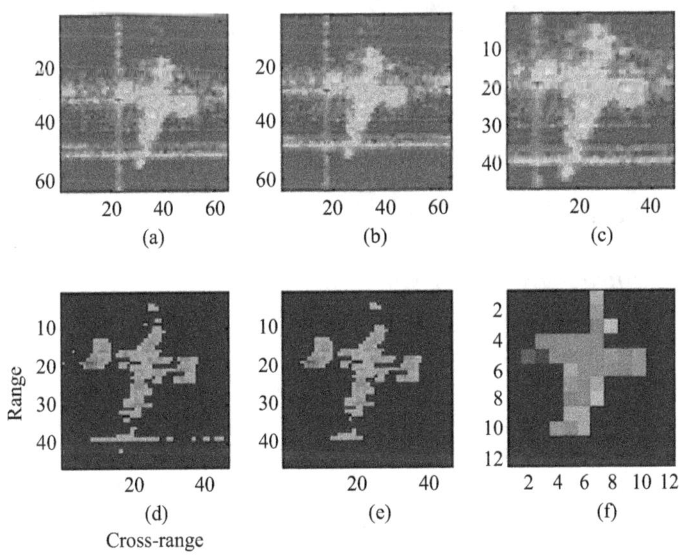

Figure 6.9 Image pre-processing (arbitrary units)

(see Fig. 6.9(a)) could serve as feature of the target under consideration (for the following discussions we will always consider 2D ISAR images as the raw data set). However, it shows some drawbacks: first of all for real measurements (not turntable data) it is very unlikely that the test data (measured data) are centred in accordance with the database used for comparison and classification. Besides that, the image may exhibit a cross-range scaling different from that of the database images. This problem of variations in cross-range scaling has been addressed above (see Fig. 6.6). In addition to scaling, an image rotation may also be present. So obviously there is a demand for features being invariant against affine transformations.

To achieve features that are invariant against translation in range and cross-range a centre of mass (COM) can be calculated from the images and for all images (test and reference data) the COM is shifted to the origin (Fig. 6.9(b)). Unfortunately this step often does not guarantee a sufficiently accurate result, e.g. changes in amplitudes or hidden parts of the target can change the COM considerably. As a result, further methods to achieve feature sets showing invariance against translation and (if possible) scaling will be discussed in the next section. In many cases the whole image plane is only partly occupied by the image under consideration, so a cutting step is implemented leading to a reduction in the number of pixels (see Fig. 6.9(c)).

As a next step, a segmentation step is introduced where connected areas composed of n connected pixels are identified (see Fig. 6.9(d)). By defining a threshold for the minimum number of pixels in such a cluster a 'cleaning' of the image is achieved where isolated clusters that may not belong to the image can be removed. Fig. 6.9(e) shows as an example an image after this pre-processing chain. This image has a resolution of 25×25 cm^2 in the range/cross-range plane. To reduce the dimension of our problem and at the same time to become less sensitive to residual errors in range/cross-range translation, scaling and rotation, a data reduction in the form of low-pass filtering of the image is advantageous. Fig. 6.9(f) shows the result of this process, where the smoothing to a reduced resolution of 1×1 m^2 is applied following our experience with aircraft data. But the advantageous effects of smoothing have to be balanced with the reduction of information generally encountered during this process. Besides that, the resolution has of course to be adapted to the identification task.

6.5.2 Feature extraction/reduction

We have already agreed on the use of images (1D and 2D images) of the target as basic features for the classification/identification task, but we still have the chance to derive effective information from these basic features for a faster, more reliable, more robust classification stage. As we had no idea (and still do not know exactly) about the origin of the different peaks and the structure of the HRR profiles, we decided to use the whole profile as an HRR feature for the following classification approach.

The same is true for the 2D ISAR images; after the pre-processing, normalisation and segmentation steps we ended up with a 2D image which we used in a first step for classification.

6.5.3 Choosing a classifier

A classification rule consists of a mapping from the feature space into the decision space giving evidence of, e.g. a target identity. Equivalently, a classification rule may be viewed as a decomposition of the feature space into non-overlapping subsets representing the different classes. If classes are well separated in the feature space, the classification task can be solved, at least in principle, without error. Otherwise, the Bayes error provides a lower limit to the classification error of any rule. In order to use the Bayes estimator [6], the underlying distribution of the data must be known to the designer, a prerequisite which is rarely met in practice. To be applicable to real-world problems, classification rules are usually based on a limited number of (independent) data samples, originating from the underlying distribution. Such a rule is called (universally) consistent if, for any distribution, with the number of samples tending to infinity, the associated error tends to the respective Bayes error. The first classification rule which was discovered to possess this property was the k-nearest neighbours (k-NN) rule, where k is somewhat related to the number of samples [7]. Ten years earlier, Cover and Hart [8] had already found that the limiting error of the NN rule, under mild restrictions concerning the underlying distribution, is bounded above by twice the Bayes error. Subsequently, many universally consistent rules have been designed. However, although being a desirable property from a theoretical point of view, the necessarily limited amount of data makes consistency a questionable concept in real-world problems. This pessimistic view may be confirmed by the fact that universally good classification rules do not exist. In contrast, any classification rule (based on a limited number of data samples) can have arbitrarily bad performance [9]. Moreover, the convergence towards the limiting error can be arbitrarily slow [10, Theorem 7.2]. Accordingly, in real applications the decision between several concurrent rules is usually decided in favour of the rule which performs best on a series of test samples, independently drawn from the same distribution. But the pragmatic question is: what are the differences between different classifiers, and is the choice of a classifier the really important part of the overall classification process?

For a practical application the question as to which classifier to choose is dependent on different constraints:

1. quality of the decision,
2. time consumption in training phase,
3. time consumption in decision phase,
4. adaptivity to changes in scenario, i.e. target types and number of classes,
5. memory needs.

While points 2 to 5 are rather different for different classifiers it is not clear from the beginning which classifier performs best. Besides that, the question of 'transparency' of a classifier should not be underestimated in the testing and developing phase. While, for example, an NN classifier easily allows for explanation of its behaviour, neural networks make it very complicated to follow the rationale of their decisions.

6.5.4 Test of classifiers

To find out about optimal classifiers in the problem of NCTI of air targets and based on sets of real data from different flight trials of NATO RSG 12 and follow-on groups, extensive tests of classifiers were performed at Fraunhofer FHR. Different classifiers were investigated and their performances as a function of signal-to-noise ratio, signal bandwidth, etc. were reported [11–14]. In the following, as one example, we discuss a comparison of nine classifiers using real data of a set of six target classes which were present during three different trials. The test and training data were strictly independent in the sense that they resulted from different trials representing targets of the individual target classes but with different tail numbers. One of the results is shown in Fig. 6.10: here for the nine different classifiers representing neural networks (Bayesian neural network (BNN), Gaussian mixture model (GMM), multi-layer perceptron (MLP), radial basis function (RBF), probabilistic neural network (PNN)), nearest neighbour classifier (NNC, k-NN), support vector machines (support vector machine linear classifier (SVML), support vector machine polynomial classifier (SVMP), support vector machine radial basis function classifier (SVMR)), and our own 'sub-space classifier' (based on an statistical, repeated selection of bins out of HRR profiles [15]) the probability of correct identification is shown. Tuning has been applied for each rule to optimise the respective free parameters. The aspect of the targets in this case was from nose-on to 14°; no rejection class was installed (i.e. forced decision classification).

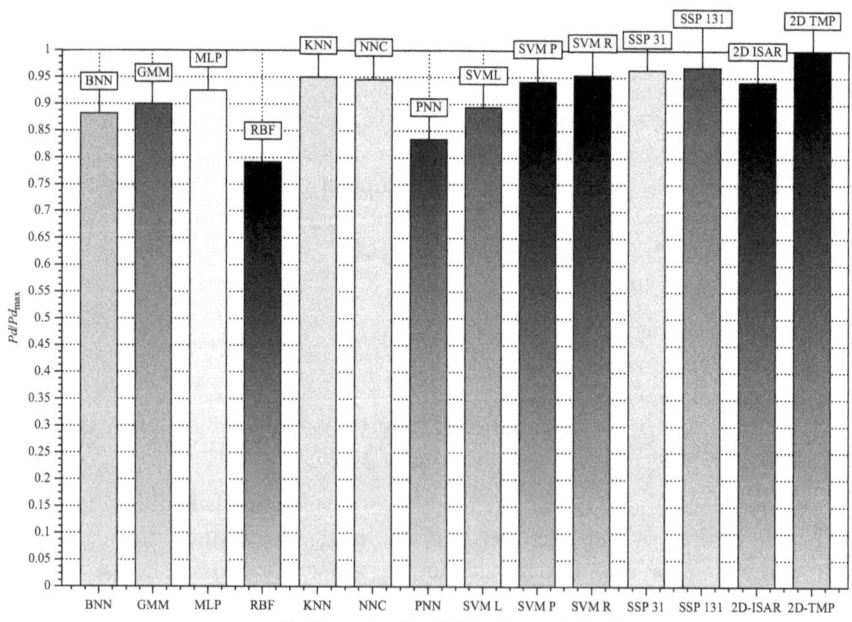

Figure 6.10 Result of application of different classifiers

All these classifiers were applied to 1D data, the HRR profiles of the targets. For comparison the last two columns show the results of an NNC classifier applied to 2D ISAR images, the last column representing the 2D template approach as described in section 6.4.3.

It should be noted that repeating the tests with different data sets or using different aspects lead to results differing slightly from the one shown here in detail, but the overall result stays the same. There are minor differences between the performances of different classifiers, but for most of them they appear rather small. In the example of Fig. 6.10, on average a relative identification rate of 0.9 is achieved with nearly all classifiers lying in a band of ± 0.05 around this value. This value more reflects the variance of the classifiers under different data conditions than quality differences of the classifier itself. While the first nine classifiers presented in Fig. 6.10 are based on HRR profiles of the targets, the last two columns show results of application of an NN classifier based on 2 D ISAR images as feature set.

In principle one expects more information in a 2D image compared to a 1D feature such as an HRR profile, but the 2D ISAR classifier does not reflect this expectation. As discussed in section 6.4.3, we believe that the stretch effect resulting from an uncertainty in the cross-range scaling more than balances (negatively) the advantage of the higher information content. So it is not surprising that the best result is achieved for the 2D template classifier which was introduced in section 6.4.3 and which, by design, should minimise the effect of unknown stretch factor.

The overall result of our investigations is that there exist a lot of classifiers that lead to very similar results in real applications. So the question which classifier to use should be more based on practical questions such as time consumption in the training phase, adaptivity to new scenarios, time consumption in the decision phase and memory consumption rather than on a 'search for best classifier' philosophy. In the overall classification process much more important than the special selection of one out of a set of already well-proved classifiers is the quality of the raw data, its pre-processing and the feature extraction stage.

6.6 Feature extraction

Images after pre-processing steps as shown in Fig. 6.9(f) will now be the basis for feature extraction for both the one- and two-dimensional features. This is done to assure that we really use the same information. From the many feature sets that could be taken from this image only some will be discussed in the rest of this chapter.

1. Collapsed range profile: By adding, for each range-cell, the amplitudes of all associated cross-range cells a kind of averaged HRR profile is formed. By construction, this feature vector is invariant against cross-range translations and cross-range scaling, not against range translation. In real applications for aircraft NCTI this feature vector leads to very promising results [11].
2. Collapsed cross-range profile: In this case, for each cross-range cell the amplitudes are averaged along the range direction. This feature vector is invariant against translation in range and range-scaling (very unlikely for well

working radar systems). All investigations have shown an inferior behaviour of this feature vector as compared to the collapsed range profile [11]. One of the reasons may be found in the missing invariance against cross-range scaling.

3. 2D ISAR: The whole 2D ISAR image is used as a feature for the classification process. This feature vector shows no invariance, neither against translation in range nor in cross-range nor in cross-range scaling.
4. Range ACF (R-ACF): From the image, a new feature vector is calculated that is independent of the absolute position in the 2D image plane. This feature vector corresponds to the autocorrelation function of the collapsed image in range direction (see feature 1). The R-ACF vector is invariant against translations of the image in range, in cross-range and against cross-range scaling.
5. Cross-range ACF (CR-ACF): By analogy to R-ACF the CR-ACF feature corresponds to the autocorrelation function of the collapsed image in cross-range direction (see feature 2). This feature set is invariant against image shifts in range, cross-range and against range scaling.
6. 2D autocorrelation function (2D ACF): We propose as a new method the use of the 2D ACF of the 2D image as feature set. The ACF is also invariant against translation in range and cross-range, but this transform conserves the structure of the image, i.e. a more 'peaky' image leads to a 'peaky' ACF. Our experience shows that if targets exhibit a more smooth amplitude distribution over the image plane then the use of the ACF as feature set is not adequate.
7. 2D FFT: A method often proposed is the use of the 2D FFT of the image. By that a data matrix is achieved that (if taking the magnitude of each pixel) is invariant against range and cross-range translation. However, unfortunately this transformation is not well suited for the problem under consideration: aircraft identification is based on the expectation of a target which shows scatterers being typically concentrated in a limited number of resolution cells. In addition to these scatterers, a continuum of smaller amplitudes fills out the target area. Applying a Fourier transform to such an image accordingly destroys the processing gain of the previous image formation process. Thus, it can be anticipated that this feature set will not yield results comparable to other feature sets.
8. Cross-range sort: Here a new image is formed where, for each range cell, a data vector is established consisting out of the amplitudes of the cross-range data belonging to this range bin arranged in amplitude order. This new feature set is invariant concerning cross-range translation and cross-range scaling.
9. 2D template: This feature differs from feature 3 (see above) by the special selection of information using only small templates of the whole image at a time but achieving by this method a considerable invariance against cross-range scaling errors (see section 6.4.3).

6.6.1 Classification results using different feature sets

For the different feature sets presented above a classification test was run. For the classifier, the NN classifier was used with no rejection class, i.e. as a forced decision classifier. As basic test and reference raw data 2D ISAR images gained from real measurements were used. The test and reference data were independent in the sense that they originated from different field trials. Six different target classes were available for this investigation. Fig. 6.11 shows the results of this investigation.

In Fig. 6.11, for each of the aforementioned feature vectors, the classification rates as a function of image resolution are depicted. The left column stands for a resolution of 1.6 m in a 1D, or 1.6 m $\times$ 1.6 m in a 2D representation, 32 $\times$ 32 pixels (middle column) reflect 0.8 m (or 0.8 m $\times$ 0.8 m) resolution and 64 $\times$ 64 pixels (right column) represent a resolution of 0.4 m (or 0.4 m $\times$ 0.4 m).

For the latter case the computational effort is shown accordingly by the pointed columns (on the far right of each grouping) in relative numbers referring to 100% for the ACF feature set.

It is quite obvious from Fig. 6.11 that the identification results are very different for different feature sets. For the same data set relative Pd/Pd_{max} results vary by a factor of 3, thus showing the big influence of the feature set on the overall classification result. Second, the resolution has a minor influence on the identification rate. In one case (2D FFT) the result is decreasing considerably with increasing resolution, a similar trend, though not as pronounced, results if the 2D ISAR image (marked as '2D picture') or the 2D template is used as feature set. In agreement with our experience, a beneficial choice for resolution is a resolution of 32 $\times$ 32 pixels corresponding to a resolution in range/cross-range of

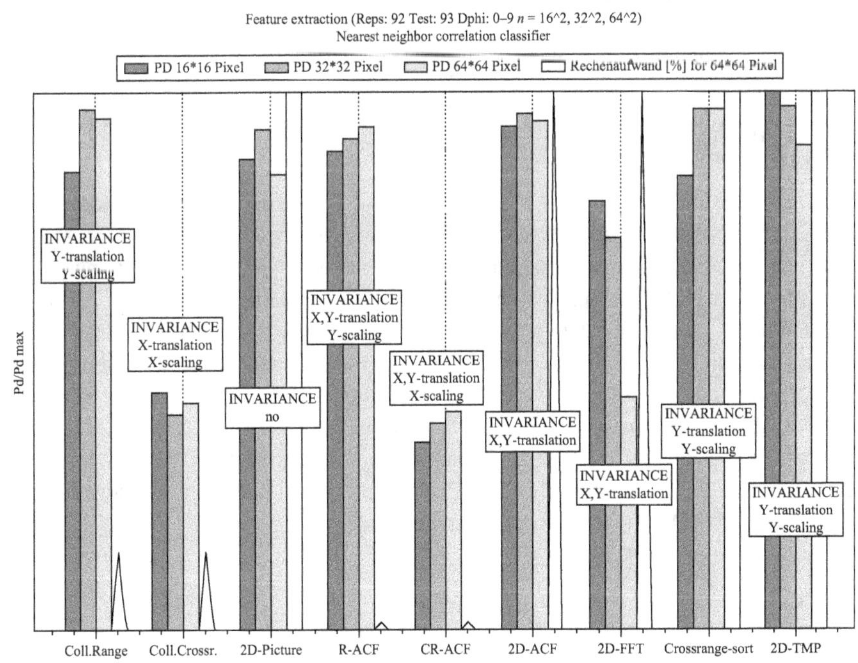

Figure 6.11 Comparison of identification rates using different feature sets (nose-on aspect, six classes)

0.8 m $\times$ 0.8 m for the task of identification of really similar aircraft in a length-interval of 17–19 m. Best results are reached by four feature sets:

1. 2D template,
2. collapsed range (corresponding to averaged HRR profile),
3. 2D ACF,
4. cross-range sort.

In all four cases very high relative identification rates are achieved.

Next in line is the 2D ISAR image itself and the R-ACF feature-set, both displaying a relative identification rate in the area of 90% $Pd/Pd_{\max}$. If the 2D FFT is used as a feature set, as expected the classification rate drops to around 80% $Pd/Pd_{\max}$ in the case of the lowest resolution and otherwise decreases even further. In the last places in this competition, the collapsed cross-range and the CR-ACF feature sets (very similar sets) are found.

As far as computational effort is concerned important differences between the different approaches are obvious: By far the highest computational burden results from use of the 2D ISAR picture feature set. The reason can be found in the image size and the missing invariance against any affine transformation leading to many additional calculations. Also computationally very costly is the 2D TMP approach, but it delivers the best result! Next in line is the cross-range sort feature set, because of a missing invariance against range-shifts. Moderate computational effort is needed for both the 2D ACF feature set and the 2D FFT features. In combination with the classification result the most efficient feature sets are the collapsed range (HRR) and (somewhat related) the R-ACF sets.

6.7 Conclusion

The choice for a particular type of classifier should be more dependent on questions of ease of implementation than on classifier type. Even then, in cases where a classifier type is looked upon as being numerically too expensive in the decision phase, as for example the NN classifier, there are ways out of the dilemma by transformation of the feature set and by further reduction of the size of the database by clustering, ranking and editing algorithms. However, as shown on the basis of real measurements, far more important than the choice of a particular classifier is the intelligent selection of robust and discriminative target describing features. Best results are achieved with feature sets that are invariant against translation and scaling operations. Interesting feature sets are the (averaged) 1D HRR profiles and if 2D images are investigated, the winners are the 2D template approach or the 2D ACF.

References

[1] P. Lacomme, *Air and Spaceborne Radar Systems: An Introduction*, Institution of Engineering and Technology, Stevenage, May 2001.

[2] R. Croci, 'Spaceborn (sic) Radar-Altimetry', http://www.alphalpha.org/radar/altim_e.html (Last updated 6 October 2002).

[3] D. R. Wehner, *High Resolution Radar*, Artech House, Stevenage, 1995.

[4] M. Martorella, 'Novel Approach for ISAR Image Cross-Range Scaling', *IEEE Trans. Aerosp. Electron. Syst.*, Vol. 44, No. 1, January 2008.

[5] M. I. Skolnik, *Introduction to Radar Systems*, McGraw-Hill, Stevenage, 1962.

[6] S.-T. Bow, *Pattern Recognition and Image Preprocessing*, M. Dekker, Stevenage, 1992.

[7] R. Fraiman, A. Justel and M. Svarc, 'Pattern recognition via projection-based *k* NN rules', *Computational Statistics & Data Analysis*, Volume 54, No. 5, pp. 1390–1403, 1 May 2010.

[8] T. M. Cover and P.E. Hart, 'Nearest Neighbour Pattern Classification', *IEEE Trans. Inform. Theory*, Vol. IT-13, No. 1, pp. 21–27, January 1967.

[9] L. Devroye, 'Any Discrimination Rule Can Have an Arbitrarily Bad Probability of Error for Finite Sample Size', *IEEE Trans. Pattern Anal.*, Vol. PAMI-4, No. 2, pp. 154–157, March 1982.

[10] L. Devroye, L. Györfi and G. Lugosi, *A Probabilistic Theory of Pattern Recognition*, Springer, Stevenage, 1996.

[11] K. Rosenbach and J. Schiller, 'Identification of Aircraft Based on 2-D-Radar Images', *IEEE International Radar Conference*, Alexandria, VA, USA, 8–11 May, 1995.

[12] K. Rosenbach and J. Schiller, 'Construction and Test of a Classifier for Non-co-operative Air-Target Identification Based on 2D-ISAR Radar Images', *International Radar Symposium IRS 98*, Munich, September 1998.

[13] K. Rosenbach and J. Schiller, 'Non Co-operative Air Target Identification Using Radar Imagery: Identification Rate as a Function of Signal Bandwidth', *IEEE International Radar Conference*, Alexandria, VA, USA, 7–12 May 2000.

[14] S. Gelf and J. Schiller, 'NCTI using Radar: An Alternative Classifier Concept', *5th International Military Sensing Symposium*, Gaithersburg, MD, USA, December 2002.

[15] K. Rosenbach and J. Schiller, 'Increasing the Classification Performance of a Classifier via Sub-space Based Classifications', *Signal Processing Symposium SPS-2007*, Wilga, Poland, 24–26 May 2007.

Chapter 7

Biologically inspired and multi-perspective target recognition

Hugh Griffiths, Alessio Balleri and Chris Baker

7.1 Introduction

In this chapter we aim to exploit experience of the natural world in which echolocating mammals are able to detect and classify objects with apparent ease. These observations suggest that waveform diversity and orientation strategies play an important role. It is this hypothesis that we test and show to be valid, as confirmed through real-world radar experiments. Specifically, there is additional information contained in different perspectives of a target that can help classify it, thus boosting performance. There is also a law of diminishing returns, in that the opportunity to extract additional new information reduces as the number of perspectives increases.

The ability to detect and locate targets on a 24-h basis, over wide areas, regardless of weather conditions has long made radar a key sensor in many military and civil radar applications. It is well recognised that the utility of the information supplied by a radar system would be hugely enhanced if targets could additionally be classified. This might be by type (e.g. to differentiate helicopters from civilian airliners from military jets) or even better, within type (e.g. a Chieftain tank from a T32 tank). It would be even more advantageous if this could be automated, at least partially. For example, huge quantities of imagery are produced by SAR and place unrealistic demands on a human image interpreter, and algorithmic aids are necessary if the data is ever to be fully exploited.

However, target and object recognition has proved far from straightforward, such that robust and reliable classifiers able to cope with wide target and clutter sets do not yet exist. Most of the approaches to classification have concentrated on exploiting higher and higher spatial resolutions both in one [1] and two [2] dimensions. Here, the aim has been to reduce the variability of the target signatures so that features cluster more tightly in feature space and classification performance is enhanced. There has also been research examining the role of polarimetric information [2,3]. However, this has not resulted in a level of classification performance enabling routine take up in operational sensing systems and hence we examine how classification proceeds in echolocating mammals, principally the bat.

First though, we examine the background to traditional forms of radar classification based upon echo signatures. As discussed in Chapter 1, the definitions of the terms ‘classification’ and ‘recognition’ are not universally accepted. Here, these terms are used interchangeably and refer to the ability of the radar system and subsequent processing to determine the class to which a target belongs. To perform automatic target recognition (ATR) in a useful manner it is necessary to achieve a sufficiently high degree of confidence that a target has been correctly classified and hence it is extremely important to minimise the probability of misclassification. This is because in both military and civilian applications there are likely to be safety issues that demand appropriately high levels of performance and a full understanding of its limits, variability and robustness.

ATR can be based either on 1D signatures, i.e. features derived from a range profile of the target, or on 2D imagery. 1D signatures are often employed due to their simplicity in terms of implementation and signal processing, although they can lead to reduced classification performance due to signature variability arising from (i) the interaction of the illuminated waveform with a complex target and (ii) the effects of radar signal processing (including the matched filter). This variability may be further exacerbated by any echo contributions from clutter and multipath. 2D imagery is less likely to contain clutter (although there will still be a significant multipath or forward component) and hence would seem to guarantee a more detailed and authentic representation of target backscattering [4]. However, target signatures in SAR imagery still exhibit wide variability as a function of small parametric changes such as geometry. Imaging requires more sophisticated signal processing and, if the data is collected over an extended period of time (as in a stepped frequency radar system), corrections for irregular target movements. Such corrections can be far from straightforward especially because the detailed form of the motion corruption is commonly unknown.

When high range resolution profiles (HRRPs) are used, slightly different target orientations due to rotational or translational motion of the target (assuming no range migration occurs) result in differences in relative range between scatterers, which lead to interference, and subsequent fluctuations in the range profile. This is known as speckle. This is further enhanced by the limited angular scattering extent exhibited by many targets. In addition, HRRPs collected at different aspect angles are likely to present even further variability due to shadowing; this occurs when one or more scatterers are occluded by other parts of the target (local shadowing) or by other targets (global shadowing) and therefore not illuminated by the incident waveform. As a consequence, a portion of the target is unavoidably masked and the signature can rapidly fluctuate. For good classification, features extracted from HRRPs should be both stable and tightly clustered so that they do not overlap with those of other targets. However, the effect described above may result in increased variability in feature vectors and hence they are spread in ‘feature space’, thus also adversely affecting classification performance.

There has been relatively little research reported that examines the utility of angular diversity for improved classification performance [5–7], although it is clear that:

(i) there is additional information in multiple perspectives,
(ii) it is extremely easy to acquire such data.

In fact, to date, the variation of target signature with angle has been more a source of performance degradation in single perspective classification approaches. Multi-perspective (MP) classification can be exploited by networks of co-operating radar systems [8]. This approach has the added advantage of improved system sensitivity, tailored coverage and reduced vulnerability to electronic and physical attack. Alternatively, a single radar system taking up different orientation angles with respect to an object can acquire multiple perspectives.

This chapter begins by examining the natural world in order to investigate what lessons may be learned. In particular it will be shown how bats classify moving and static targets, such as insects or flowers of bat-pollinated plants. Following this, micro-Doppler and its application to target classification is briefly examined and applied to human recognition and motion classification.

Finally, the concept of MP target classification is examined for radar. Using full-scale target signature measurements as the source data, it is seen how multiple perspectives enhance classification performance. Classification performance is assessed both as a function of the number of perspectives and of the signal-to-noise ratio (SNR).

7.2 Biologically inspired NCTR

In the natural world, echolocating mammals use waveform diversity as an inherent component of their normal behaviour. The constantly changing time and frequency structure, and location and direction of their transmitted signals represent a pro-active approach to interrogation of the surrounding environment. In addition, a multiplicity of processing streams collectively extracts information from received echoes to build up an accurate picture that is supplemented by long-term 'experiential' memory. In this way bats, whales and dolphins are able to 'perceive' their environment, that is to say, they can 'see with sound'. Further, bats, for example, are able to do this such that they can autonomously navigate, feed, socialise and otherwise conduct their lives. Here we examine the use of diversity in natural sensing systems with a view to exploitation in synthetic sensors, with particular emphasis on target classification.

7.2.1 Waveform design

Echolocating mammals such as bats, whales and dolphins have been using waveform diversity for over 50 million years. In contrast, synthetic systems such as sonar and radar have been in existence for not much more than 100 years. Echolocating mammals vary parameters including the pulse repetition frequency (PRF), and power and frequency content of their transmitted waveforms. Recent developments in technology mean that it is now possible to replicate such diversity in radar and sonar systems. With modern digital technology, high-precision, wide-bandwidth waveforms can be generated and varied even on a pulse-to-pulse basis.

Echolocating mammals potentially offer valuable insights that may enable capabilities such as autonomous navigation and automatic target classification that have hitherto proved elusive. If we understand how bats exploit echolocation for autonomous navigation, collision avoidance and recognition, perhaps we can build this understanding into synthetic systems. Potentially, this can provide a step function change in the utility of unmanned systems that can take advantage of the attributes of radar and sonar. Reference 9 provides further insight on this topic.

7.2.2 Nectar-feeding bats and bat-pollinated plants

Nectar-feeding bats play an important role in the process of pollination of plants, which, with their flowers, represent a very interesting class of organisms for the study of target classification. First, flowers are motionless and silent so that bats cannot rely on Doppler information or passive location based on target sounds, and second their habitat is often a densely cluttered environment. Finding and approaching a flower is a gradual process that involves all the bats' senses. While bats can only use vision at dusk, they can rely on their sense of smell and on echolocation to search for flowers in darkness. Nectar-feeding bats have a highly developed sense of smell and they are attracted by the scent of sulphur compounds which are produced by many plants pollinated by bats [10]. Although the sense of smell certainly plays an important role, and in particular for long-range attraction, latest research results support the theory that nectar-feeding bats largely rely on echolocation to plan their approach flight and to detect and select the flowers in the proximity of the plant. Indeed, it has been shown that bats are capable of detecting and selecting objects by echolocation only.

The flowers of the bat-pollinated vine *Mucuna holtonii* plant release the greatest amount of nectar reward only on the first visit by the bat. By landing into the flower corolla, the bat ignites an explosion of the flower's vexillum which functions as a trigger for the pollen and nectar release. Behavioural experiments have shown that bats can successfully select unvisited flowers within an inflorescence in darkness and this is a task that can only be done by echolocation. The scent of the flower, in fact, remains unchanged before and after the explosion. It has further been shown that by modifying the echo-acoustic properties of these flowers, via insertion of a small pad of cotton in the cavity of the vexillum (i.e. without altering the scent and visual aspect of the flower), there is a resulting drop in the rate of successful classification [11,12].

Choosing the most suitable flowers to be visited within an inflorescence is a task for which echolocation certainly plays a fundamental role and, although classification of flowers in such an environment is demanding, nectar-feeding bats succeed in their feeding and pollination task. Because it is in the interest of both the plant and the bat that pollination takes place, it is widely hypothesised that co-evolution has shaped such flowers in order to ease classification by bats so that the chances of successful pollination are increased [13]. Understanding how features derived from echoes of the flowers aid the bat in its search for nectar is key to understanding strategies that underpin the task of classification. Thus this can offer interesting insights that potentially can be applied to radar systems such that they are better able to perform target classification and recognition. Indeed, there are a number of similarities with the radar and sonar cases. For example, the types of waveforms deployed by bats are modulated in a fashion similar to linear and hyperbolic chirps commonly used by radar and sonar systems [9]. Also, the ratio between the flower size and the range resolution used by bats is similar to the ratio of target size to range resolution which appears to give best performance using radar or sonar sensors. This provides still further cause for optimism that methods employed by natural systems can be usefully exploited in synthetic counterparts.

7.2.3 Classification of flowers

Here, we analyse HRRPs of two *Cobaea scandens* flowers taken at different stages of maturity and then manually modified. A photo of a *C. scandens* is shown in Fig. 7.1.

The classification performance of a *k*-NN classifier trained to distinguish between these flowers is assessed. This enables us to investigate and quantify the information contained in their echo-acoustic signatures to see if it can be used to perform recognition. In particular, we place emphasis on the differences in echoes that are due to the flower wilting and also due to manual removal of the flower petals. In other words, can the bat distinguish between version of the same flower that may or may not provide a suitable source of nourishment?

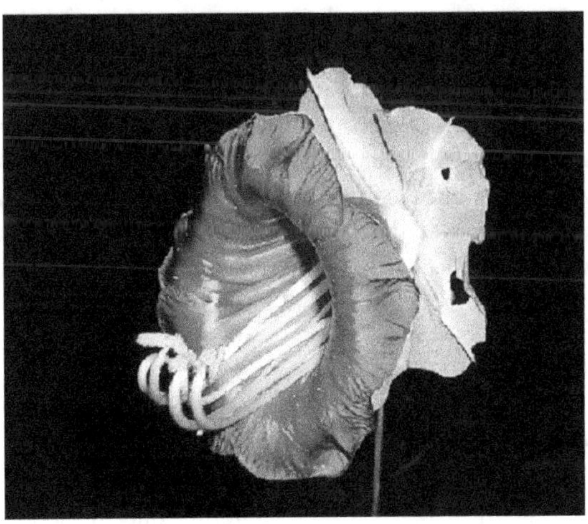

Figure 7.1 Photo of a desiccated Cobaea scandens *flower with protruding anthers, a bell-shaped corolla of largely merged petals whose unmerged petal ends fold back, and a ring of partly merged sepals at its base*

7.2.3.1 Data collection

The data consists of HRRPs of a few individual flowers of *C. scandens*. These were collected at the School of Biological Sciences of the University of Bristol in October 2008. The flowers to be ensonified were impaled on a thin pin that was placed at the centre of a rotating turntable with an angular resolution of 1°. A linear down-chirp spanning the frequencies between 50 kHz and 250 kHz (the range of frequencies used by nectar-feeding bats) was transmitted towards the flower with a custom-built loudspeaker [14]. The echo, recorded with an ultrasound microphone (G.R.A.S. type 40 BF), was sampled at a rate of 500 kHz using a National Instrument PCIe-6251 card. During the experiments, both the loudspeaker and the microphone were placed into a custom-built artificial bat head, capable of hosting one loudspeaker and two microphones in order to reproduce the spatial arrangement of the mouth and the two ears of the bat. The setup allowed measurements with a range resolution of about 1.5 mm. Three different horizontal images were recorded, each one representing HRRPs of a *C. scandens* in three different cases:

(i) a flower in ideal condition for pollination,
(ii) a desiccated flower,
(iii) a flower whose petals were removed by hand.

7.2.3.2 Data pre-processing and results

For each angular perspective the target HRRP $\mathbf{x}_j$ consisting of a series of consecutive range cells was formed from a starting range of 18 cm to finishing range of 32 cm (i.e. those range bins that contained the echo from the targets). Similarly, a noise profile $\mathbf{n}_j$ was formed with the elements corresponding to the range bins that contained noise only and no echo returns from targets or clutter.

The images showing the magnitude of the HRRPs for each angular perspective are presented in Figs. 7.2–7.4. These are related to the angular perspectives ranging between $-90°$ and $+90°$, where the 0° line corresponds to the case when the ‘centre line’ of the flower was aligned with that of the microphone and the loudspeaker. The images were normalised to the same value in order to allow a fair comparison. The magnitude of the HRRPs of the individual of *C. scandens* that was ready for pollination is shown in Fig. 7.2. It exhibits strong reflections at a distance of between 21 cm and 23 cm from the microphone. These strong reflections originate from the petals of the corolla. There are weaker reflections from between 20 cm and 21 cm that are due to the anthers protruding from the corolla. There is also a weaker scattering that is caused by the bell-shaped inner part of the flower’s corolla from between 25 cm and 27 cm. This is the part of the corolla that contains nectar and from which the pistil grows. Weaker scattering due to the echo generated by the ring of sepals that cover the external back side of the flower is also visible from about 28.5 cm to 29.5 cm. It is interesting to observe how the petals can scatter a considerable amount of power over a wide range of angles that goes between $-60°$ and $+60°$. There are changes to the corolla that arise as the flower wilts (and stops producing nectar, hence losing attractiveness to bats) that tell us about which features are responsible for recognition. The HRRPs of the desiccated *C. scandens* that

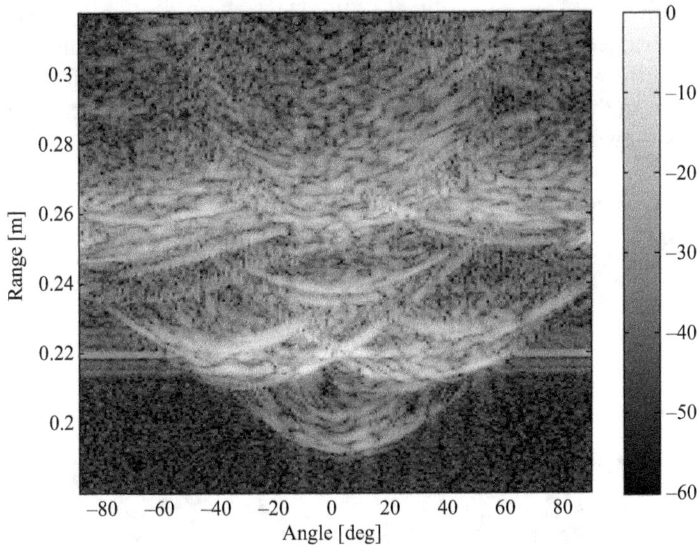

Figure 7.2 Magnitude of HRRPs over 180° for a C. scandens *ready for pollination. 0° is to the front of the flower. HRRP's direction is horizontal, i.e. from left to right with respect to the flower's natural orientation. Scaled in relative decibel*

had started to wilt due to an overnight frost are displayed in Fig. 7.3. The time plot shows that the scattering due to the petals from between ranges of 21 cm and 23 cm are less complex than those shown in Fig. 7.2 where there is no wilting of the flower. The scattering, though, covers a broad set of illumination angles between −60° and 0° and is still present at a range of approximately 22 cm. This is probably due to one petal that had remained in good condition at the time of the measurement. However, most of the scattering from the petals is no longer visible, leading to a marked loss of information. This is because the petals tend to fold back during the wilting process and the mirror effect that occurs when they are in the upright position diminishes. The loss in turgor might also affect reflective properties of the plant tissue. The scattering from inside the corolla does not show any loss of complexity, but it is weaker than before with respect to the maximum value of the image. However, it is clear that the wilting of the flower produces a very different echo signature to that of a flower that has yet to wilt. These differences are used by bats to determine which flowers are suitable sources of food and therefore we conclude that they contain clues to successful classification.

To investigate what the image of the flower would be without petals at the corolla, the front ends of the petals from the same individual were removed manually using a pair of scissors. Fig. 7.4 shows the HRRP image obtained from this modified flower. In particular it is noted that the scattering from the petals disappears over all angles. A fraction of power is still scattered at 0°, probably due to some residuals of petal that might have not been removed properly or simply by the line delimiting the aperture of the corolla. The fact that the image is very similar to the one obtained from the wilting flower suggests that the scattering from petals tends to significantly diminish when the flower is no longer suitable for pollination. Scattering from the inside of the flower is now more visible when compared with the HRRPs shown in Fig. 7.3.

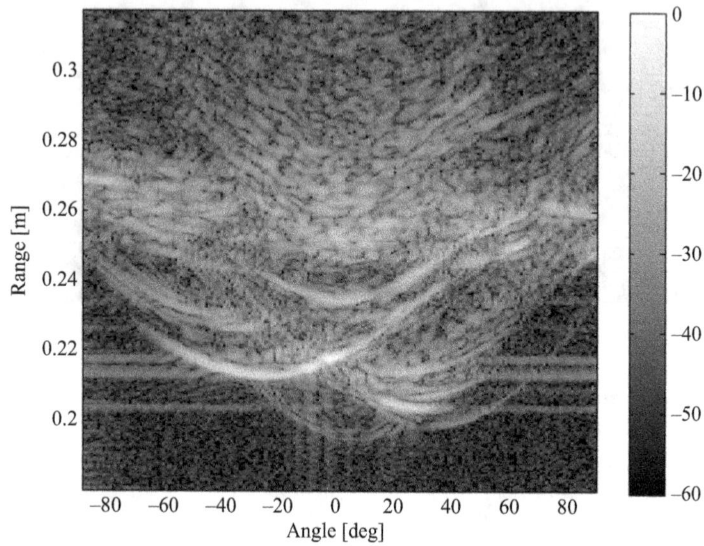

Figure 7.3 Magnitude of HRRPs over 180° for a desiccated C. scandens

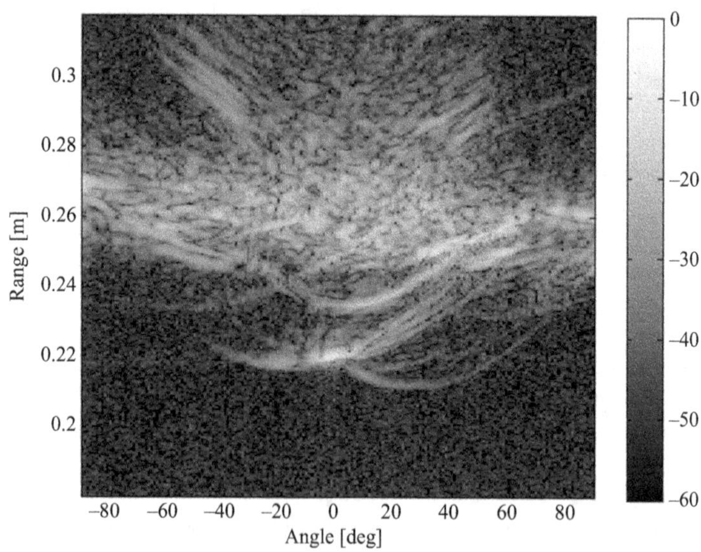

Figure 7.4 Magnitude of HRRPs over 180° for a modified C. scandens

Differences in the echo-acoustic structure were assessed using a k-NN classifier trained with HRRPs extracted from each image with an angular step of 2° to 5°. The remaining HRRPs for each image were used to test classification performance. For each HRRP the principal components were extracted with a principal component analysis algorithm [15]. The maximum value of the cross-covariance was used as a measure to identify the k-nearest neighbours (k-NNs) to each test. The profile was assigned to the class owning the highest number of neighbours between the k selected. When the k-NN classifier was set to make a decision with $k=3$ nearest neighbours, 77% correct classification was achieved. This was obtained with an angular separation between consecutive training profiles equal to 3°, corresponding to a total number of 60 training profiles and 120 test profiles. Dimensionality reduction was performed through the PCA algorithms by selecting the principal 20 features for each profile. Classification performance improved significantly up to values over 90% correct classification when the classifier was trained with profiles separated by 2° and similarly performance dropped to 68% when the decisions were made on $k=5$ neighbours while keeping the angular separation between the training profiles equal to 3°.

Results support the hypothesis that the echo-acoustic signatures of bat-pollinated flowers contain information that changes with the stage of maturity of the flower and can therefore be used to achieve higher levels of classification performance. In particular, scattering from the petals and the way in which this changes as a function of the state of the flower could be a way of letting bats know which flower is most suitable for pollination, i.e. holds a nectar reward. We can conclude that this plays an important role for recognition and subsequent selection of flowers. An additional analysis of power reflectivity shows that differences in the backscattered power can also be used to aid recognition or indicate the presence of a flower at longer ranges [15].

The trajectory of a *Choeronycteris mexicana* nectar-feeding bat approaching an artificial feeder containing nectar was recorded during a set of behavioural experiments carried out in Germany in 2008 [16,17]. An analysis of the position of the bat, when emitting a call, showed that the bat emitted about 10 calls before moving towards the feeder, and that it proceeded on a straight line at constant vertical angle of 40° with respect to the horizontal plane. Within these 10 calls the bat exploited a set of angular perspectives that covered about 40° on the vertical plane and 15° on the horizontal plane. On completion of its mission the bat had emitted calls over a number of angular perspectives covering about 30° on the horizontal plane. Although the information on the orientation of the echolocation call beam with respect to the feeder was not available, these results indicate that, in the proximity of the feeder, the bat explores only a limited number of angular perspectives. This suggests that bats (a) may choose a suitable angular sector that contains key features to make a decision and (b) that if classification is based on the presence of largely merged petals then a number of HRRPs over a wider angular sector are used to improve classification performance. Although these hypotheses can only be speculative, the bats' strategy is consistent with one that seeks to maximise the success rate of classification while minimising energy expenditure. This provides motivation to investigate the implication that these speculations may have on classification performance. For this reason, here, the classification performance analysis presented above is repeated and assessed as a function of an angular sector composed of a limited number of perspectives. This enables a determination as to which perspectives contain key features for flower classification and how critical the choice of the trajectory can be for the success of the mission.

Classification performance was assessed using a test set of HRRPs covering a limited number of perspectives. The classifier was trained with the same training set as previously used. To investigate the impact of the physical features of the flower the classification performance was examined for two classes, one representing the desiccated flower and the other the modified flower. Because the modified flower was obtained by removing petals from the desiccated individual, the only differences between the two classes were due to the removed features. The results show that a high level classification performance can be achieved over the sectors that contained the perspectives where the contribution of the petals that were removed was most evident. The performance of the k-NN classifier indicates that HRRPs that contain the petals remain highly correlated over wide angular sectors. On the other hand, the performance of the Naive Bayesian classifier indicates that the presence of the petals can change the parameter of the HRRP distribution significantly. These results corroborate the hypothesis that the petals contain the physical parts of the flower that play an important role in determining the echo signature and therefore correspondingly in aiding the bats to assess when the flower is at its most suitable for feeding. It will only be possible to draw firm conclusions about the actual bats' strategies when the neural processing that takes place in the brain of the bats is sufficiently well understood. However, it is suggestive of the new ideas and concepts that can be exploited to improve the classification performance of radar and sonar systems.

7.2.4 *Classification of insects*

In this section we examine classification of insects by bats where the bat is emitting a continuous frequency frequency (CF) signal. Fig. 7.5 shows an example of a CF waveform composed of three main harmonics at frequencies of approximately 30 kHz, 60 kHz and 120 kHz. The spectrogram of Fig. 7.5 reveals that each component begins and ends with small downward frequency sweeps. It has been argued that the portions at the beginning and end of the waveform are used for target localisation, while the long constant frequency component aids classification of targets that exhibit bulk and micro motions [18]. Multi-harmonic waveforms of this type are not used in radar systems. However, the fact that mammals perform at such a high level utilising these structures suggests that it is worth investigating whether or not such waveforms might aid radar target classification. Thus here we examine in more detail these waveforms and the nature of information subsequently imparted on the received echoes.

In flight most species of CF bats, such as horseshoe bats, lower their call frequency in relation to their flight speed in order to compensate for Doppler shifts

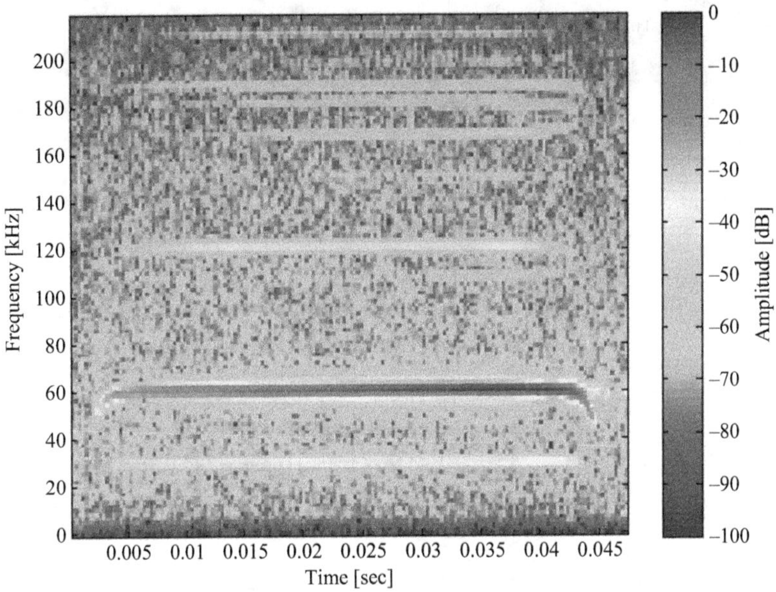

Figure 7.5 Example of a multi-harmonic CF signal with initial and final down-sweep

induced by their own forward motion, so that echoes always return at a frequency best matched to the peak of the bat's hearing sensitivity [19]. Bats that emit CF signals can detect and classify fluttering insects from the amplitude and frequency modulations of the echoes that result from the movement of the insect's wing [20,21]. These modulations, called 'acoustical glints', provide information about wing beat rate and flight angle and are very specific to each species of insect. Glints turn out to be useful not only for the detection of the prey but also for its identification. Amplitude modulations are the result of the fact that echo strength is strongest when the insect wings are perpendicular to the sound source and gets weaker as the insect wing moves away from the perpendicular position. The precise timing of the glint depends on the angular orientation of the insect [22]. The amplitude modulations give information about the target elevation and are related to the timing of the glint production [23].

The wing movement towards and away from the receiver induces Doppler shifts in the echoes, providing information of wing beat movement that are important signatures for species identification, since insect wing beat frequency scales with body size [24]. Because the way in which a given insect species moves its wings is highly specific, the structure of glints varies across different species and even insects with the same wing beat frequency may provide different spectral signature in the echo [25].

Fig. 7.6 shows the spectrogram of four echoes from four different insects: *Deilephila elpenor*, *Scotia exclamationes*, *Melolontha melolontha* and *Tipula*

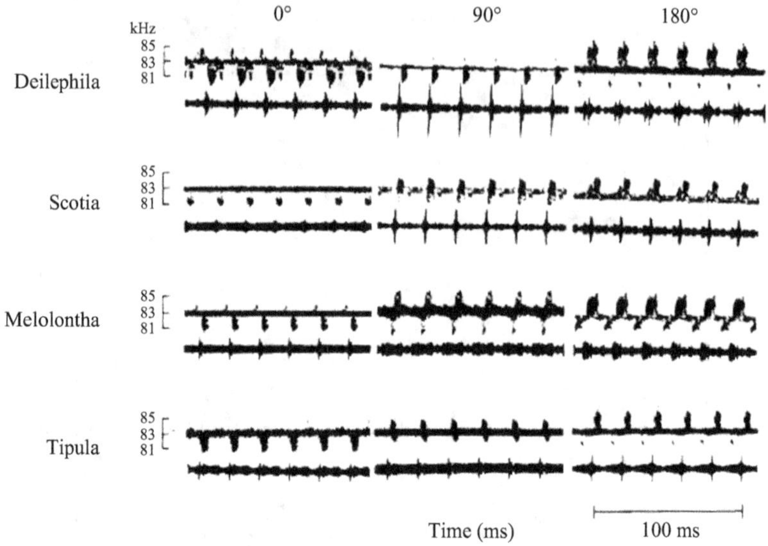

Figure 7.6 Echoes from four insect species all fluttering at 50 Hz but with different spectral patterns in echoes from their wing beats. For each insect, the upper trace represents the spectrogram of the echo and the lower trace the time oscillation [25]

oleracea. These were flying with the same wing beat rate and were illuminated from 0°, 90° and 180° with respect to the sound source. The figure shows that spectrograms for each species present different characteristics, although the flutter frequency is the same. It also shows that, for the same insect, the properties of the spectrogram are also dependent on the angle from which the sound source emanates, confirming that spectral cues provide information on angular position [23,26].

The way CF bats classify fluttering insects has a very close parallel with that used in radar and sonar systems. The approach is to classify targets by using micro-Doppler information in the echo spectrogram. A big difference, though, is that bats are able to combine information from both amplitude modulations and frequency modulations and subsequently manage to obtain a remarkable level of performance. For example, bats have shown a peculiar ability to distinguish between insects that present the same wing beat and similar dimensions with a level of accuracy that is beyond any demonstrated using radar or sonar systems. It is therefore evident that micro-Doppler signatures of a moving target contain both FM and AM structures that might lead to improved classification performance.

7.3 Acoustic micro-Doppler

The echo produced by a moving target that is illuminated by a radar or ensonified by a sonar system contains frequency modulations caused by the time-varying delay occurring between the target and the sensor. The main bulk translation of the target towards or away from the sensor induces a frequency or Doppler shift of the echo as a result of the Doppler effect [27]. Additional movements of small parts of the target contribute with frequency modulations around the main Doppler shift and these are commonly called micro-Doppler modulations. Micro-Doppler modulations contain a signature of the target that can be used for target recognition.

Classification of targets by micro-Doppler signatures has been widely investigated for the radio frequency (RF) regime and in particular for applications related to radar systems. V.C. Chen *et al.* have modelled the radar micro-Doppler phenomenon and simulated micro-Doppler signatures for various targets such as rotating cylinders, vibrating scatterers and personnel targets [28–30]. The authors also show that a time-frequency analysis of the radar return can be used to extract micro-Doppler signatures from the received signal, leading to additional information on the target that can be used for classification and recognition. An easy way to perceive target micro-Doppler modulations is to listen to an appropriate audible version of the radar return. Indeed, this method has been used to perform classification of targets but it has the disadvantage of requiring the presence of a human operator. Various studies have looked at classifying targets automatically through the analysis of their micro-Doppler signatures [31–39], but have yet to yield sufficiently robust methodologies.

Although target classification by micro-Doppler signatures has been widely exploited in the RF regime, there has been relatively little research done on the same topic in the acoustic regime [40–43]. Previous work has resulted in the development of acoustic systems capable of collecting micro-Doppler signatures of frequencies of up to 40 kHz. However, it is desirable for an active surveillance system to operate discreetly, ideally be undetectable at all times and have minimal impact on the environmental surroundings including on any animals other than humans. This requires an acoustic radar to operate at frequencies outside the hearing sensitivity of, at least, most common non-human targets and hence implies frequencies of above about 60 kHz. The use of higher frequencies leads to an additional advantage. As the Doppler shift is directly proportional to the carrier frequency, higher frequencies are more sensitive to small movements and hence induce a wider frequency shift leading exaggeration of the micro-Doppler signature. On the other hand, the increased acoustic attenuation at higher frequencies lowers the operational range of the system. The range of frequencies deployed by bats that forage on insects represents a potentially attractive option despite the range limitations of the higher frequencies.

Here, a coherent acoustic radar (i.e. one that measures amplitude and phase) operating at 80 kHz is used to gather experimental micro-Doppler signatures of human subjects performing various differing actions. The data was collected at University College London between 2010 and 2011. Classification performance of a k-NN classifier and a Naive Bayesian classifier trained to distinguish between different human motions and different personnel targets are then presented. We begin by describing the acoustic radar system.

7.3.1 Description of the acoustic radar

The acoustic radar was developed at UCL and is composed of a signal generator, a loudspeaker, a microphone, two pre-amplifiers that amplify the signals before transmission and after recording and a Data Acquisition (DAQ) card that digitises the recorded signal. The waveforms are generated by a National Instrument NI PXI-6733 card capable of generating up to 500 MSamples/s (32 bit resolution) per channel. Each card can transmit on eight channels and therefore up to eight different waveforms can be transmitted simultaneously. The generated waveform is input to the pre-amplifier (Ultra Sound Advice S55A) which is capable of generating a maximum output of 140 V peak to peak. The loudspeaker provides a monitor output which is 1/100 of the actual output of the amplifier. This allows monitoring of the output and of the gain of the amplifier. The signal is transmitted by a loudspeaker (Ultra Sound Advice) which can nominally operate between 20 kHz and over 200 kHz. It can generate an output level greater than 105 dB SPL (Sound Pressure Level) between 20 kHz and 50 kHz (measured at 0.25 m) and greater than 85 dB SPL up to 150 kHz. The active area of the loudspeaker is circular and has a diameter of 50 mm. In the receive section, echoes are captured by a microphone capsule (Ultra Sound Advice UM3 capsule). The sensitive element of the capsule is a 1.8 cm diameter disk made of a very thin aluminised polyester film. Nominally its sensitivity is better than −57 dB from 20 kHz to 120 kHz and better than −70 dB at 180 kHz. The maximum gain obtainable by the amplifier is nominally × 220 at 100 kHz. The amplified received signal is digitised by a DAQ card (NI PXI-6133) that is capable of receiving eight channels simultaneously with a sampling frequency up to 500 kHz per channel. The digitised data is stored in a file and the signal processing is performed offline.

7.3.2 Experimentation

The data containing micro-Doppler signatures were gathered at UCL on various trials between 2010 and 2011. The microphone and the loudspeaker were arranged in a pseudo-monostatic configuration next to each other as shown in Fig. 7.7. Their active areas were contained on the same plane perpendicularly to the ground floor with their phase centres positioned 12 cm from each other and both were at a height of 105 cm with respect to the floor. During the experiments the human subject targets were placed facing forwards towards the microphone and the loudspeaker at a distance of 2 m. A 10 s long CW tone at 80 kHz with 0.3 V amplitude was generated with the NI PXI-6733 card used to ensonify the targets. The gain of the loudspeaker pre-amplifier, measured by using the monitor output, was about 20 dB. Echoes were recorded with the microphone and then sampled at a rate of 500 kHz. Micro-Doppler signatures were gathered for three different personnel targets undertaking various actions:

(i) walking,
(ii) running,
(iii) walking while carrying an object in one hand,

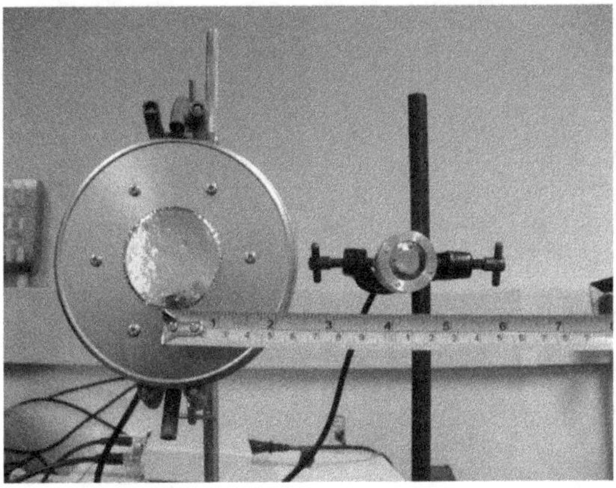

Figure 7.7 Front and lateral view of the space arrangement of the microphone and the loudspeaker during the experiment

(iv) walking while carrying an object with both hands,
(v) walking with a heavy backpack on the shoulders,
(vi) shadow boxing,
(vii) picking up an object from the floor,
(viii) oscillating with the torso while standing on the spot.

When walking or running the targets were moving on a Pro Fitness manual treadmill and were ensonified from the back to avoid any masking effects from the treadmill itself. This was deployed to remove the main Doppler shift contribution from the Doppler signatures, and to keep the SNR of the received signal as constant as possible during the recording. Also the treadmill removed any movement constraint on the targets and therefore allowed longer duration recordings.

A recording of the background clutter was removed from the raw data for each target recording. The difference signal was mixed with the complex conjugate of the transmitted tone to obtain a baseband signal containing the micro-Doppler signatures. The data analysed were obtained by applying the Matlab *Hilbert.m* function to the original signals. It was observed experimentally that the micro-Doppler signatures had no frequency components above 2.5 kHz. Thus the mixer output was filtered with an ideal low-pass filter cutting off all the frequencies over 2.5 kHz. The resulting signal was down-sampled of a factor 50 leading to a final sampling frequency of 10 kHz. This was done to reduce the processing load without altering the base information. A short-time Fourier transform was applied to all the recordings in order to visualise the micro-Doppler signatures. This was formed by using 30 ms long, 50% overlapping windows weighted with a Hanning function. The measured micro-Doppler signatures for one of the two human subject targets when walking and when running are shown in Figs. 7.8 and 7.9 respectively.

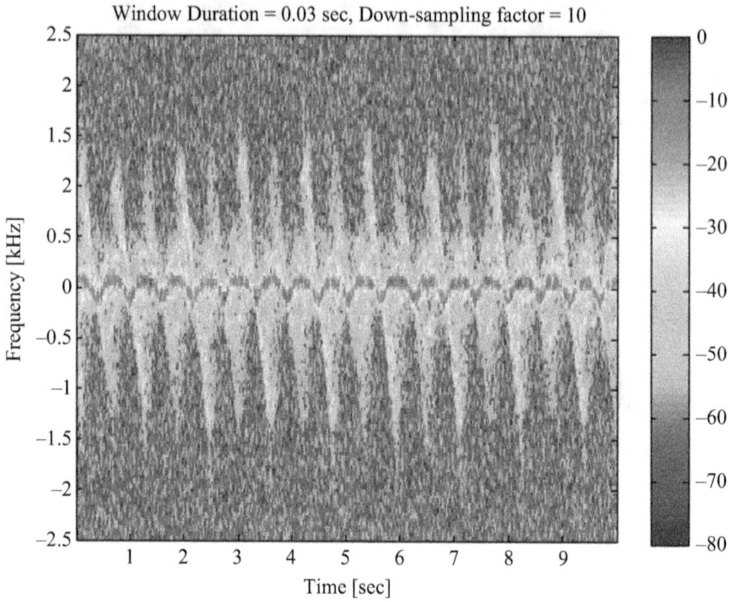

Figure 7.8 Micro-Doppler signature of Target A *walking on a treadmill, facing the acoustic radar at a distance of 2 m*

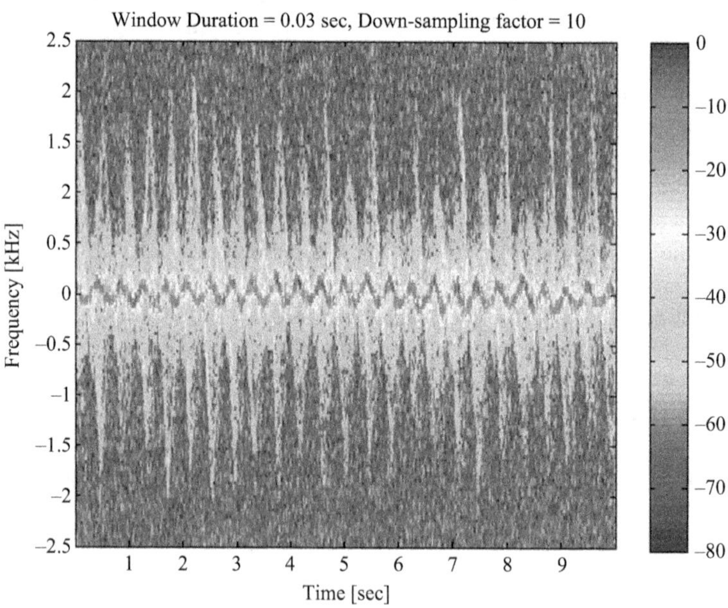

Figure 7.9 Micro-Doppler signature of Target A *running on a treadmill*

7.3.3 *Classification performance results*

The baseband data obtained after pre-processing contains the micro-Doppler signatures and it is these that are used to assess and compare the performance of (i) a Bayesian classifier and (ii) a *k*-NN classifier. The classification goal is to distinguish between different human targets undertaking the same action or to distinguish between different actions undertaken by the same person. The time sequences corresponding to each class were divided in windows of duration T_w with a 50% overlap. The analysis was repeated for three different window durations; $T_w = 30$ ms, $T_w = 100$ ms and $T_w = 150$ ms. This resulted in a total number N_w of windows for each class. The first N_{Train} windows for each class correspond to the first 1.5 s of the recorded signal and were used to train the classifiers. This results in the information used for training remaining the same in all three cases.

Principal Component Analysis (PCA), Cepstrum and MEL Cepstrum algorithms were used to extract N_f features from each window to reduce the dimensionality of the data and reduce the computational load of the classifiers [44–46].

Table 7.1 shows the classification performance that was obtained from the comparison between the walking gait of Target A and that of Target C for $T_w = 30$ ms. This is divided in a sub-table per classifier, each one containing the three confusion matrices related to the three feature extraction algorithms that were used. In each confusion matrix, the percentage number is the ratio between the number of assignments and the total number of windows under test for a particular class. The averaged probability of correct classification (P_{cc}) is given below each confusion matrix. This was calculated under the assumption each class had the same *a priori* probability. The parameter *k* of the *k*-NN classifier was set equal to 3. Results show that high level classification performance can be obtained. In particular, for this case, the combination of PCA plus Naive Bayesian classifier leads to a rate of correct classification over 90%. The lowest rate is given by PCA followed by the *k*-NN classifier presenting a high number of wrong decisions when Target A is being assessed. Overall, in all other cases, correct classification performance is better than 80% [45].

The confusion matrices obtained from the comparison of the walking gaits of all three targets, for $T_w = 30$ ms, is given in Table 7.2. Here the parameter *k* of the *k*-NN classifier was set to 5 and all cases that could not univocally be assigned to any of the three classes were treated as unknowns. Classification performance drops for both the classifiers and all the feature extraction algorithms. All present a high rate of wrong decisions when having to choose between Target B and Target C. The highest rate of correct classification is achieved by the PCA algorithm followed by the Naive Bayesian classifier, and this is equal to 64%. The drop in performance may be due to differences in the gait of Target B and Target C that cannot be resolved by the feature extraction algorithms. Classification performance improves when the duration of the windows increases. For example, when classification performance is assessed on the same data set for a window duration $T_w = 150$ ms, correct classification improves to 71% for the combination of the PCA algorithm

Table 7.1 Confusion matrices reporting the performance of the Naive Bayesian classifier and the k-NN classifier (k = 3) testing the walking gait of Target A *and that of* Target C *for all feature extraction algorithms* ($T_w = 30$ *ms)*

	PCA		Cepstrum		MEL Cepstrum	
Naive Bayesian	*Target A*	*Target C*	*Target A*	*Target C*	*Target A*	*Target C*
Target A	97%	3%	91.7%	8.3%	85%	15%
Target C	11.9%	88.1%	27.6%	72.4%	24%	76%
	$P_{cc} = 92.57\%$	$P_{cc} = 82.03\%$	$P_{cc} = 80.52\%$			
***k*-NN**	*Target A*	*Target C*	*Target A*	*Target C*	*Target A*	*Target C*
Target A	47.2%	52.8%	89.8%	10.2%	80.3%	19.7%
Target C	1%	99%	18.8%	81.2%	12.3%	87.7%
	$P_{cc} = 73.1\%$	$P_{cc} = 85.52\%$	$P_{cc} = 84\%$			

followed by the Naive Bayesian classifier. The same behaviour is observed for all feature extraction algorithms.

Overall the results show that performance varies depending on the target and the type of actions under test. In particular the results indicate that the closer the classes under test, the higher the drop in classification performance. For example, correctly classifying a target walking normally versus the same target walking with an object being held in one hand proves to be highly challenging. It shows, on average, a drop in performance to about 60% for $T_w = 30$ ms. Overall, the classification performance tends to increase as the window duration T_w becomes longer.

There is a trade-off between time duration and classification performance. In general, longer window durations allow a beneficial characterisation of the micro-Doppler signature. For example, the Cepstral coefficients vary as a function of time. Longer durations allow extraction of additional time-varying features, such as the signature period and maximum bandwidth. This must be balanced against the disadvantage of longer time gaps between decisions, which can lead to a reduction in classification performance.

7.4 Multi-aspect NCTR

Inspired by observations that bats appear to exploit multiple perspectives as part of their strategies for classifying targets, we now examine how this may aid radar target classification. Thus in this section we examine MP target classification by comparing the classification performance as a function of numbers of perspectives (including the traditional mono-perspective case). Real radar data is used to compute classification performance and consists of HRRPs of full-sized vehicle targets. Performance is assessed under the assumption that targets are already detected and tracked by the illuminating radar system and that the signature to be recognised is a 1D HRRP. The complete database used is collected at all azimuth angles (i.e. over

Table 7.2 Confusion matrices reporting the performance of the Naive Bayesian classifier and the k-NN classifier (k = 3) testing the walking gait of Target A, Target B, *and that of* Target C *for all feature extraction algorithms* (T_w = *30 ms)*

	PCA			Cepstrum			MEL Cepstrum		
Naive Bayesian	*Target A*	*Target B*	*Target C*	*Target A*	*Target B*	*Target C*	*Target A*	*Target B*	*Target C*
Target A	85.2%	85.2%	85.2%	89%	6%	5.1%	80.7%	12.2%	7.2%
Target B	15.7%	39.8%	44.5	36%	29.8%	34.1%	32.5%	23.8%	43.8%
Target C	3%	30%	67%	23.7%	28.5%	47.8%	21.8%	21.3%	57%
	P_{cc} = ***64.04%***	P_{cc} = ***55.53%***	P_{cc} = ***53.78%***						
***k*-NN**	*Target A*	*Target B*	*Target C*	*Target A*	*Target B*	*Target C*	*Target A*	*Target B*	*Target C*
Target A	30%	24.6%	27.6%	78.2%	9%	4%	70.5%	12.7%	8%
Target B	1.3%	19.4%	74%	21%	30.1%	36.6%	13.6%	31.1%	40.2%
Target C	0.2%	10.6%	87%	11.6%	20%	57.4%	5.3%	26.5%	54%
	P_{cc} = ***44.80%***	P_{cc} = ***55.23%***	P_{cc} = ***51.86%***						

360°) for an elevation angle of 0°. The radar system is stationary and receives echoes from a target mounted on a rotating platform.

During the period of collection and processing, a number of factors can contribute to the echo strength of the range profiles causing variability, even though the orientation of the target and system parameters remain the same. These factors include measurement noise, rotational and translational range migration, clutter and forward scatter or multipath. Here, the effects of noise, clutter, forward scatter and any other external interference have been minimised. Although this is not wholly realistic, the purpose of the research is to determine whether or not multiple viewing perspectives improve classification performance, i.e. there is an in-built assumption that any improvement observed would also be apparent for targets in a wider variety of environments. However, this assumption is currently untested and will require careful verification.

Similar to the approach used earlier in this chapter HRRPs of radar targets are used to form feature vectors for subsequent classification. HRR data of full-scale real vehicle targets measured using the Thales airborne data acquisition system (ADAS) provides our signature set from which MP target classification is examined. The data collection geometry is depicted in Fig. 7.10. Two stationary trihedral targets are located in front and behind a turntable, and provide a means for computation of the radar point spread function. For collection of the signatures for classification complex vehicle targets replace the two rotating corner reflectors. In the figure two moving reflectors are located on opposite sides of the rotating turntable. During the 360° target rotation, HRR profiles are collected at an angular interval of 2 min of arc. HRR is achieved by transmitting a linear frequency-modulated (LFM) chirp pulsed waveform [47]. On reception pulse compression is utilised to generate an HRRP. For example for a pulse duration t_p, the chirp waveform is given by

$$t_p = \frac{1}{\sqrt{t_p}} \text{rect}\left(\frac{1}{t_p}\right) e^{jt(\omega_c + \pi\gamma t)} \tag{7.1}$$

where:

ω_c = carrier frequency
γ = rate of frequency variation with time

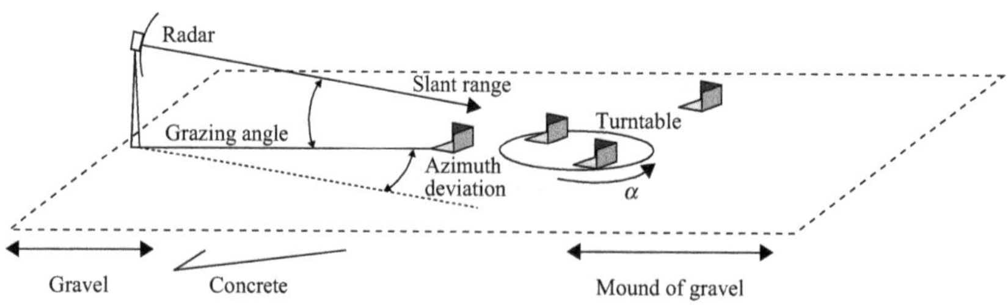

Figure 7.10 ISAR geometry: two stationary corner reflectors are located in front and behind the turntable, while two rotating ones are placed on the turntable

The instantaneous frequency is

$$f(t) = \frac{1}{2\pi}\frac{d}{dt}(\omega_c t + \pi\gamma t^2) = \frac{\omega_c}{2\pi} + \gamma t \tag{7.2}$$

Therefore, the bandwidth is $B = \gamma t_p$. The pulse compression ratio (PCR) is equal to the time-bandwidth product Bt_p and represents the range compression factor (and is equal to the ratio of the pulse length before time compression to its length after compression). The PCR is a gauge of the range resolution improvement given by pulse modulation.

The ADAS data chirp parameters are shown in Table 7.3. These provide HRRPs with a range resolution of approximately 39 cm after weighting. Fig. 7.11 shows an example of the form of the radar data. It depicts the intensity-modulated time history of HRRPs from the two stationary and the two rotating trihedral reflectors for the geometry depicted in Fig. 7.10. The two stationary trihedrals show a constant response at near and far range as expected. For the two rotating trihedral targets, when the line-of-sight is on the trihedral bisector, a peak of reflection

Table 7.3 ADAS single chirp parameters

Centre frequency	f_c	9.25 GHz
Chirp rate	γ	$1.46 \cdot 10^{15}$
Pulse compression ratio	*PCR*	170.5
Transmitted bandwidth	B_{tx}	500 MHz
Sampling frequency	f_s	500 MHz
Range resolution	Δr	30 cm

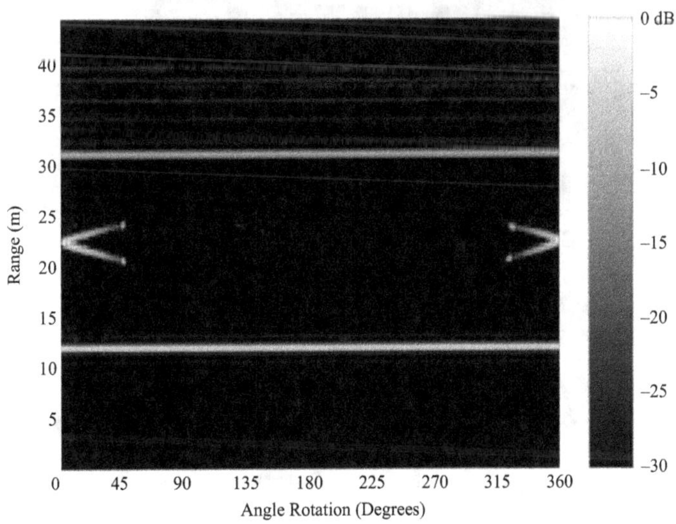

Figure 7.11 History of HRRPs (30 cm range resolution) from four corner reflectors, two rotating and two stationary

occurs. This is consistent with the expected theoretical response [48]. As the trihedral targets rotate, the backscattered field decreases progressively until a point is reached where there is a local peak of reflection. This is a specular reflection from one of the sides making up the trihedral which is orthogonal to the illuminating radar system (i.e. it faces the radar beam and looks like a flat plate reflector). At increasing rotation angles the Radar Cross Section (RCS) of the target drops rapidly because the orientation of the trihedral is such that it tends to reflect incident radiation away from the radar. This angular dependency of the RCS of a known reflector, such as a trihedral, begins to illustrate how the backscattering properties of real targets may vary with the orientation of observation. For example, if a target has part of its structure that mimics a trihedral, it will only show this feature over a limited angular range. In an MP environment, different angular samples of a target signature should improve the likelihood of observing a corner or corner-like reflector. Indeed, such shapes are common on many man-made structures and are often quite dominant features and hence may prove useful for classification.

In Fig. 7.12 the complete angular ambit of range profiles spanning 360° are shown for a Land Rover vehicle. This highlights a number of different scattering behaviours: the strong peaks from specular reflections (0°, 90°, 180°,...) appear over a very limited range of angles and obscure nearby point-like backscattering. Possible corner-like returns can be observed at a far range (6 m) for two range angular spans ((10°–60°) and (130°–180°)). These returns appear to correspond to the trihedral-like structures formed at the rear of the Land Rover. This is a vehicle without the rear soft top and has a metallic bench seat that makes a corner where it

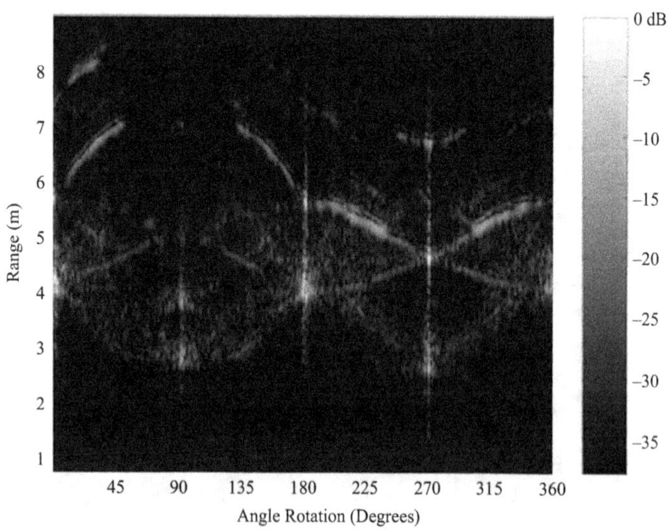

Figure 7.12 History of HRRPs (8 cm range resolution) from a series of X-band stepped frequency chirps illuminating a ground vehicle as it rotates over 360° with the turntable

joins the rear bulkhead of the driver's cabin. At 8 m range, well beyond the physical extent of the Land Rover, there is a double bounce return corresponding to a multiple bounce in the back of the pick-up. Scatter effects of this type can increase the information that can be extracted (i.e. we can infer that it is a vehicle with an open back) and its observation is more likely as the number of perspectives increases.

Classification has been performed on a subpopulation of the total data set consisting of a total of three ground vehicles (A, B and C) exploiting multiple perspectives of HRRPs from each of the vehicles. For each, there are 10,500 profiles spanning 360°. The data are divided into contiguous 10° sectors, and 36 signatures randomly selected from each of the 10° sectors are given to the classifier to build the internal representation of the class. These signatures are then removed completely from the data set. The remaining profiles then provide the input to the classifiers.

7.4.1 Data preparation

After pulse compression and weighting the signal is Fourier transformed to form a full-range profile in which the target lies. Subsequently, the HRR profile of the target only is cut out and consists of 52 range bins, each with a sampled resolution of 30 cm, and an actual resolution of approximately 39 cm (because of the window applied). However, the data exhibit a low level contribution from zero Doppler clutter (ZDC). ZDC is stationary or quasi-stationary clutter contained in the beam, making a contribution to the measured HRRPs and hence impacting the classification performance adversely. Thus the ZDC has been estimated and subtracted. This has been performed in the frequency domain following the method outlined in Reference 49. The mean value of ZDC over about 50° of turntable rotation is subtracted from those frequency signatures, and then the final HRRP is extracted. After ZDC subtraction, a mean filter over a neighbourhood of three echoes is applied to a sequence of range profiles. Since two consecutive HRRPs are spaced by only 2 min of arc of target rotation, the non-coherent averaging partially averages out those phenomena that are independent of the geometry. In addition, the SNR is increased as demonstrated in Reference 50 by a factor $\sqrt{N}$, where N is the number of target echoes. Prior to non-coherent averaging, it is usually necessary to align the range profiles. However, here, because the input patterns for recognition are collected from the turntable measurements, no additional alignment is needed between training and test profiles. Finally, the signatures are normalised following Reference 51 to make the target area in the HRRP more prominent. A target mask is generated to select the target only as part of the range profile by measuring the noise mean value μ and standard deviation σ. A threshold $\mu + 3\sigma$ is applied to eliminate the non-target zone. The target masks are used so that target-only data is presented to the classifier.

7.4.2 Feature extraction

In a typical pattern recognition problem it is often necessary to reduce the data dimension of the input of the classifier. This is mainly because of an intrinsic degree of redundancy within the data where there are regions where no information is contained. It also helps to reduce the very high quantities of data that otherwise have to be processed. The range profiles are used to derive feature vectors that aim to contain the key target information enabling successful classification. After focusing on those range resolution cells whose intensities are significant and hence depict the target backscattering, the number of elements representing the back-scatter can be reduced with an information loss that is assumed negligible. Dimensional reduction of this type also attempts to emphasise the differences between patterns and hence enhance classification performance. PCA [15,45] is a statistical method that enables the data to be represented in a different vector basis such that it is possible to remove similarities (which therefore do not contribute to the classification process). After subtracting the mean $\bar{f}$ from each of the vectors of the training set F and producing a zero-mean set of data, the covariance matrix $\boldsymbol{Cov}$ can be formed

$$\boldsymbol{Cov}(f) = \frac{1}{N}\sum_{n}(f_n - \bar{f})(f_n - \bar{f})^T \tag{7.3}$$

After calculating the Eigenvectors of $\boldsymbol{Cov}(f)$, the P most significant Eigenvectors with the largest Eigenvalues are selected and form a new basis vector $V = (v_1, v_2, \ldots, v_P)$. The test and training feature vectors can then be transformed as follows:

$$f' = V^T(f_n - \bar{f}) \tag{7.4}$$

The number of principal components PC $= P$ is chosen as a function of the classification rate achieved. This usually becomes stable once the PCs necessary to fully describe the data have been selected. After testing the classifiers, their mean value CCRs are plotted against the number of principal components representing the feature vectors. For the data used here the dimension of the feature vector is 52. However, as can be seen in Fig. 7.13, the probability of correct classification P_{cc} is almost unaltered until fewer than approximately 12 components. At this point they are no longer representative of the different target classes. Therefore, in order to depict the target with sufficient robustness, the feature vectors used in the following sections are obtained using the first 21 principal components.

7.4.3 Multi-perspective classifiers

In order to avoid any bias introduced by a single algorithm, a total of three different classifiers have been implemented. These are (1) a Naive Bayesian classifier, (2) an NN classifier and (3) a neural network classifier. All three are described in detail in Reference 52. A traditional single-perspective classifier, after training the ATR algorithm with a set of templates, is tested with profiles collected from all orientations of the target. In an N-perspective scenario, the parameter that distinguishes the perspective topology is the vector $\boldsymbol{\Phi} = \{\Delta\phi_{i,j} : i,j = 1, \ldots, N\}$, where the element $\Delta\phi_{i,j} = \phi_i - \phi_j$ represents the angular displacement between the nodes (perspectives) i and j. For example in a two-perspective (2P) scenario, the

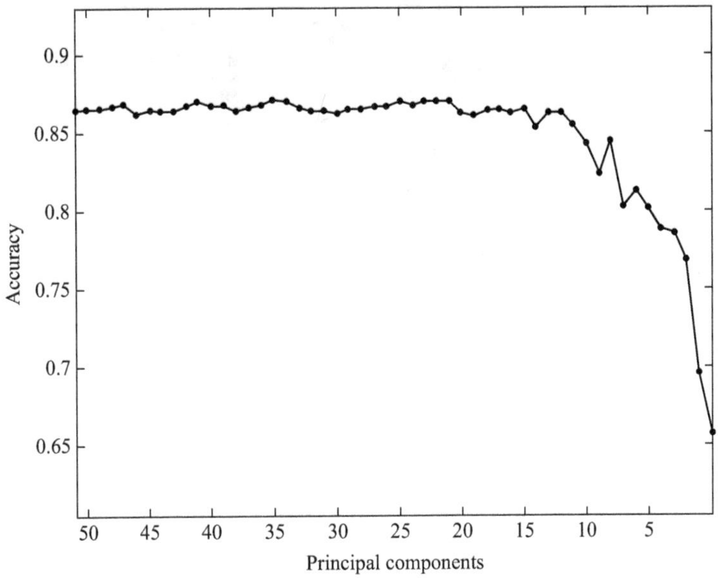

Figure 7.13 Classification accuracy versus number of principal components obtained with the PCA algorithm

parameter that distinguishes the perspective node locations is their relative angular displacement $\Delta\phi_{1,2} = \phi_2 - \phi_1$. Hence, after fixing $\Delta\phi_{1,2}$, the 2P classifier is tested with all possible pairs of HRR profiles displaced by that angle covering all the possible orientations of the target. Having a test set consisting of N profiles, the same number of pairs can be formed to test the 2P classifier. As shown in Fig. 7.14, to test the MP classifier, the sequence of all the possible pairs of HRR profiles displaced by $\Delta\phi_{1,2}$ is given as an input, and the accuracy for a 2P classifier is subsequently computed. Thus to evaluate the 3P recognition performance, after choosing the set $\Phi = \{\Delta\phi_{1,2}, \Delta\phi_{1,3}\}$, the classification accuracy is measured as the average correct classification over all the possible HRR profiles triples that could be formed from the MP topology Φ. This procedure is extended for networks having four and five perspectives. The training set of representative vectors for each class is made up of 36 range profiles, taken approximately for every 10° of target rotation. The 'testing set' of each class consists of the remaining range profiles neglecting the templates used for training. The angular displacement between nodes is not processed as information by the MP classifier. Note, this is only one of the possible approaches to MP classification and is used as it is simple to implement and the computational burden is reduced to a minimum. Although the target has already been assumed to have been detected and tracked, the angular displacements F of the nodes are unknown to the network. Therefore, in case of a single sensor collecting multiple perspectives from a moving target, it is assumed that to combine the signatures, it is not necessary to estimate the motion parameters

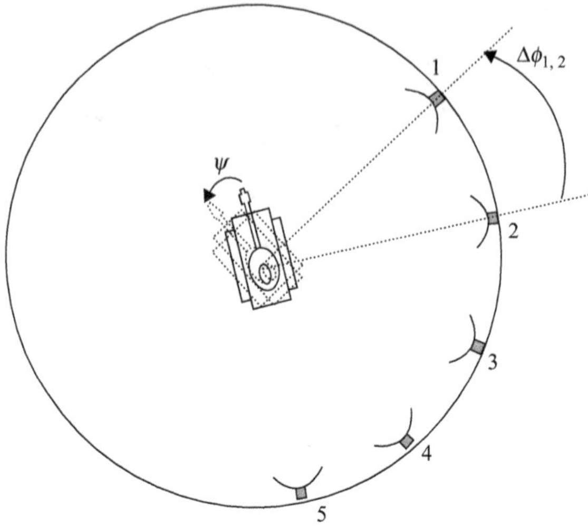

Figure 7.14 Multi-perspective environment approximation

and the centre of rotation of the target. Here, we consider a network of radars consisting of N nodes, each single perspective j of the target is represented by the HRRP X_j. Consequently, the set of signatures collected by the network is represented by $\{X_j : j = 1, \ldots, N\}$ and is directly related to the vector of angular node displacements $\mathbf{\Phi}$. To test classification performance, the signatures $\{X_j : j = 1, \ldots, N\}$ correspond to the HRRPs after applying PCA.

7.4.4 Multi-perspective classification performance

Classification performance is evaluated in a forced decision environment using confusion matrices (e.g. see Reference 53) where each column represents the instances in a predicted class, whereas the rows represent the output distribution for an actual class. In Table 7.4, the confusion matrices of the three classifiers are shown with respect to the number of perspectives involved in the decision processing. The single-perspective results represent the more usual single perspective classifier performance. It should be noted that even with a single perspective the classification performance is very good. This is a function of using turntable data together with a small number of sub-populations. However, it does allow the trends with a number of perspectives to be explored. Table 7.4 clearly shows an improvement in classification performance, as the number of perspectives is increased. This trend is observed for all three classifiers and hence we can conclude that it is not a function of the classification process but due to the additional perspectives. Interestingly, the biggest increase in performance occurs when just one additional perspective is employed. A parameter to measure the goodness of a classifier is the CCR (i.e. the correct classification rates). It is the mean value of the elements on the diagonal of the confusion matrix. In Table 7.5, the CCR of the three implemented classifiers are displayed using different numbers of

Table 7.4 Multi-perspective confusion matrices for three classifiers (k-NN, Naive Bayesian and FANNs) using different numbers of perspectives

Perspective	Classifier	*k*-NN (%)			Bayesian (%)			FANNs (%)		
	Class	*A*	*B*	*C*	*A*	*B*	*C*	*A*	*B*	*C*
1	*A*	79.37	1.72	18.91	78.39	6.90	14.71	89.73	1.4	8.87
	B	2.92	85.82	11.26	2.10	67.99	29.91	2.58	82.06	15.36
	C									
2	*A*	91.61	0.48	7.91	90.38	1.74	7.88	96.77	0.15	3.08
	B	1.11	94.92	3.97	2.04	77.82	20.12	0.39	94.58	5.03
	C	0.25	15.97	83.78	3.83	8.20	87.97	0.45	5.97	93.58
3	*A*	92.26	0.07	7.67	95.27	0.71	4.02	98.12	0.05	1.83
	B	0	97.88	2.12	1.84	84.52	13.64	0.19	96.41	3.40
	C	0.04	7.59	92.37	2.58	6.68	90.74	0	4.03	95.97
4	*A*	96.88	0.01	3.11	98.24	0.19	1.57	99.44	0	0.56
	B	0	99.65	0.35	1.55	89.86	8.59	0.05	98.66	1.29
	C	0	7.05	92.95	1.78	5.66	92.56	0	2.14	97.86
5	*A*	97.34	0	2.66	98.93	0.09	0.98	99.75	0	0.25
	B	0	99.82	0.18	0.94	95.18	3.88	0	99.36	0.64
	C	0	3.5	96.5	1.09	4.99	93.92	0	1.20	98.80

perspectives. Again all three show the same trend of improvement in performance as the number of perspectives is increased. Fig. 7.15 shows this graphically and again highlights that the increase in classification performance is independent of the particular classifier approach used. Furthermore, it emphasises that the MP improvements are not linear with the number of perspectives involved. On average an 8.8% improvement in the correct classification rate, with respect to the single perspective classifier, is reached when two perspectives are used, whereas the benefit in relation to the 2P classifier is 3.6% when three perspectives are used. It is 2.51% when going from three to four perspectives and finally 1.5% when going from four to five perspectives. This is a consequence of those signatures that relate to particular target orientations where specular reflections occur and the profile is thus dominated by only a very few scattering centres. These patterns are common to all the target classes and consequently are more easily misclassified. Nevertheless, for these target orientations, the decision confidence of the classifier is low and is therefore

Table 7.5 Multi-perspective correct classification rates (CCR) for the three classifiers using a progressive number of perspectives

	***k*-NN (%)**	**Bayesian (%)**	**FANNs (%)**
1	80.81	77.06	86.22
2	90.10	85.39	94.97
3	94.17	90.17	96.83
4	96.49	93.55	98.65
5	97.88	96.01	99.30

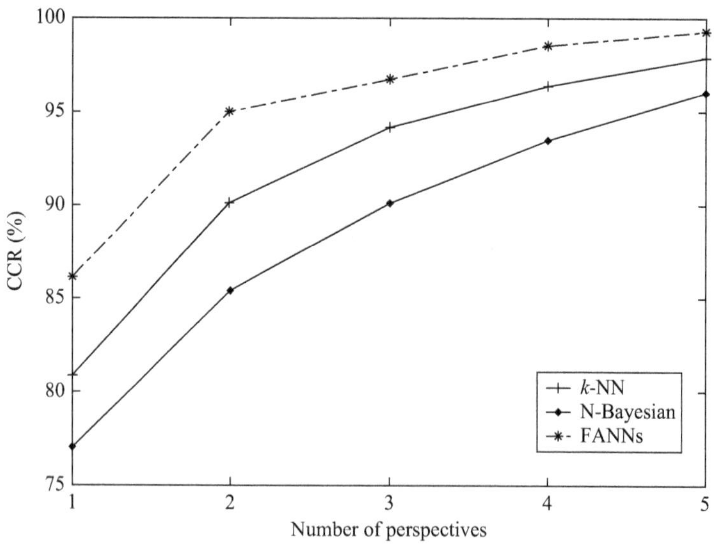

Figure 7.15 Correct classification rates of three classifiers using different numbers of perspectives

significantly aided by a second perspective. On the other hand, if both the first and second perspectives correspond to those with reduced information signatures, a third perspective is less effective in improving performance. A third perspective has less influence to correct the weights of the others and thus the improvement from subsequent perspectives is progressively attenuated. This trend may be different if the perspective displacement of the nodes of the network is processed by the classifier as useful information. Nonetheless, the nonlinear relationship between CCR and number of radar perspectives suggests that only the employment of a small number of nodes in the network is necessary in order to achieve the best trade-off between system complexity and improvement in classification performance.

The benefits of using multiple perspectives to improve classification performance are now examined as a function of different SNR levels. From each target class i, the noise level is measured in the non-target area before target normalisation. The target signal level is estimated as the mean value of the maximum target backscattering over the 360° range profile history and the $SNR_{\text{class-i}}$ using:

$$SNR_{\text{class-i}} = \frac{1}{KN}\sum_{n=1}^{N}\sum_{k=1}^{K} 20\log\left(\frac{r_{n,k}}{\mu_n}\right) \tag{7.5}$$

where:

μ_n = mean noise intensity for the nth, k-dimensional profile
$r_{n,k} = (r_{n,1}, r_{n,2}, \ldots, r_{n,K})$
N = number of profiles collected for the ith target class.

The result of applying this procedure to the original data leads to

$$SNR = (1/n_c)\sum_i SNR_{\text{class-i}} = 28.4\text{dB} \tag{7.6}$$

The profiles are subsequently progressively corrupted with different levels of Gaussian noise applied to the I and Q channels, before matched filtering. Then, the SNR is measured as above, and classification attempted on PCs extracted from the 1D signatures. The original training dataset is preserved. As there was no bias introduced by any of the three classifiers, only the Feed-forward Artificial Neural Networks (FANNs) classifier is used to evaluate performance as a function of SNR. In Fig. 7.16, the CCR results for just the FANNs classifier are depicted against the number of perspectives used and the SNR levels. In all cases increasing the number of perspectives improves classification performance. In comparison to a single-perspective classifier, the CCRs remain stable above an SNR threshold of 17 dB, and then a very rapid deterioration of the classification performance is observed. This is because of the information loss caused by the increasing noise that corrupts the range profile. The relatively high SNRs at which this occurs can be partly explained by the fact that it is a combination of smaller magnitude and larger magnitude scatterers that provide the classification information. The smaller

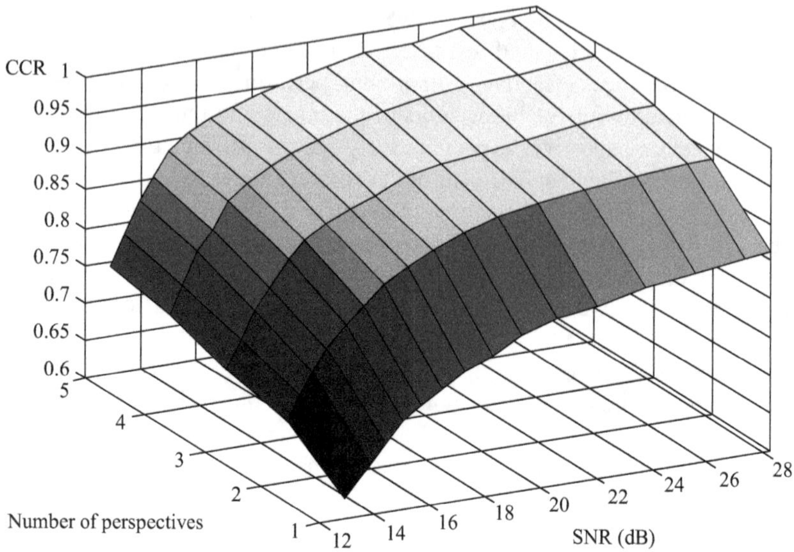

Figure 7.16 Correct classification rates for an MP FANNs classifier with different SNR test sets

magnitude scatterers are more severely affected by the increasing noise and hence quickly have an impact on reducing performance. As the number of network nodes increases, the information is partially recovered with the help of other perspectives because the noise is not correlated and hence modifies the profiles differently. This result does highlight the importance of having adequate SNR such that small magnitude scatterers are preserved as an input to the classification process.

7.5 Summary

Multiple perspectives appear to be available to the echolocating bat and may play a role in the way in which bats classify targets. In this section multiple perspectives have been shown to consistently enhance classification performance of an HRR radar system. The use of multiple perspectives introduces more information, unique to a particular target. This explains both the improvement in classification performance and also the fact the improvement is subject to a law of diminishing returns. Noise has been shown to have a significant impact of achievable performance regardless of the number of perspectives used. The average SNR required for reliable classification is higher than that of detection and may be related to the SNR of the smallest echo that contains key classification information.

An acoustic radar operating at 80 kHz was developed for collection of micro-Doppler signatures of moving targets, doubling the frequency used in previous works. The acoustic radar was deployed in a set of experimental trials in which micro-Doppler signatures of various personnel targets undertaking a number of actions were collected in a highly indoor cluttered environment. These were used to test a *k*-NN and a Naive Bayesian classifier trained to distinguish between walking gaits of different targets or between different actions undertaken by the same targets. Results show that the acoustic radar can be successfully deployed at short ranges to collect micro-Doppler signatures of moving targets. Classification performance results show that the information contained in these acoustic signatures can be used to perform identification and recognition of personnel targets. Data has been gathered and analysed for a range of personnel target motions and good levels of classification have been achieved. Although the acoustic radar can only survey short ranges due to the high attenuation of sound waves in air, it can be successfully deployed in indoor environments such as airports, to monitor, for example, the flow of passengers through doors or security checks with the potential to identify suspicious behaviour. In addition to this, the acoustic radar offers an easy and inexpensive way to collect micro-Doppler signatures of a wide range of targets, often difficult to obtain, that can be used to analyse human walking gaits or other type of behaviour in a number of applications. Future work will look at understanding how additional features such as geometrical features (i.e. the signature period or the signatures maximum Doppler shift) or how the history of the current features (i.e. the time variation of the Cepstrum or MEL-Cepstrum coefficients) can be deployed to enhance classification performance. Extending the work presented in this chapter to the underwater regime is also under consideration.

Multi-perspective radar target classification based on that used by bats shows how radar target classification performance can be further improved. The method is based on the processing of real, full-scale target signatures collected at different perspectives in a network of monostatic radars. In a first single perspective stage, depending on the methodology adopted, the 1D imageries are processed in parallel, providing the following MP stage with partial outputs eventually used for the final decision. The benefits of the MP classifier implementation have been analysed and show a nonlinear but very clear CCR improvement with the number of perspectives. As the number of radar nodes increases, the MP CCR stabilises, suggesting that most benefit accrues when using a small network of only two or three nodes and hence avoids any significant increase of system implementation complexity. For example, this enables a simple approach to be adopted for a single monostatic radar that obtains multiple perspectives as it traverses an object or area of interest (provided sufficient angular diversity is obtained). The performance has also been investigated as a function of different noise levels. The incidence of multiple perspectives on correct classification performance improvement is observable for any value of the SNR range of values examined. In addition, the MP benefits are more pronounced for low SNR levels that are for those SNR values that are typical of not controlled real-world experiments. Further research is required to investigate the effects of the separation of perspectives used to evaluate any relationship between angles of perspectives selected and the resulting classification performance. This is particularly important if single monostatic radar systems are to be used as it could minimise the time at which targets are illuminated.

References

[1] Williams R., Westerkamp J., Gross D., Palomino A., Williams R. and Fister T., 'Automatic target recognition of time critical moving targets using 1D high range resolution (HRR) radar', *IEEE AES Magazine*, Vol. 15, no.4, pp. 37–43, 2000.

[2] Novak L.M., Halversen S.D., Owirka G. and Hiett M., 'Effects of Polarization and Resolution on SAR ATR', *IEEE Trans. Aerosp. Electron. Syst.*, Vol. 33, pp. 102–116, 1997.

[3] Sadjadi F., 'Improved Target Classification using Optimum Polarimetric SAR Signatures', *IEEE Trans. Aerosp. Electron. Syst.*, Vol. 38, pp. 38–49, 2002.

[4] Novak L.M., 'A Comparison of 1-D and 2-D Algorithms for Radar Target Classification', *IEEE International Conference on Systems Engineering*, pp. 6–12, August 1991.

[5] Shihao J., Xuejun, L. and Carin L., 'Adaptive Multiaspect Target Classification and Detection with Hidden Markov Models', *IEEE Sens. J.*, Vol. 5, pp. 1035–1042, 2005.

[6] Xuejun L., Runkle P. and Carin L., 'Identification of Ground Targets from Sequential High-Range-Resolution Radar Signatures', *IEEE Trans. Aerosp. Electron. Syst.*, Vol. 38, pp. 1230–1242, 2002.

[7] Runkle P.R., Bharadwaj P.K., Couchman L. and Carin L., 'Hidden Markov Models for Multiaspect Target Classification', *IEEE Trans. Signal Proces.*, Vol. 47, pp. 2035–2040, 1999.

[8] Baker C.J. and Hume A.L., 'Netted Radar Sensing', *IEEE AES Magazine*, Vol. 18, pp. 3–6, 2003.

[9] Baker C.J., Griffiths H.D. and Balleri A., 'Biologically Inspired Waveform Diversity', in *Waveform Design and Diversity for Advanced Radar Systems* (F. Gini, A. De Maio and L. Patton, editors), *IET Radar Sonar Navigat.* Series 22, pp. 149–172, 2012.

[10] von Helversen O., Winker L. and Bestmann H., 'Sulphur-Containing "Perfumes" Attract Flower-Visiting Bats', *J. Comp. Physiol. A*, Vol. 186, pp. 143–153, 2000.

[11] von Helversen D. and von Helversen O., 'Acoustic Guide in Bat-Pollinated Flower', *Nature*, Vol. 398, pp. 759–760, 1999.

[12] von Helversen D., von Helversen O., 'Object Recognition by Echolocation: A Nectar-Feeding Bat Exploiting the Flowers of a Rain Forest Vine', *J. Comp. Physiol. A*, Vol. 189, pp. 327–336, 2003.

[13] von Helversen D. Holderied M.W. and von Helversen O., 'Echoes of Bat-Pollinated Bell-Shaped Flowers: Conspicuous for Nectar-Feeding Bats?' *J. Exp. Biol.*, Vol. 206, No. 6, pp. 1025–1034, 2003.

[14] Streicher A., Muller R., Peremans H. and Lerch R., 'Broadband Ultrasonic Transducer for an Artificial Bat Head', *IEEE Symposium on Ultrasonics*, Vol. 2, pp. 1364–1367, 2003, Honolulu, HI.

[15] Balleri A., Griffiths H.D., Baker C.J., Woodbridge K. and Holderied M.W., 'Analysis of Acoustic Echoes from a Bat-Pollinated Plant Species: Insight into Strategies for Radar and Sonar Target Classification', *IET Radar Sonar Navigat.*, Vol. 6, No. 6, pp. 536–544, July 2012.

[16] Volz A., *Echolocation and Flight Behaviour of Neo-tropical Nectar-Feeding Bats (Chiroptera, Glossophaginae) during Flower Approach*, Masters thesis, University of Erlangen, 2006.

[17] Balleri A., Griffiths H.D., Woodbridge K., Baker C.J. and Holderied M., 'Impact of Flight Trajectory on the Detection and Selection of Flowers by Nectar-Feeding Bats', *2nd International Workshop on Cognitive Information Processing (CIP)*, pp. 350–353, 14–16 June 2010, Elba Island, Italy.

[18] Tian B. and Schnitzler H.-U., 'Echolocation Signals of the Greater Horseshoe Bat (*Rhinolophus ferrumequinum*) in transfer flight and during landing', *J. Acoust. Soc. Am*, Vol. 101, No. 4, pp. 2347–2364, 1997.

[19] Schuller G. and Pollak G.D., 'Disproportionate Frequency Representation in the Inferior Colliculus of Doppler Compensating Greater Horseshoe Bats. Evidence for an Acoustic Fovea', *J. Comp. Physiol.*, Vol. 132, pp. 47–54, 1979.

[20] Ostwald J., 'Tonotopical Organization and Pure Tone Response Characteristics of Single Units in the Auditory Cortex of the Greater Horseshoe Bat', *J. Comp. Physiol. A*, Vol. 155, pp. 821–834, 1984.

[21] von der Emde G. and Schnitzler H.-U., 'Fluttering Target Detection in Hipposiderid Bats', *J. Comp. Physiol. A*, Vol. 159, pp. 765–772, 1986.

[22] Schnitzler H.-U., Menne D., Kober R. and Heblich K., 'The Acoustical Image of Fluttering Insects in Echolocating Bats', *Neuroethology and Behavioral Physiology: roots and growing points*, pp. 235–250, 1983.

[23] Kober R. and Schnitzler H.-U., 'Information in Sonar Echoes of Uttering Insects Available for Echolocating Bats', *J. Acoust. Soc. Am.*, Vol. 87, No. 2, pp. 882–896, 1990.

[24] Sotavalta O., 'The Flight Tone (Wing Stroke Frequency) of Insects', *Acta Entomol. Fenn.*, Vol. 4, pp. 5–117, 1947.

[25] von der Emde G. and Schnitzler H.-U., 'Classification of Insects by Echolocating Greater Horseshoe Bats', *J. Comp. Physiol. A*, Vol. 167, pp. 423–430, 1990.

[26] Schnitzler H.-U., *Echoes of Fluttering Insects: Information for Echolocating Bats*, Recent Advances in the Study of Bats, pp. 226–243, Cambridge University Press, Cambridge, 1987.

[27] Gill T.P., *The Doppler Effect, an Introduction to the Theory of the Effect*, Logos Press, 1965, London.

[28] Chen V.C., Li F., Ho S.-S. and Wechsler H., 'Analysis of Micro-Doppler Signatures', *IEE Proc. Radar Sonar Navigat.*, Vol. 150, No. 4, pp. 271–276, 2003.

[29] Chen V.C., Li F., Ho S.-S. and Wechsler H., 'Micro-Doppler Effect in Radar: Phenomenon, Model, and Simulation Study', *IEEE Trans. Aerosp. Electron. Syst.*, Vol. 42, No. 1, pp. 2–21, January 2006.

[30] Chen V.C., 'Analysis of Radar Micro-Doppler with Time-Frequency Transform', *Proceedings of the Tenth IEEE Workshop on Statistical Signal and Array Processing*, pp. 463–466, 2000, Pocono Manor, PA, USA.

[31] Smith G.E., Woodbridge K. and Baker C.J., 'Template Based Micro-Doppler Signature Classification', *IET Seminar on High Resolution Imaging and Target Classification*, pp. 127–144, 2006, London, UK.

[32] Smith G.E., Woodbridge K. and Baker C.J., 'Micro-Doppler Signature Classification', *CIE International Radar Conference*, Shanghai, 2006.

[33] Smith G.E., Woodbridge K. and Baker C.J., 'Naive Bayesian Radar Micro-Doppler Recognition', *International Radar Conference*, pp. 111–116, 2008, Adelaide, Australia.

[34] Tahmoush D. and Silvious J., 'Radar Micro-Doppler for long range front-view gait recognition', *IEEE 3rd International Conference on Biometrics: Theory, Applications, and Systems*, pp. 1–6, 2009.

[35] Tivive F., Bouzerdoum A. and Amin M., 'Automatic Human Motion Classification from Doppler Spectrograms', *2nd International Workshop on Cognitive Information Processing (CIP)*, pp. 237–242, 14–16 June 2010, Elba Island, Italy.

[36] Vignaud L., Ghaleb A., Le Kernec J. and Nicolas J.-M., 'Radar High Resolution Range & Micro-Doppler Analysis of Human Motions', *SEE International Radar Conference*, Bordeaux, France, 12–16 October 2009.

[37] Ghaleb A., Vignaud L. and Nicolas J., 'Micro-Doppler Analysis of Wheels and Pedestrians in ISAR Imaging', *IET Signal Proces.*, Vol. 2, No. 3, pp. 301–311, September 2008.

[38] Chen V.C., Miceli W. and Himed B., 'Micro-Doppler Analysis in ISAR—Review and Perspectives', *SEE International Radar Conference*, Bordeaux, France, 12–16 October 2009.

[39] Sammartino P. and Fortuny-Guash J., 'Space and Frequency Diversity for Moving Personnel Spectrogram Estimation', *IEEE International Radar Conference*, Washington DC, 2010.

[40] Kalgaonkar K. and Raj B., 'Acoustic Doppler Sonar for Gait Recognition', *IEEE Conference on Advanced Video and Signal Based Surveillance*. AVSS, pp. 27–32, 2007, 5–7 Sept. 2007, London.

[41] Zhang Z., Pouliquen P.O., Waxman A. and Andreou A.G., 'Acoustic Micro-Doppler Radar for Human Gait Imaging', *J. Acoust. Soc. Am. Expr. Lett.*, Vol. 121, No. 3, pp. 110–113, 2007.

[42] Zhang Z., Pouliquen P.O., Waxman A. and Andreou A.G., 'Acoustic Micro-Doppler Gait Signatures of Humans and Animals', *41st Annual Conference on Information Sciences and Systems*, pp. 627–630, 2007, 14–16 March 2007, Baltimore, MD, USA.

[43] Zhang Z. and Andreou A., *Human Identification Experiments using Acoustic Micro-Doppler Signatures, Proceedings of the Argentine School of Micro-Nanoelectronics, Technology and Applications*, pp. 81–86, 18–19 Sept. 2008, Buenos Aires. pp. 81–86, 2008.

[44] Duda R., Hart P. and Stork D., *Pattern Classification*, Wiley, 2001, New York.

[45] Theodoridis S. and Koutroumbas K., *Pattern Recognition* (4th edition), Academic Press, 2009, Amsterdam/London.

[46] Balleri A., Chetty K. and Woodbridge K., 'Classification of Personnel Targets by Acoustic Micro-Doppler Signatures', *IET Radar Sonar Navigat.*, Vol. 5, No. 9, pp. 943–951, December 2011.

[47] Sullivan R.J., *Radar Foundations for Imaging and Advanced Concepts*, SciTech Publishing, 2004, Raleigh, NC.

[48] Knott E.F., Shaeffer J.F. and Tuley M.T., *Radar Cross Section*, Artech House, 1985, Boston/London.

[49] Showman G.A., Richards M.A. and Sangston K.J., 'Comparison of Two Algorithms for Correcting Zero-Doppler Clutter in Turntable ISAR Imagery', *Conference on Signals, Systems & Computers*, Vol. I, pp. 411–415, 1998, 1–4 Nov. 1998, Pacific Grove, CA, USA.

[50] Skolnik M.I., *Introduction to Radar Systems*, McGraw-Hill, 1980, New York/London.

[51] Zyweck A. and Bogner R.E., 'Radar Target Classification of Commercial Aircraft', *IEEE Trans. Aerosp. Electron. Syst.*, Vol. 32, pp. 598–606, 1996.

[52] Vespe M., Baker C.J. and Griffiths H.D., 'Radar Target Classification using Multiple Perspectives', *IET Radar Sonar Navigat.*, Vol. 1, No. 4, pp. 300–307, August 2007.

[53] Tait P., *An Introduction to Radar Target Recognition*, IEE, Stevenage, 2005, ISBN: 0863415016.

Chapter 8
Radar applications of compressive sensing

Albert G. Huizing and Reinier G. Tan

8.1 Introduction

In radar systems, there is a growing need for higher resolutions in the spatial and temporal domains to enable the detection of small targets in a complex background, to improve the tracking of multiple closely spaced targets and to support automatic target recognition. Higher resolutions in the spatial domain can be achieved by increasing the instantaneous radar bandwidth and/or enlarging the size of the radar aperture. In the temporal domain, a higher resolution can be achieved by a longer coherent integration time. As an example, consider a radar system that classifies an air target based on a relatively small set of K features that are extracted from a range profile with a high resolution (see Fig. 8.1).

The high range resolution profile may be obtained with different types of waveforms such as a very short pulse, a chirp with a large bandwidth or a stepped frequency waveform [1]. However, the common element of these waveforms is that the sampling rate in the receiver needs to satisfy the Nyquist criterion, which results in a number of samples N that is often much larger than the number of features that is used for classification. Although this approach is widely used, not only in radar but also in other sensors such as receivers for electronic support and cameras, it seems to be quite inefficient to collect so much data that is discarded in a later stage.

Compressive sensing (CS) is a novel data acquisition/processing method that allows reconstruction of a signal from measurements that are much fewer than specified by the Nyquist criterion [2,3]. CS is based on the premise that the signal is sparse in some (transform) domain and a random sampling of the signal. Fig. 8.2 shows a block diagram of a radar that uses random sampling and a signal reconstruction processes to allow the K features to be extracted from a number of samples M that is significantly smaller than N. After reconstruction of the radar signal, the relevant features can be extracted and the target can be classified. This CS approach avoids sampling at the Nyquist rate and can, e.g., reduce the dwell time of a radar system using a stepped frequency waveform for obtaining a high range resolution profile. Note that in this approach the Nyquist rate radar signal is reconstructed in the original signal domain, which still has a size N. A further step in reducing the inefficiency in the radar processing chain may be achieved by

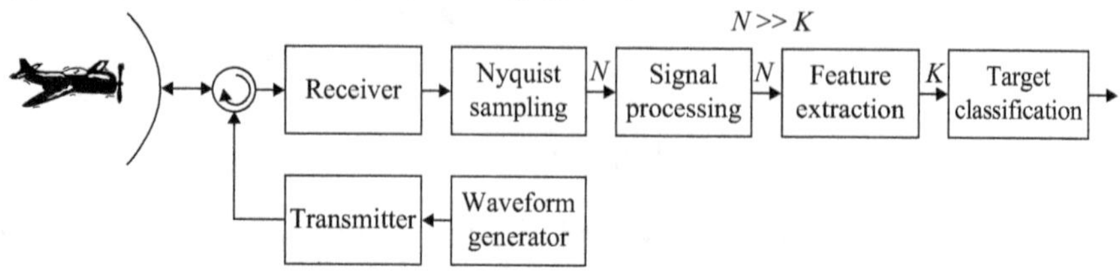

Figure 8.1 Simplified block diagram of a radar system for target recognition based on a high range resolution profile using Nyquist sampling of the radar signal

reconstruction of the signal in the feature domain directly through a combination of the sparse signal reconstruction process and the feature extraction process.

The purpose of this chapter is to give a brief overview of the principles of CS and to show how CS may be applied in a radar system to support automatic target recognition. The chapter is organised as follows. Section 8.2 gives an introduction of the basic principles of CS. Section 8.3 presents an overview of some of the main algorithms for reconstruction of sparse signals. The application of CS to target recognition based on jet engine modulation (JEM) is described in section 8.4. Section 8.5 shows how CS may be applied to high resolution imaging of targets using inverse synthetic aperture radar (ISAR). Finally, section 8.6 gives conclusions.

8.2 Principles of compressive sensing

8.2.1 Sparse and compressible signals

Radar signals just like many other sensor signals have concise representations when expressed in a convenient basis. Consider, e.g., the radar echo of an aircraft with jet engines sampled by a pulse-Doppler waveform with a pulse repetition frequency (PRF) of 40 kHz represented in the time domain in Fig. 8.3(a) and its Doppler spectrum in Fig. 8.3(b), which is computed by the Fourier transform of the time samples. Although all of the time samples have non-zero values, the Fourier coefficients offer a concise summary: most coefficients are small, and the relatively few large coefficients capture most of the information. The highest peak in the

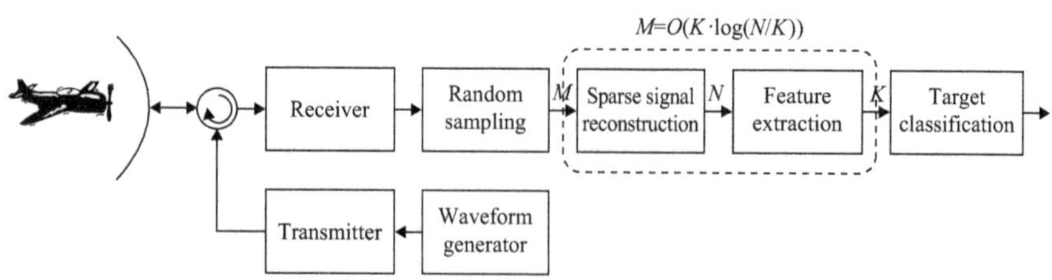

Figure 8.2 Simplified block diagram of a radar system for target recognition based on a high resolution range profile using random sampling and sparse signal reconstruction

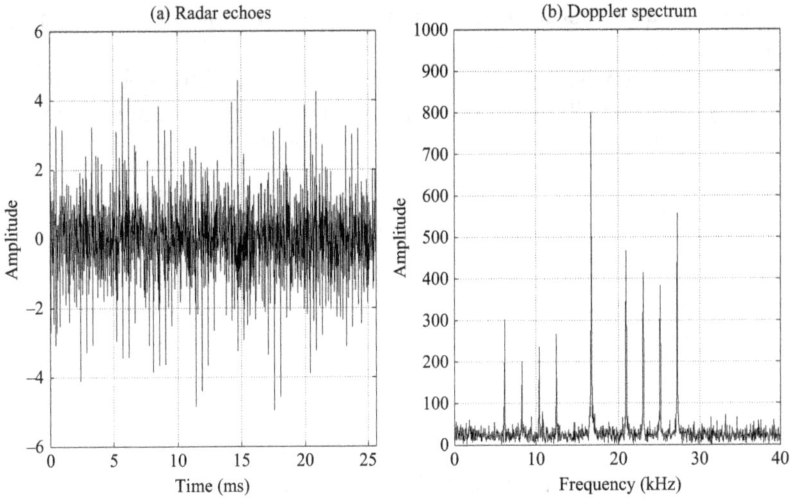

Figure 8.3 (a) Simulated radar echoes and (b) Doppler spectrum of an aircraft with a jet engine which is sparse in the frequency domain

Doppler spectrum is associated with the echo from the aircraft body while the four spectral lines on each side of the body line are caused by the reflections from the jet engines. The radar signal in the Doppler domain is sparse and the implication is that not much information is lost by discarding the Doppler spectrum coefficients with values smaller than a threshold of say 100.

Let the radar signal be denoted by an $N \times 1$ column complex-valued vector $\boldsymbol{x}$ in $\mathbb{C}^N$ with elements $x(n)$, $n = 1, 2, \ldots, N$. Using the $N \times N$ basis matrix $\Psi = [\psi_1|\psi_2|, \ldots, |\psi_N]$ with a set of complex orthonormal vectors ψ_n as columns, a radar signal $\mathbf{x}$ can be expressed as [2]

$$x = \sum_{n=1}^{N} s(n) \cdot \psi_n \text{ or } x = \Psi \mathbf{s} \tag{8.1}$$

where $\boldsymbol{s}$ is the $N \times 1$ column vector of weighting coefficients $s(n) = \langle \mathbf{x}, \psi_n \rangle = \psi^{H}{}_{n}\mathbf{x}$ and ^{H} denotes the conjugate transposition. The radar signal $\mathbf{x}$ may be represented in the time or space domain and $\boldsymbol{s}$ is a representation of the radar signal in the Ψ domain. The matrix Ψ is also referred to as a dictionary and the vectors ψ_n are also known as atoms. Examples of well-known dictionaries are the Fourier and wavelet bases.

An N-dimensional signal $\mathbf{x}$ is K-sparse with respect to the dictionary Ψ if it can be represented by a linear combination of K (or fewer) atoms ψ_n of the dictionary Ψ with K much smaller than N:

$$\|s\|_0 \leq K \ll N \tag{8.2}$$

where $||\boldsymbol{s}||_0$ is a quasi-norm, which counts the number of non-zero elements in a vector.

In practice, signals are encountered that are not exactly sparse but whose coefficients decay rapidly when sorted according to their magnitude. Mathematically, compressible signals are those signals that satisfy power law decay [4]:

$$|s^*(n)| \leq c \cdot n^{-1/q} \tag{8.3}$$

where:

$\mathbf{s}^*$ = non-increasing arrangement of $\boldsymbol{s}$
c = some positive constant and $0 < q < 1$

Compressible signals are well approximated by sparse signals, so the sparse approximation framework applies to this class of signals.

Compressible signals are well approximated by sparse representations and this is currently often employed to store and transmit high resolution images and videos at much lower data rates after using a compression algorithm, such as JPEG or H.124 advanced video encoding. Since radar signals with a high spatial and/or temporal resolution are compressible too, the CS principle can also be used to avoid the high sampling rates in radars that are often needed to support automatic target recognition. In CS, a compressed signal representation is directly obtained without the need for the acquisition at the Nyquist sampling rate.

To illustrate the principles and challenges of CS, consider a general linear measurement process $\mathbf{\Phi}$ that acquires a vector $\boldsymbol{y}$ containing M samples of the original N-dimensional signal $\mathbf{x}$ plus additive noise $\mathbf{n}$:

$$y = \mathbf{\Phi x} + \mathbf{n} = \mathbf{\Phi}\Psi s + \mathbf{n} = \Theta s + \mathbf{n} \tag{8.4}$$

where:

$\mathbf{\Phi} = M \times N$ measurement matrix
$\Theta = M \times N$ matrix with $M < N$

The measurement matrix $\mathbf{\Phi}$ is fixed and does not depend on the signal $\mathbf{x}$. The two main challenges in CS are (i) to design a measurement matrix $\mathbf{\Phi}$ that makes sure that the salient information in the compressible signal $\mathbf{x}$ is not lost by the dimensionality reduction from $\mathbf{x} \in \mathbb{C}^N$ to $\mathbf{y} \in \mathbb{C}^M$ and (ii) to develop a reconstruction algorithm that recovers $\mathbf{x}$ from only $M \approx K$ measurements [2].

8.2.2 *Restricted isometric property and coherence*

The measurement matrix $\mathbf{\Phi}$ must allow the reconstruction of the K-sparse signal $\boldsymbol{s}$ with length N, or equivalently the signal $\mathbf{x}$, from a measurement vector $\boldsymbol{y}$ with $M < N$ measurements. This problem is ill-conditioned since there are more unknowns than measurements available. However, if $\boldsymbol{s}$ is K-sparse, then the problem can be solved provided $M \geq K$. A necessary and sufficient condition for this problem to be well conditioned is that, for any K-sparse vector $\boldsymbol{s}$ and for some small $\varepsilon > 0$, the matrix $\Theta = \boldsymbol{\Phi}\Psi$ satisfies the restricted isometry property (RIP) [2]:

$$1 - \varepsilon \leq \frac{\|\Theta \mathbf{s}\|_2}{\|s\|_2} \leq 1 + \varepsilon \tag{8.5}$$

The RIP implies that the matrix Θ must preserve the lengths of K-sparse vectors. An equivalent description of the RIP is that all subsets of K columns taken from Θ are in fact nearly orthogonal (the columns of Θ cannot be exactly orthogonal since there are more columns than rows) [4].

A property that is related to RIP is referred to as coherence. The coherence between the measurement matrix $\boldsymbol{\Phi}$ and the representation matrix Ψ is defined as [3]

$$\mu(\Phi, \Psi) = \sqrt{N} \cdot \max_{1 \leq k,j \leq N} |\langle \phi_j, \psi_k \rangle| \tag{8.6}$$

The coherence measures the largest correlation between any two column vectors of $\boldsymbol{\Phi}$ and Ψ. The coherence is large if $\boldsymbol{\Phi}$ and Ψ contain correlated vectors; otherwise, $\boldsymbol{\Phi}$ and Ψ have a low coherence and are incoherent.

The design of a measurement matrix $\boldsymbol{\Phi}$ such that $\Theta = \boldsymbol{\Phi}\Psi$ has the RIP requires that all possible combinations of K non-zero entries in the vector $\boldsymbol{s}$ of length N have to satisfy (8.5), which is an NP-complete problem. However, both the RIP and the incoherence can be achieved with high probability simply by designing $\boldsymbol{\Phi}$ as a random matrix [3]. An example of a random matrix is a Gaussian matrix with elements that are independent and identically distributed (i.i.d.) random variables from a Gaussian probability density function with mean zero and variance $1/N$. A Gaussian measurement matrix has the RIP with high probability if $M \geq cK\log(N/K)$, with c a small constant [2]. Therefore, K-sparse and compressible signals of length N can be recovered from only $M \geq cK\log(N/K) \ll N$ random Gaussian measurements. Furthermore, a Gaussian measurement matrix has the RIP with high probability regardless of the choice of orthonormal basis Ψ, which makes it almost universally applicable.

Other examples of random measurement matrices with the RIP are the Bernoulli matrix with element values that are equal to $+1/\sqrt{M}$ and $-1/\sqrt{M}$ with equal probability and a partial Fourier matrix with K randomly selected rows [5].

8.2.3 Signal reconstruction

In addition to the challenge of designing a measurement matrix with the RIP, CS has to provide an algorithm that can reconstruct the signal $\boldsymbol{s}$ from the measurements $\boldsymbol{y}$ given the assumption that the signal is K-sparse. Since the number of measurements, i.e. the dimension of the vector $\boldsymbol{y}$, is smaller than the dimension of the signal $\boldsymbol{s}$, the problem is underdetermined, which means that additional constraints must be applied to obtain useful solutions. In this section, three methods based on minimisation of the ℓ_p norm with $p = 0$, 1 or 2 will be discussed. The ℓ_p norm of a vector $\boldsymbol{s}$ is defined as [4]

$$||s||_p := \left(\sum_{n=1}^{N} |s(n)|^p\right)^{1/p} \tag{8.7}$$

for $0 \leq p < \infty$ where $||s||_0 = |\text{supp}(\bar{s})|$ and $||s||_\infty = \max |s(n)|$.

8.2.3.1 Minimum ℓ_2 norm reconstruction

An approach that is often used to estimate a vector $\boldsymbol{s}$ from a set of measurements $\boldsymbol{y}$ is by minimising the ℓ_2 norm of the vector $\boldsymbol{s}$ under the condition that $\boldsymbol{y} = \Theta \boldsymbol{s}$:

$$\hat{s} = \arg \min_{y=\Theta s} ||s||_2 \tag{8.8}$$

A solution to this problem is given by the Moore–Penrose pseudo-inverse:

$$\hat{s} = (\Theta^{\text{H}}\Theta)^{-1}\Theta^{\text{H}}y \tag{8.9}$$

However, since large values are penalised by the square operation in the minimisation of the ℓ_2 norm, the solution provided by the Moore–Penrose pseudo-inverse will almost always be a vector with many non-zero elements and will therefore not be K-sparse.

8.2.3.2 Minimum ℓ_0 norm reconstruction

Instead of penalising large values in the minimisation of the ℓ_2 norm, a better alternative is to minimise the number of non-zero entries in $\boldsymbol{s}$, which is given by the ℓ_0 norm of $\boldsymbol{s}$. The optimisation procedure that minimises the ℓ_0 norm

$$\hat{s} = \arg \min_{y=\Theta s} ||s||_0 \tag{8.10}$$

recovers a K-sparse signal exactly with high probability using only $M = K + 1$ i.i.d. Gaussian measurements. Unfortunately, the minimisation of the ℓ_0 norm is both numerically unstable and NP-complete because it requires an exhaustive enumeration of all possible $\binom{N}{K}$ locations of the non-zero entries in $\boldsymbol{s}$ [2].

8.2.3.3 Minimum ℓ_1 norm reconstruction

By minimising the ℓ_1 norm instead of the ℓ_0 norm, the combinatorial problem can be relaxed to a convex optimisation problem:

$$\hat{s} = \arg \min_{y=\Theta s} ||s||_1 \tag{8.11}$$

A K-sparse signal can be exactly recovered by a convex optimisation procedure with high probability using only $M \geq O(K\log(N/K))$ random measurements. This procedure conveniently reduces to a linear program known as basis pursuit (BP) whose computational complexity is about $O(N^3)$ [2]. Section 8.3 gives an overview of these reconstruction algorithms and other related algorithms.

8.2.3.4 Example of ℓ_1 norm versus ℓ_2 norm reconstruction

This section gives an example of the results of ℓ_1 norm versus ℓ_2 norm reconstruction of a sparse range profile of a target that consists of 100 range bins with 10 range bins containing complex point scatterers, while the rest of the range bins are empty (see Fig. 8.4(a)). Fig. 8.4(b) shows a simulated complex radar signal (without noise) received from this target using a stepped frequency waveform with 100 frequency steps.

Fig. 8.5(a) and (b) shows the results of the algorithms that reconstruct the sparse range profile by minimising the ℓ_1 norm and ℓ_2 norm, respectively, based on only 25 random samples (indicated with dots in Fig. 8.4(b)) of the original 100 frequency samples. The BP approach described in section 8.3.1.1 has been used for minimising the ℓ_1 norm of the range profile while the Moore–Penrose pseudo-inverse of the matrix has been used to obtain the minimum ℓ_2 norm reconstruction. The BP approach provides an exact reconstruction of the range profile based on 25 measurements, which is only slightly higher than the minimum number given by the RIP, i.e. $K\log(N/K) = 23$. The reconstruction of the range profile by minimising the ℓ_2 norm using the Moore–Penrose pseudo-inverse does not give a sparse solution at all and does not appear to be very useful.

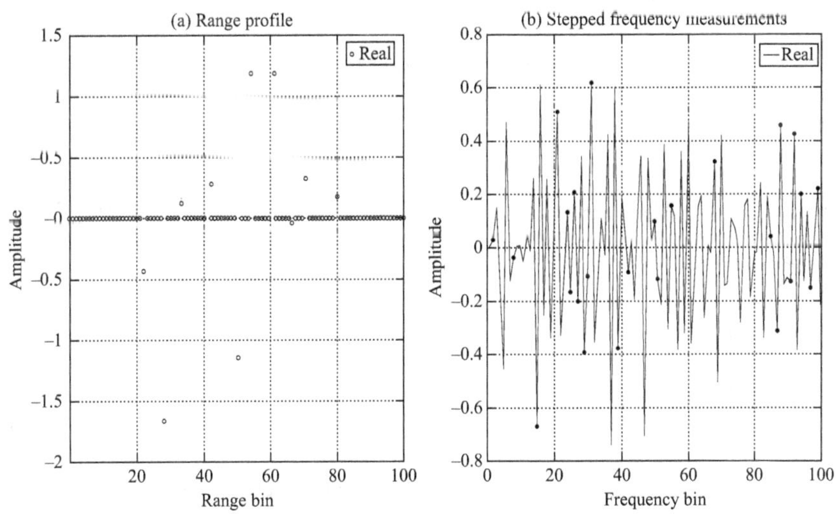

Figure 8.4 (a) Range profile of a target with 10 complex point scatterers and (b) 100 stepped frequency measurements. The solid dots in (b) indicate the 25 random samples taken from the 100 available measurements for sparse reconstruction

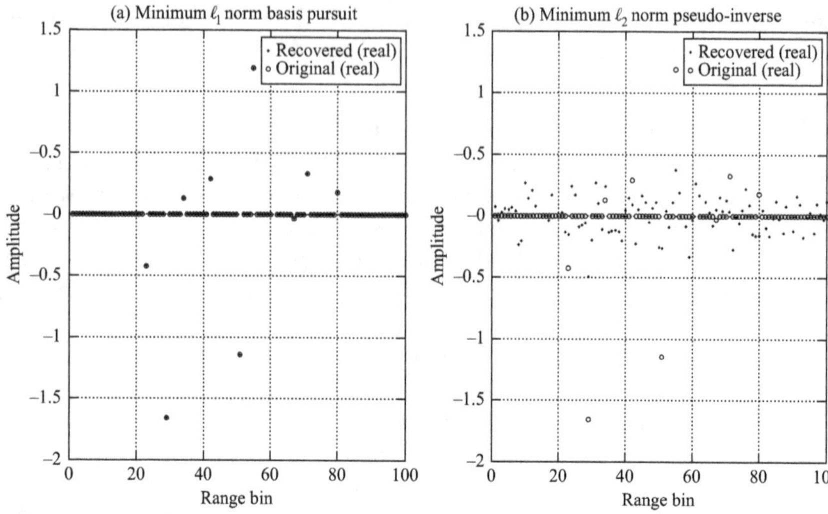

Figure 8.5 (a) Reconstruction of the range profile by minimising the ℓ_1 norm with the basis pursuit algorithm and (b) reconstruction of the range profile by minimising the ℓ_2 norm using the pseudo-inverse

8.3 Reconstruction algorithms

In the past decades, many classes of algorithms for solving sparse reconstruction problems have been investigated. In this section, examples of algorithms from three classes of sparse reconstruction methods will be briefly reviewed.

8.3.1 Convex optimisation

8.3.1.1 Basis pursuit

The combinatorial problem of finding a signal with the minimum ℓ_0 norm is replaced with a convex optimisation problem that minimises the ℓ_1 norm, which is a convex function that is closest to the ℓ_0 norm. The BP approach for finding a solution to a convex optimisation problem minimises the ℓ_1 norm of the sparse signal $\boldsymbol{s}$ given the measurements $\boldsymbol{y}$ [6,8]:

$$\hat{s} = \arg\min_{s} ||s||_1 \text{ subject to } y = \Theta s \tag{8.12}$$

For real-valued signals, this problem can be recast as a linear programming (LP) problem, while for complex-valued signals it is equivalent to a second-order cone program (SOCP) [8].

A well-known method for solving LP problems is the simplex algorithm, which was first developed by Dantzig and Thapa [7]. The simplex algorithm finds the minimum of an objective function subject to a number of constraints. The name of the algorithm is derived from the simplex that describes all possible solutions to this problem. Although the simplex algorithm is guaranteed to find a solution in a finite number of steps (except for some theoretical cases in which cycles occur), the computational complexity of the simplex algorithm is $O(N^3)$, which makes it unsuitable for signals with high dimensions.

As an alternative to the simplex algorithm, Karmarkar proposed an interior point projective method, which traverses the interior of the simplex. The lowest computational complexity currently attained by an interior point method is $O(M^2 \cdot N^{1.5})$ [3].

8.3.1.2 Basis pursuit de-noising

The BP approach using a simplex algorithm is guaranteed to find a reconstruction of a *K*-sparse signal if there is no measurement noise. However, every sensor measurement contains noise and this needs to be addressed in the reconstruction algorithm. The influence of noise on the signal reconstruction can be minimised by applying the Dantzig and Thapa (BPDN) approach, which finds a solution to the following problem [8]:

$$\hat{s} = \arg \min_{\mathbf{s}} \|s\|_1 \text{ subject to } \|y - \Theta s\|_2 \leq \sigma \tag{8.13}$$

where the positive parameter σ is an estimate of the noise level in the data. The case $\sigma = 0$ corresponds to the BP problem. The BPDN problem can be solved by algorithms that are used in LP for real-valued signals or SOCP for complex-valued signals.

8.3.1.3 Least absolute shrinkage and selection operator

Another formulation of the convex relaxation problem is referred as the least absolute shrinkage and selection operator (LASSO) [8]:

$$\hat{s} = \arg \min_{\mathbf{s}} \|y - \Theta s\|_2^2 \text{ subject to } \|s\|_1 \leq \tau \tag{8.14}$$

An alternative expression of this minimisation problem is given by [9]

$$\hat{s} = \arg \min_{\mathbf{s}} \frac{1}{2} \|y - \Theta s\|_2^2 + \lambda \|s\|_1 \tag{8.15}$$

where λ is a Lagrangian multiplier that is related to the constraint τ in (8.14). These optimisation problems described by (8.14) and (8.15) are in some sense equivalent and their solutions coincide for appropriate parameter choices of λ and t. However, the parameters that make these problems equivalent cannot be known a priori except for special cases, such as an orthogonal matrix Θ [8].

A spectral-projected gradient (SPG) algorithm has been developed by van den Berg and Friedlander to solve the minimisation problem in (8.14) [8]. The SPG algorithm is suitable for large-scale problems and for problems involving complex signals that often occur in the radar domain. One of the key issues of the SPG algorithm is the choice of the parameter τ, which allows a trade-off between the ℓ_1 norm and the fit of the reconstructed signal with the measurements. In practice, different values of τ can be tried and the reconstructed signal with the smallest error can then be selected. Another recent solution is the adaptive CAMP algorithm, which solves the complex LASSO problem and adaptively estimates the optimal value of τ from the input data [9–11].

8.3.2 Greedy constructive algorithms

Greedy constructive techniques start with the observation that the measurement vector $\boldsymbol{y}$ consists of the weighted sum of K columns of the matrix $\Theta = [\theta_1, \theta_2, \ldots, \theta_N]$. The approach taken by constructive algorithms is to identify those K columns and the corresponding weights by correlating the column vectors of the matrix Θ with the measurement vector $\boldsymbol{y}$. The estimate of the sparse vector $\boldsymbol{s}$ is iteratively constructed by selecting the atoms with the highest correlation, i.e. a greedy selection.

8.3.2.1 Matching pursuit

The matching pursuit (MP) algorithm developed by Mallat and Zhang is one of the first greedy algorithms that have been used to solve sparse reconstruction problems [12]. The MP algorithm iteratively selects the column θ_j of Θ that is most strongly correlated with the current residue, where the residue is the difference between the measurement vector and the measurement vector predicted by the reconstructed signal. In other words, the MP algorithm selects the atom of the dictionary that on its own leads to the largest reduction in the approximation error. In the first iteration, the sparse signal estimate is equal to the null vector and the residue is equal to the measured vector:

$$\hat{s}_0 = \mathbf{0} \quad \mathbf{r}_0 = y \tag{8.16}$$

In each iteration the column of Θ with the highest correlation with the current residue $\mathbf{r}_t$ is selected:

$$j_t = \arg\max_j |\theta_j{}^{\mathrm{H}} \mathbf{r}_t| \tag{8.17}$$

The appropriate element of the current sparse signal estimate is then updated with the value of the correlation:

$$\hat{s}_t(j_t) = \hat{s}_{t-1}(j_t) + \theta_{j_t}^{\mathrm{H}} \cdot \mathbf{r}_t \tag{8.18}$$

and the new residue is calculated by subtracting the selected column weighted by the correlation value from the current residue:

$$\mathbf{r}_{t+1} = \mathbf{r}_t - (\theta_{j_t}^{\mathrm{H}} \cdot \mathbf{r}_t) \cdot \theta_{j_t} \tag{8.19}$$

After a sufficient number of iterations, the MP algorithm will have identified the correct set of columns. There are several criteria possible for halting the iterations:

- Stop after a fixed number of iterations.
- Stop when the residual has small magnitude.
- Stop when no column vector explains a significant amount of energy in the residual.

An intrinsic feature of the MP algorithm is that when stopped after a few iterations, it yields an approximation using only a few column vectors. When the matrix Θ is orthogonal, the MP algorithm recovers a K-sparse signal perfectly after K iterations [6]. However, in CS the matrix Θ is not orthogonal and the MP algorithm converges only very slowly due to the correlations between the columns of the matrix Θ.

8.3.2.2 Orthogonal matching pursuit

The orthogonal matching pursuit (OMP) algorithm avoids the slow convergence of the MP algorithm by updating the coefficients of all previously selected column vectors so that the newly derived residual is orthogonal to not only the immediately selected column vector, but also all the column vectors selected at previous iterations. As a consequence, once a column vector is selected, it is never selected again in subsequent iterations [13,14]. Fig. 8.6 shows a schematic diagram of the OMP algorithm in which the selected columns of the matrix Θ in the set I_t are orthogonalised at each step. In fact, the estimate of the sparse vector $\boldsymbol{s}$ at the t-th iteration is the least-squares estimate using the columns that have been selected so far:

$$\hat{s}_t = (\Theta_{I_t}^{H}\Theta_{I_t})^{-1}\Theta_{I_t}^{H}y \tag{8.20}$$

The identification of the column with the largest correlation is computationally the most demanding step in the OMP algorithm. The most direct approach to finding the column with the largest correlation computes the maximum inner product via the matrix–vector multiplication, which costs $O(MN)$ for an unstructured dense matrix. If the OMP algorithm is stopped after K iterations, the computational complexity is $O(KMN)$ [15].

8.3.2.3 Stage-wise orthogonal matching pursuit

For some applications, the OMP algorithm is not fast or accurate enough and alternatives based on the same greedy construction principles have been investigated. For example the stage-wise orthogonal matching pursuit (StOMP) algorithm selects multiple columns at each step [16]. Fig. 8.7 shows a schematic diagram of the StOMP algorithm, which is equal to the flow diagram of the OMP algorithm

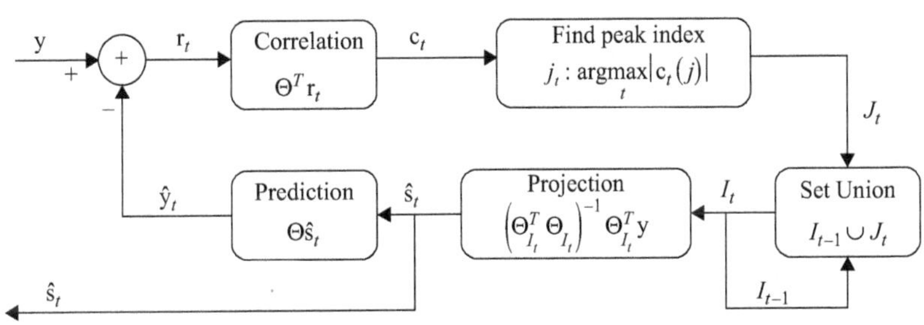

Figure 8.6 Schematic diagram of the OMP algorithm

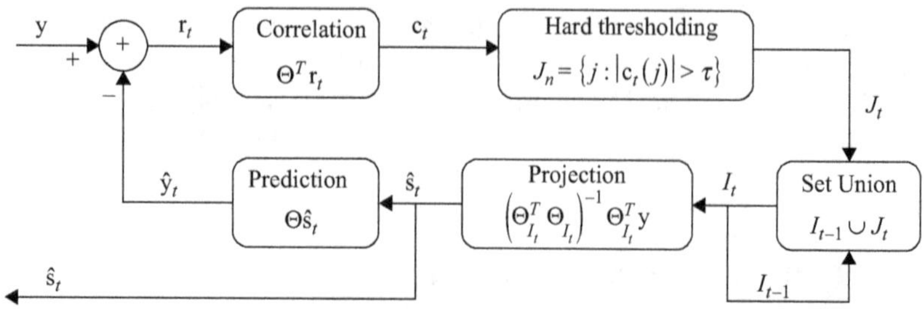

Figure 8.7 Schematic diagram of the StOMP algorithm

except that multiple atoms of the dictionary are selected by hard thresholding of the correlation coefficients with a threshold τ.

The StOMP algorithm iterates only through a fixed number of stages and then terminates, whereas OMP requires K iterations, where K is the sparsity level. Although the structure of StOMP is similar to that of OMP, the computation time is significantly reduced because StOMP selects multiple columns at each iteration. If iterative methods are used to solve the least-squares problem, a runtime bound of $cTKN + O(N)$ is achieved, where T is the fixed number of iterations run by StOMP and c is a constant that depends only on the accuracy level of the least-squares problem [4].

8.3.3 Iterative thresholding algorithms

Greedy algorithms such as MP, OMP and StOMP are relatively fast iterative procedures that have been used extensively in practical applications. The performance of these methods is, however, not guaranteed in general and only under very strict conditions can they be shown to optimise the ℓ_1 norm. Another class of algorithms is based on iterative thresholding that solves the following formulation of the ℓ_1 norm minimisation problem:

$$\hat{s} = \arg\min_{\mathbf{s}} \frac{1}{2}\|y - \Theta s\|_2^2 + \tau\|s\|_1 \tag{8.21}$$

The objective function to be minimised consists of the least-squares fit of the signal reconstruction and the ℓ_1 norm of the reconstructed signal with the Lagrangian parameter τ controlling the relative weights of the two contributions [5].

A sparse solution to the objective function in (8.21) can be found via the iteration [17,18]:

$$\hat{s}_{t+1} = D_\tau\left(\hat{s}_t + \frac{1}{\alpha}\mathbf{g}_t\right) = D_\tau\left(\hat{s}_t + \frac{1}{\alpha}\Theta^{\mathrm{H}}(y - \Theta\hat{s}_t)\right) \tag{8.22}$$

where D_τ is a thresholding operator, $\mathbf{g}_t$ is a gradient, which is equal to the correlation of the residue with the columns of Θ, and α is a suitable step size. The basic idea of an iterative thresholding algorithm is that it improves the current estimate

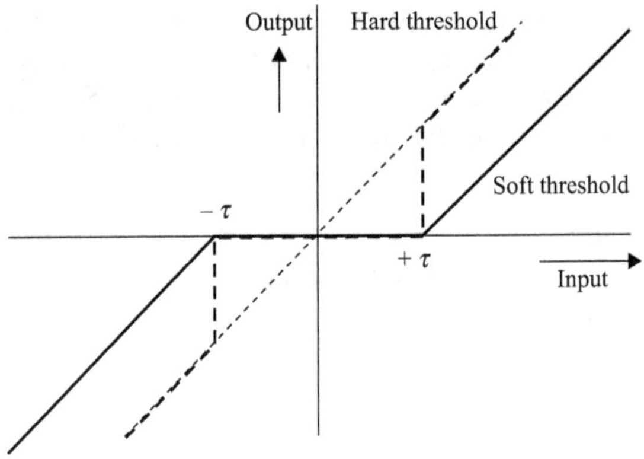

Figure 8.8 Hard and soft thresholding functions

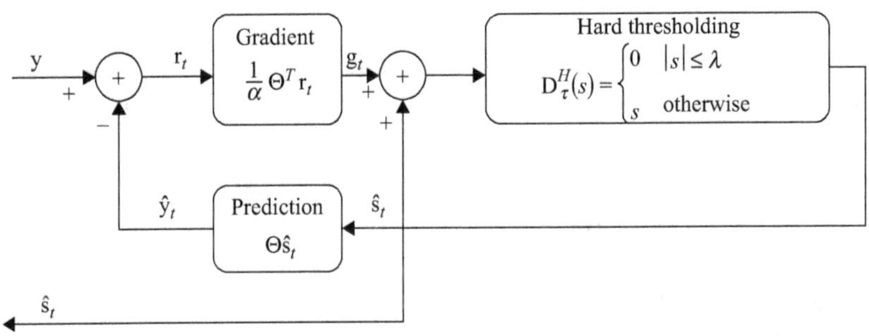

Figure 8.9 Schematic diagram of the IHT algorithm

of the sparse signal by adding the weighted gradient of the measurement residue, and then applying a thresholding and/or shrinkage operation to increase the sparsity.

8.3.3.1 Iterative hard thresholding

Iterative hard thresholding (IHT) is a simple iterative algorithm that applies a hard threshold to a new estimate of a sparse signal [19]:

$$D_\tau^{\mathrm{H}}(s) = \begin{cases} 0 \; |s| \le \tau \\ s \text{ otherwise} \end{cases} \tag{8.23}$$

Fig. 8.8 shows the hard threshold function, which sets small signal values equal to 0 and leaves large signal values intact.

Fig. 8.9 shows a flow diagram of the IHT algorithm. It involves only the computation of the gradient $\boldsymbol{g}_t$, the hard thresholding operation and the prediction of the measurements. The computation of the gradient $\boldsymbol{g}_t$ is the same as the computation of the correlation between the residue and the columns of the matrix in the MP, MP and StOMP algorithms.

The computational complexity of the IHT algorithm is $O(TMN)$, where T is the number of iterations. There is empirical evidence that IHT is reasonably effective for solving sparse approximation problems in practice. However, there are also indications that simple thresholding techniques do not perform well in the presence of noise [15].

8.3.3.2 Iterative shrinkage and thresholding

The iterative shrinkage and thresholding (IST) algorithm applies a soft threshold to the estimate of a complex signal **s**, which sets small signal values equal to 0 and shrinks large signal values (see Fig. 8.8) [18].

$$D_\tau^S(s) = \begin{cases} 0 & |s| \le \tau \\ (|s| - \tau) \cdot s/|s| & \text{otherwise} \end{cases} \tag{8.24}$$

Fig. 8.10 shows a schematic diagram of the IST algorithm, which is equivalent to the IHT algorithm except for the application of a soft instead of a hard threshold.

Like the IHT algorithm, the IST algorithm is simple but it is generally slow to converge, in particular if the matrix Θ is ill-conditioned and τ is small. Faster algorithms based on IST have been developed such as FISTA, TWIST and SpaRSA [15,18].

8.4 Jet engine modulation

8.4.1 Introduction

The radar signal reflected by an aircraft with jet engines is modulated in phase and amplitude by the rotation of the compressor and/or turbine blades [1]. In the Doppler spectrum, these JEMs manifest themselves as spectral lines with a constant frequency spacing that is given by the blade chop frequency, which is the product of the rotation rate of the engine(s) and the number of rotor blades. A radar system can use the JEM effect to support automatic target recognition by measuring the blade chop frequency and inferring the jet engine type of the aircraft from this measurement.

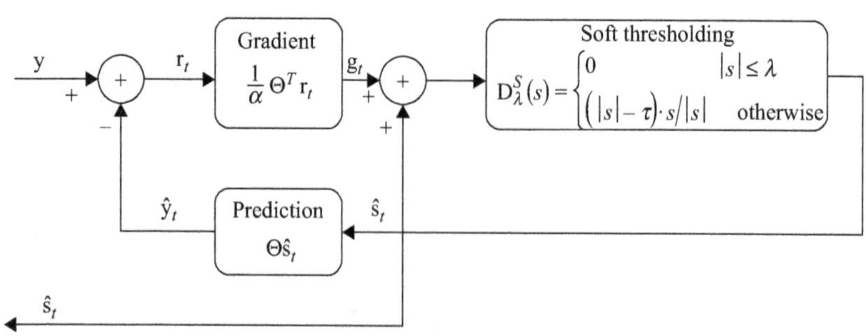

Figure 8.10 Schematic diagram of the IST algorithm

To measure the JEM spectrum unambiguously for all types of jet engines, the sampling rate in the Doppler domain (PRF for pulse-Doppler radars) needs to be high [1]. In addition, the coherent integration time needs to be long to achieve a Doppler resolution that is sufficient to discriminate JEM lines from the rotor blades at the minimum engine speed. Both the sampling rate requirement and the coherent integration time requirement for JEM-based automatic target recognition impose severe constraints on a radar system. Since the Doppler spectrum due to JEM is sparse, CS may be used to alleviate these constraints by reducing the (average) sampling rate and/or the coherent integration time. Some results of the application of CS to JEM will be discussed in this chapter.

8.4.2 Jet engine model

A simplified model of the signal received by a radar system from the rotor of a jet engine is described by [20]

$$x\,(t) = A \cdot e^{j2\pi f_d t \cdot \frac{4\pi R(t)}{\lambda}} \cdot [1 + V(\phi)\sum_{b=1}^{B} x_b(t)] + n(t) \tag{8.25}$$

where:

$x\,(t)$ = Echo received at time t
$n(t)$ = Additive complex white Gaussian noise
A = Amplitude of the echo of the body of the aircraft
f_{d} = Doppler frequency of the body of the aircraft
λ = Radar wavelength
$R(t)$ = Range of the aircraft to the radar
ϕ = Aspect angle from the radar to the aircraft (0° equals a nose-on aspect)
B = Number of rotor blades
$V(\phi)$ = Fan visibility function

The simplest visibility function $V(\phi)$ is equal to 1 for aspect angles less than V_e and zero for aspect angles larger than V_e.

For each blade the signal $x_b(t)$ is described by

$$x_b(t) = \left[\alpha + \beta \cos\left(\omega_{\mathrm{r}} t + \frac{2\pi b}{B}\right)\right] \cdot \sin\mathrm{c}\left[\frac{4\pi}{\lambda} \cdot \frac{(L_2 - L_1)}{2} \cdot \cos\phi \cdot \sin\left(\omega_{\mathrm{r}} t + \frac{2\pi b}{B}\right)\right] \cdot \\ \cdot e^{-j2\pi \frac{L_2+L_1}{\lambda} \cdot \cos\phi \cdot \cos\left(\omega_{\mathrm{r}} t + \frac{2\pi b}{B}\right)} \tag{8.26}$$

where:

ω_{r} = Rotation rate of the rotor
L_1 = Radius of the engine hub
L_2 = Radius of the main rotor (i.e. the length of the blades plus the hub radius)

and the variables α and β are given by

$$\alpha = \sin(|\phi| + \rho) + \sin(|\phi| - \rho) \tag{8.27}$$

$$\beta = -\text{sgn}(\phi) \cdot [\sin(|\phi| + \rho) - \sin(|\phi| - \rho)] \tag{8.28}$$

with ρ, the pitch angle of the blades.

8.4.3 Simulation results of JEM compressive sensing

This section shows some results of CS applied to a simulated radar signal that originates from an aircraft with a single jet engine. Fig. 8.11 shows the Doppler spectrum of the echoes from a pulse train consisting of 1024 pulses with a constant PRF of 40 kHz that is emitted by a pulse-Doppler radar with a carrier frequency of 10 GHz. The rotor has 38 blades and rotates at 3300 rpm. The length of the rotor blades (L_2) is 1.1 m and the radius of the engine hub (L_1) is 0.35 m. The pitch of the blades is 35° and the aspect angle is 65°. The aircraft has a velocity of 250 m/s, which leads to a Doppler frequency of the body line of $2 \times 250/0.03 = 16.7$ kHz. Fig. 8.11 shows that on each side of the body line there are four JEM lines visible with a constant frequency spacing of approximately 2.1 kHz, i.e. the product of the number of rotor blades (38) and the rotor rotation rate (3300/60 Hz). The difference in amplitude of the JEM lines on the left- and the right-hand side of the body line is caused by the pitch of the rotor blades. Gaussian complex noise has been added to the JEM signal with a signal-to-noise ratio of the body line of 15.5 dB.

To illustrate the application of CS to JEM, 256 samples of the original 1024 samples of the JEM signal are randomly selected and the JEM spectrum is then

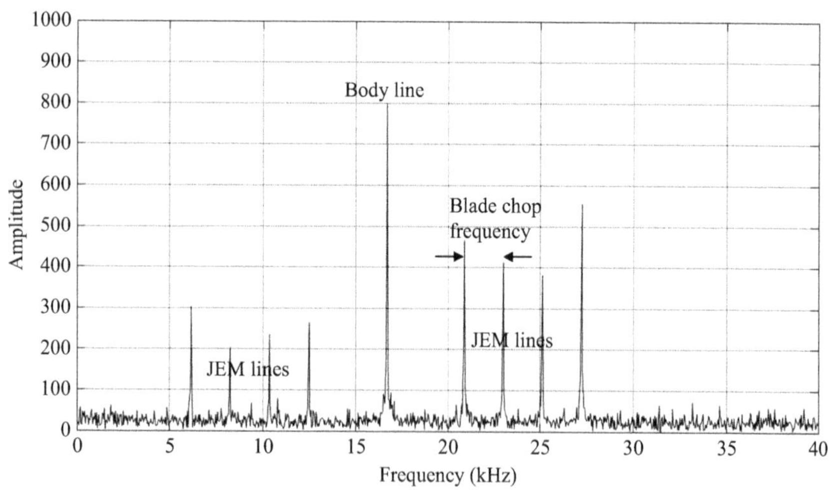

Figure 8.11 Simulated Doppler spectrum of a jet engine rotor with 38 blades rotating at 3300 rpm sampled by an X-band radar with a PRF of 40 kHz

reconstructed, i.e. the matrix Θ is a partial Fourier matrix with randomly selected rows from a complete Fourier matrix. The SPGL1 Matlab solver for large-scale one-norm regularised least squares, made available by Friedlander and van den Berg [21], has been used to reconstruct the JEM spectrum with the LASSO method (see section 8.3.1.3). Fig. 8.12 illustrates the reconstruction of the JEM spectrum for three different values of the parameter τ that constrains the ℓ_1 norm of the reconstructed spectrum as defined in equation (8.14). Fig. 8.12(a) shows that for a value of $\tau = 1000$, only the body line and some of JEM lines are reconstructed with a computation time of 0.022 s. For a value of $\tau = 4000$, the body line and all JEM lines are reconstructed in a computation time of 0.0286 s, but the amplitude of the lines is not very accurate (see Fig. 8.12(b)). Fig. 8.12(c) shows that the best reconstruction of the JEM spectrum is achieved with a value of $\tau = 10{,}000$ but the computation time is now much longer (2.24 s). It can be concluded that for the purpose of jet engine recognition, a value of $\tau = 4000$ would be optimal since the blade chop frequency can be determined accurately enough from the JEM spectrum and the time needed to reconstruct the JEM spectrum is short.

To determine how much the number of random measurements can be reduced, the JEM spectrum has been reconstructed with the LASSO method using a value of $\tau = 3500$ for three different numbers of random measurements: 256, 128 and 64 (see Fig. 8.13). Fig. 8.13(b) shows that by using only 128 samples, the blade chop frequency can still be extracted from the reconstructed JEM spectrum, although the reconstruction time has increased from 0.0273 s to 0.178 s. If only 64 samples are used, the reconstructed JEM spectrum becomes less useful for jet engine recognition and the reconstruction time has increased to 6.62 s.

The conclusion from the CS experiments using simulated JEM signals is that the blade chop frequency of a jet engine can be determined with an average PRF of

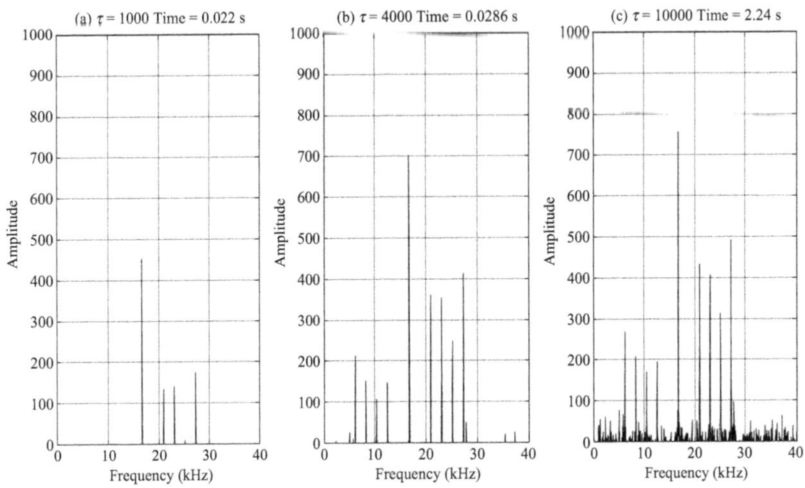

Figure 8.12 Reconstruction of the JEM spectrum by using the LASSO method for three different values of the ℓ_1 norm constraint τ

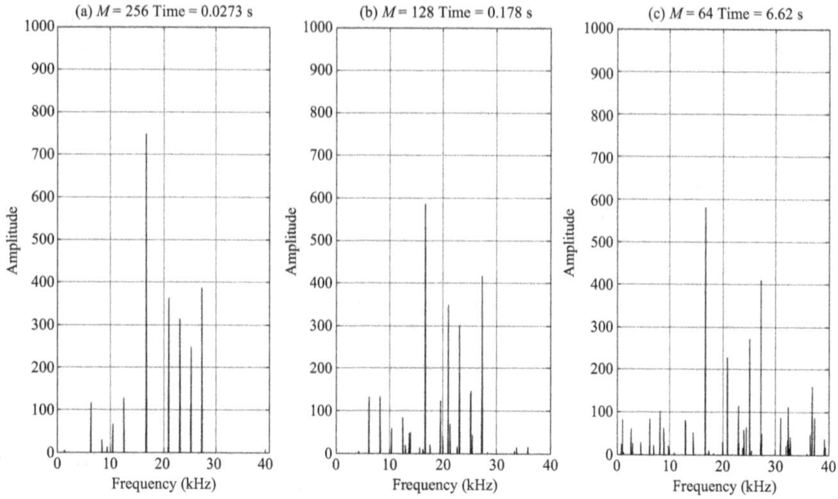

Figure 8.13 Reconstruction of the JEM spectrum by using the LASSO method with different numbers of random measurements M*: 256, 128 and 64*

a random pulse-Doppler waveform that is a factor of 8 lower than the Nyquist rate. The CS experiments also showed that for the LASSO method the value of the ℓ_1 norm constraint τ has a large impact on the accuracy and computation time of the reconstruction. It is recommended to compare different values of τ and select the value with the smallest approximation error within the available computation time. Finally, the results of the simulations have shown that the reconstruction time of the LASSO method increases significantly when the number of measurements decreases.

8.5 Inverse synthetic aperture radar

8.5.1 Introduction

Inverse synthetic aperture radar (ISAR) is a radar imaging technique that employs different Doppler shifts of various parts of a moving radar target to obtain a high cross-range resolution while the resolution in down-range is achieved with a large radar bandwidth. Two-dimensional high resolution ISAR images can be used to automatically recognise targets by extracting relevant features from the ISAR image such as the length and width of the target or the positions of dominant scatterers.

The down-range resolution Δr_{d} of a radar waveform with a bandwidth B is given by

$$\Delta r_{\mathrm{d}} = \frac{c}{2 \cdot B} \tag{8.29}$$

where c is speed of light (3.10^8 m/s). If a stepped frequency waveform is used to synthesise the bandwidth B with N frequency steps and a frequency step size Δf, the unambiguous range interval is computed from:

$$R_\mathrm{d} = \frac{c}{2 \cdot \Delta f} \tag{8.30}$$

As an example, a down-range resolution of 12.5 cm and an unambiguous range interval of 25 m can be provided with a stepped frequency waveform using 200 frequency steps and a step size of 6 MHz.

The cross-range resolution ΔR_c that can be achieved with ISAR is computed from [1]

$$\Delta R_\mathrm{c} = \frac{\lambda}{2 \cdot \Delta \phi} \tag{8.31}$$

where λ is the radar wavelength and $\Delta\phi$ is the change in the target aspect angle during the ISAR coherent integration time. As an example, a resolution of 12.5 cm in cross-range can be achieved with a radar wavelength of 3.2 cm and a change in aspect angle of 7.2°.

The bandwidth and change in aspect angle that are required for high resolution ISAR images impose significant constraints on a radar. If the radar echo of a target is assumed to be sparse in the spatial domain, i.e. it consists of point scatterers, CS may be used to alleviate these constraints by reducing the ISAR sampling requirements for a pulse-Doppler radar using a stepped frequency waveform. Some results of the application of CS to ISAR will be discussed in this chapter.

8.5.2 Simulation model

To simulate radar signals of a target for the application of CS to ISAR, a software program called RAPPORT (Radar signature Analysis and Prediction by Physical Optics and Ray Tracing) is used [22]. RAPPORT, which has been developed by TNO, uses a combination of geometrical optics (GO) and physical optics (PO) methods to calculate the radar return of a specified object [23]. The target is represented by a 3D computer aided design (CAD) model, consisting of a large number of triangular facets. RAPPORT uses GO to determine which facets of the CAD model are hit by rays from the radar's EM radiation. The PO calculating method is then used to determine the amount of reflected radiation for each facet, which is hit by the radar's EM radiation. This is done for all facets and the returning radiation is summed over all facets, resulting in one amplitude and phase. This can be done for all frequencies in the radar waveform and all aspect angles with respect to the target. Therefore, when a stepped frequency waveform and a number of aspect angles are defined, ISAR measurements can be generated with RAPPORT.

The target that is considered in this CS experiment is a 3D CAD representation of an unmanned aerial vehicle (UAV) (see Fig. 8.14). This particular CAD model

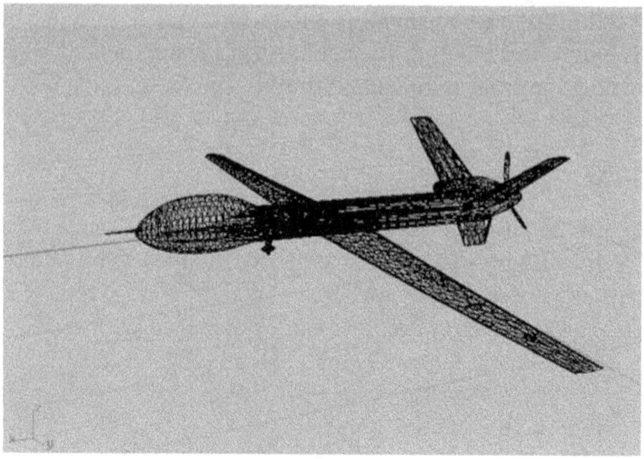

Figure 8.14 CAD model of an unmanned aerial vehicle

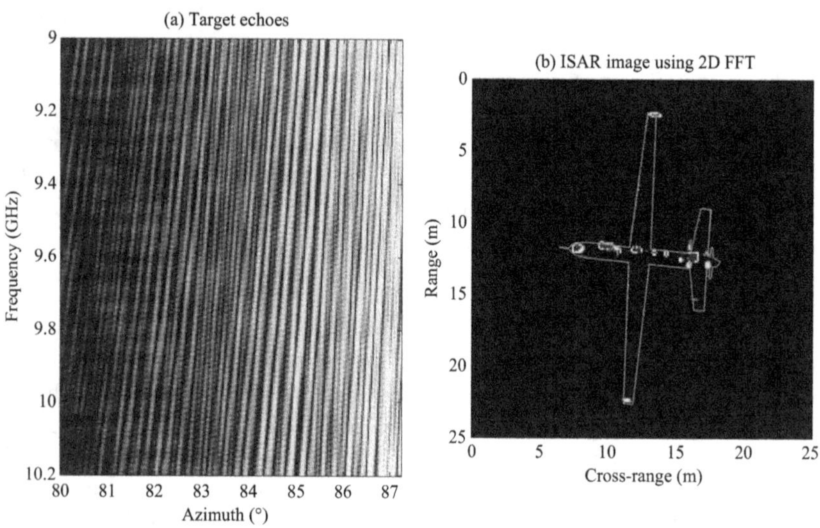

Figure 8.15 (a) Echoes of a UAV as a function of frequency and azimuth and (b) ISAR image using a 2D FTT and a 2D Hamming taper

consists of approximately 5000 facets and the length and span of the UAV are approximately 10 m and 20 m, respectively.

With RAPPORT the complex echoes are calculated of a pulse-Doppler radar that transmits a stepped frequency waveform with 200 frequency steps between a frequency of 9 GHz and 10.2 GHz. The UAV is illuminated with the stepped frequency waveform at a fixed elevation angle of 5° and at 210 azimuth angles between 80° and 87.2° (0° is nose-on). Fig. 8.15(a) shows the amplitude of the radar echoes of the UAV (without noise) as a function frequency and azimuth and

Fig. 8.15(b) shows the ISAR image that has been obtained with a 2D fast Fourier transform (FFT) after application of a 2D Hamming taper to the complex radar echoes. The outline of the UAV projected on the ISAR image clearly shows correspondence of large echoes with the nose, the wing tips and the tail of the UAV.

CS is now applied to ISAR by taking random samples of the radar echoes in the frequency domain for each azimuth angle (i.e. the matrix Θ is a partial Fourier matrix with randomly selected rows) [24]. After sampling in the azimuth domain with the same (randomly) stepped frequency waveform, a 1D FFT is used to achieve a high cross-range resolution for each frequency sample. Then, the range profile is reconstructed for each cross-range bin by using the BPDN method (see section 8.3.1.2) and the SPGL1 Matlab solver with a value of σ that is equal to 0.1.

Fig. 8.16(a) shows the ISAR image using all 200 frequency samples and the FFT and Fig. 8.16(b)–(d) shows the reconstructed images using 50, 25 and 10 randomly chosen frequency samples from the available 200 frequency samples, respectively. The quality of the reconstructed image for 50 and 25 frequency samples is close to the quality of the FFT image. The reconstructed ISAR image with only 10 frequency samples shows a significant degradation in the image quality due to the appearance of spikes in the down-range direction.

The preliminary conclusion of the CS application to ISAR imaging is that the average sampling rate in the frequency domain for a stepped frequency waveform can be reduced by a factor of 8. CS, thereby, provides with more flexibility in the design of an ISAR waveform. For example the angular sampling rate can be

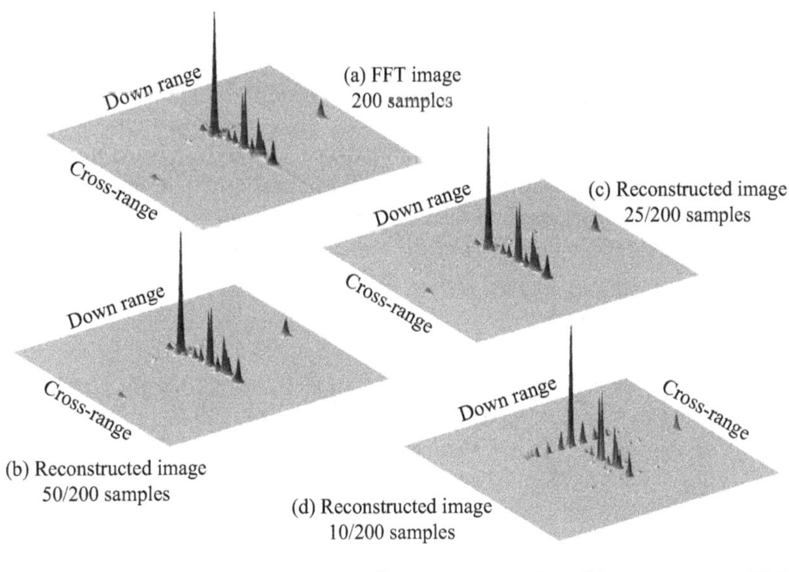

Figure 8.16 (a) FFT image using 200 frequency samples, (b) reconstructed BPDN image using 50 out of 200 frequency samples, (c) reconstructed BPDN image using 25 out of 200 frequency samples and (d) reconstructed image using 10 out of 200 frequency samples

increased with a factor of 8 to improve ISAR imaging of fast and/or manoeuvring targets. Another possibility is to transmit eight successive pulses at the same frequency before stepping to the next frequency to enable the suppression of surface clutter or the effects of JEM by Doppler filtering.

8.6 Conclusions

CS is a new paradigm in data acquisition and signal processing that allows sparse signals to be sampled at a much lower rate than the Nyquist rate. In radar systems that use wide bandwidths and/or many receiver channels for automatic target recognition, CS can provide a significant reduction in acquisition time, PRF or hardware costs. An overview of the basic principles of CS has been given and different algorithms that allow sparse signals to be reconstructed from random samples of a radar signal have been described. The application and benefits of CS in radar systems have been shown for automatic target recognition based on JEM and ISAR images.

The conclusion from the CS experiments using simulated JEM signals is that the blade chop frequency of a jet engine can be determined with a random pulse-Doppler waveform using an average PRF that is a factor of 8 lower than the Nyquist rate. The CS experiments also show that for the LASSO method the value of the ℓ_1 norm constraint τ has a large impact on the accuracy and computation time of the reconstruction. It is recommended to compare different values of τ and select the value with the smallest approximation error within the available computation time. Finally, the results of the JEM simulations have shown that the reconstruction time of the LASSO method increases significantly when the number of measurements decreases.

The preliminary conclusion of the CS application to ISAR imaging is that the average sampling rate in the frequency domain for a stepped frequency waveform can be reduced by a factor of 8. CS, thereby, provides more flexibility in the design of an ISAR waveform. For example, for a stepped frequency waveform, the angular sampling rate can be increased with a factor of 8 to improve ISAR imaging of manoeuvring targets. Another possibility is to transmit eight successive pulses at the same frequency to enable the suppression of surface clutter or the effects of JEM.

Acknowledgements

The authors acknowledge the help of Philip van Dorp from TNO for providing the Matlab software to simulate JEM signals, and Michael Friedlander and Ewout van den Berg for sharing the SPGL1 Matlab software for compressive sensing.

References

[1] P. Tait, *Introduction to Radar Target Recognition*, Institution of Electrical Engineers (IEE), London, 2005.

[2] R. G. Baraniuk, ‘Compressive Sensing’, *IEEE Signal Process. Mag.*, pp. 118–124, Vol. 4, No. 4, July 2007.

[3] E. Candès and M. Wakin, 'An Introduction to Compressive Sampling', *IEEE Signal Process. Mag.*, pp. 21–30, Vol. 25, No. 2, March 2008.

[4] D. Needell, 'Topics in Compressive Sensing', PhD Thesis, University of California, 2009.

[5] H. Rauhut, 'Compressive Sensing and Structured Random Matrices', in M. Fornasier (editor), *Theoretical Foundations and Numerical Methods for Sparse Recovery* (Vol. 9, pp. 1–92). Radon Series on Computational and Applied Mathematics, Walter de Gruyter GmbH & Co. Berlin/New York, 2010.

[6] S. S. Chen, D. L. Donoho and M. A. Saunders, 'Atomic Decomposition by Basis Pursuit', *SIAM Journal on Scientific Computing*, Vol. 20, pp. 33–61, 1998.

[7] G. B. Dantzig and M. N. Thapa, *Linear Programming*, Springer, New York, NY, 1997.

[8] E. van den Berg and M. P. Friedlander, 'Probing the Pareto Frontier for Basis Pursuit', *SIAM Journal of Scientific Computing*, Vol. 31, No. 2, pp. 890–912, November 2008.

[9] A. Maleki, L. Anitori, Z. Yang and R. Baraniuk, 'Asymptotic Analysis of Complex LASSO via Complex Approximate Message Passing (CAMP)', *IEEE Trans. Inf. Theory*, pp. 4290–4309, Vol. 9, No. 7, July 2013.

[10] L. Anitori, 'Compressive Sensing and Fast Simulations. Applications to Radar Detection', PhD Thesis, TU Delft, 2012.

[11] L. Anitori, M. Otten and P. Hoogeboom, 'CS Radar Imaging via Adaptive CAMP', *Proceedings of 9th European Conference on SAR (EUSAR)*, Nuremberg, Germany, pp. 263–266, 23–26 April 2012.

[12] S. Mallat and Z. Zhang, 'Matching Pursuit in a Time-Frequency Dictionary', *IEEE Trans. Signal Process.*, Vol. 41, pp. 3397–3415, December 1993.

[13] Y. C. Pati, R. Rezaiifar and P. S. Krishnaprasad, 'Orthogonal Matching Pursuit: Recursive Function Approximation with Applications to Wavelet Decomposition', *Proc. 27th Annular Asilomar Conf. Signals, Systems, Comput.*, Vol. 1, pp. 40–44, November 1993.

[14] J. Tropp and A. Gilbert, 'Signal Recovery from Random Measurements via Orthogonal Matching Pursuit', *IEEE Trans. Inf. Theory*, Vol. 53, pp. 4655–4666, December 2007.

[15] J. Tropp and S. J. Wright, 'Computational Methods for Sparse Solution of Linear Inverse Problems', *IEEE Proc.*, Vol. 98, No. 6, pp. 948–958, June 2010.

[16] D. L. Donoho, Y. Tsaig, I. Drori and J.-L. Starck, 'Sparse Solution of Underdetermined Linear Equations by Stagewise Orthogonal Matching Pursuit', *IEEE Trans. Inf. Theory*, Vol. 58, No. 2, pp. 1094–1121, 2012.

[17] M. Zibulevsky and M. Elda, 'L1-L2 Optimisation in Signal and Image Processing', *IEEE Signal Process. Mag.*, Vol. 27, No. 3, pp. 76–88, May 2010.

[18] A. Beck and M. Teboulle, 'A Fast Iterative Shrinkage Thresholding Algorithm for Linear Inverse Problems', *SIAM J. Imag. Sci.*, Vol. 2, No. 1, pp. 183–202, 2009.

[19] T. Blumensath and M. Davies, 'Iterative Hard Thresholding for Compressed Sensing', *Appl. Comput. Harmonic Anal.*, Vol. 27, No. 3, pp. 265–274, 2009.

[20] E. Piazza, 'Radar Signals Analysis and Modellization in the Presence of JEM Application to Civilian ATC Radars', *IEEE AES Syst. Mag.*, pp. 35–40, January 1999.

[21] M. Friedlander and E. van den Berg, 'SGPL1: A Solver for Large-Scale Sparse Reconstruction', http://www.cs.ubc.ca/labs/scl/spgl1/

[22] R. van der Heiden, L. J. van Ewijk and F. C. A. Groen, 'In-Flight Measurements and RCS-Predictions: A Comparison on broad-side radar range profiles of a Boeing 737', *International Radar Conference*, Edinburgh, pp. 444–448, 1997.

[23] M. G. E. Brand, 'Radar Signature Prediction and Analysis by Physical Optics and Ray Tracing. The RAPPORT Code for RCS Prediction', FEL-95-A097, TNO, The Hague, The Netherlands, 1995.

[24] L. Zhang, M. Xing, C.-W. Qiu, J. Li, J. Sheng, Y. Li and Z. Bao, 'Resolution Enhancement for Inversed Synthetic Aperture Radar Imaging Under Low SNR via Improved Compressive Sensing', *IEEE Trans. Geosci. Remote Sens.*, Vol. 48, No. 10, pp. 3824–3838, October 2010.

Chapter 9

Advances in SAR change detection

Leslie M. Novak

9.1 Introduction

A typical SAR image gathered by the General Dynamics data collection system (DCS) is shown in Fig. 9.1. The image size is 4096 × 4096 pixels and the resolution of the data is 1 ft by 1 ft. The box superimposed on the image shows a region of interest containing two interesting change detection scenes that will be investigated. These initial non-coherent change detection (NCCD) studies will focus on a scene containing parked vehicles (the 'vehicle scene'); the initial coherent change detection studies will focus on a scene containing a subtle man-made disturbance due to people who walked in a grassy area (the 'racetrack scene').

Figs. 9.2 and 9.3 show the 1024 × 1024 region of interest containing the change detection scenes; Fig. 9.2 shows the SAR reference image and Fig. 9.3

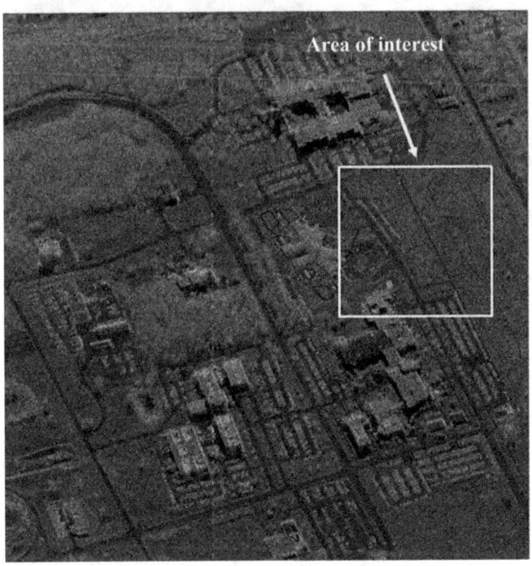

Figure 9.1 SAR reference image, size = 4096 × 4096 pixels; area of interest size = 1024 × 1024 pixels

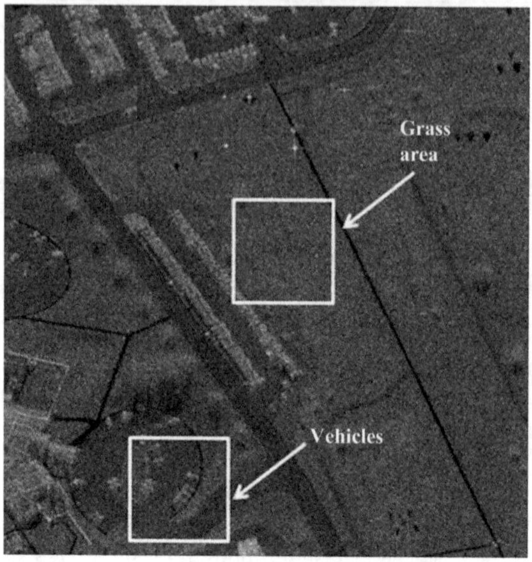

Figure 9.2 SAR reference image

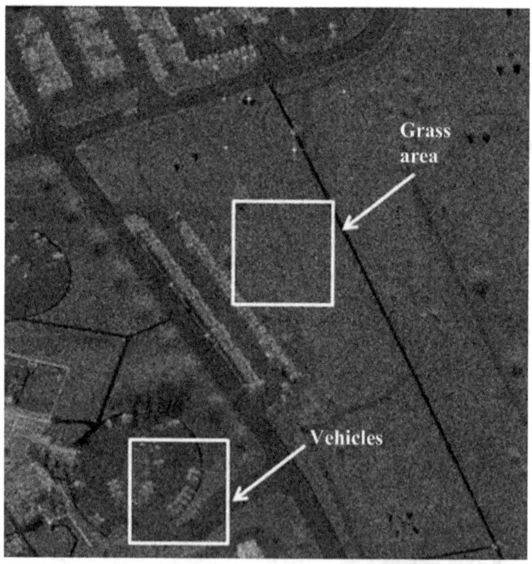

Figure 9.3 SAR test image

shows the corresponding SAR test image. These reference and test images will be processed using the coherent change detection (CCD) algorithm [1,2,3] and NCCD algorithm [4] as defined in Table 9.1.

Figs. 9.4 and 9.5 show the CCD and NCCD change images obtained from comparisons of a test image and a previously gathered reference image; several

Table 9.1 SAR change detection algorithms

Coherent change detection algorithm	Non-coherent change detection algorithm
$CCD = \dfrac{\left\lvert \sum_{k=1}^{N} x_k^* x_{k+N} \right\rvert}{\sqrt{\left(\sum_{k=1}^{N} \lvert x_k \rvert^2\right)\left(\sum_{k=1}^{N} \lvert x_{k+N} \rvert^2\right)}}$	$NCCD = 1 - \dfrac{\left(\frac{1}{N}\sum_{k=1}^{N} \lvert x_k \rvert^2\right)\left(\frac{1}{N}\sum_{k=1}^{N} \lvert x_{k+N} \rvert^2\right)}{\left(\frac{1}{2}\left(\frac{1}{N}\sum_{k=1}^{N} \lvert x_k \rvert^2 + \frac{1}{N}\sum_{k=1}^{N} \lvert x_{k+N} \rvert^2\right)\right)^2}$

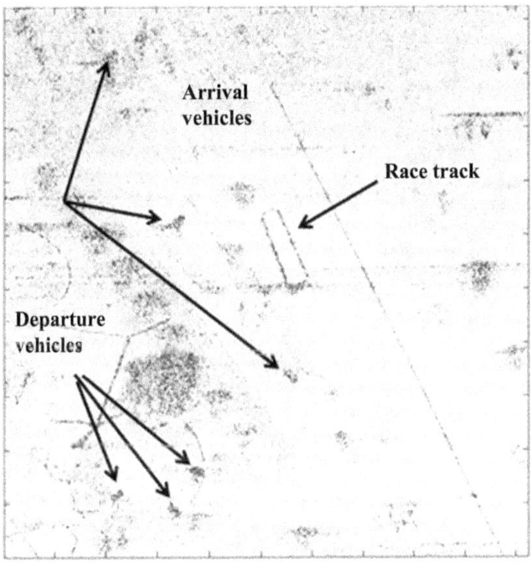

Figure 9.4 Coherent change image

detected changes are pointed out on the NCCD and CCD images. Note that only coherent change detection has detected the 'racetrack' in the grass area and, although the change in amplitude between the reference and test images is too small to be detected by the NCCD algorithm, the change in phase (i.e. the 'coherence') between the reference and test images is sufficient to permit detection of this subtle change by the CCD algorithm.

9.2 An analysis of the CCD algorithm

Fig. 9.6 presents a simplified block diagram of the baseline coherent change detection (CCD) algorithm we use in these studies. The reference and test images comprise complex pixels, denoted as images $X_{m,n}$ and $\hat{X}_{m,n}$. As indicated in the

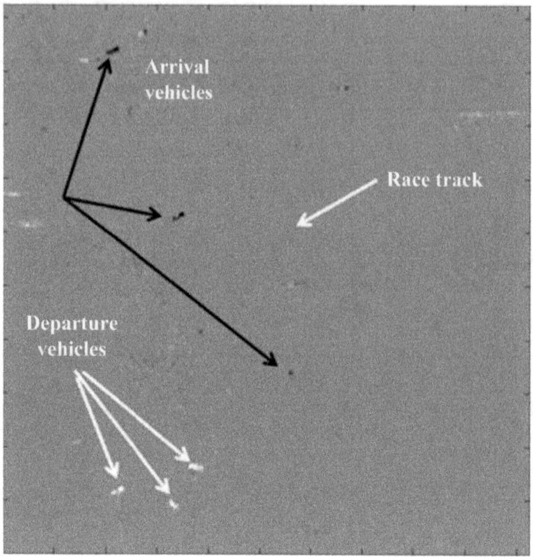

Figure 9.5 Non-coherent change image

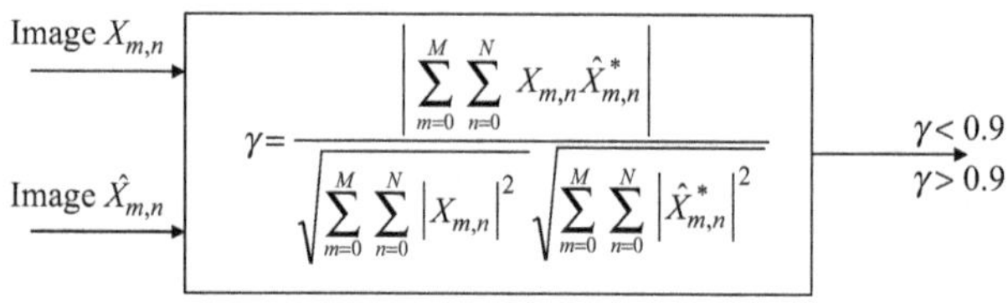

Figure 9.6 Block diagram of the CCD algorithm

figure, the algorithm calculates the coherence $'\gamma'$ between the reference and test images (i.e. the magnitude of the complex cross-correlation between the reference and test images). Also in these studies, the coherence is calculated using a 3×3 cluster of the complex image data (thus, $M = N = 2$).

An analysis of the coherent change detection algorithm is given as follows. We write the complex pixel data in amplitude and phase format:

$$X_{m,n} = |X_{m,n}|e^{j\phi_{m,n}} \text{ and } X^{*}_{m,n} = |X_{m,n}|e^{-j\phi_{m,n}}$$

The coherence equation defined in Fig. 9.6, expressed in amplitude/phase format, is given as follows:

$$\gamma = \frac{\left|\sum_{m=0}^{M}\sum_{n=o}^{N} |X_{m,n}||\hat{X}^{*}_{m,n}|\, e^{j(\phi_{m,n} - \hat{\phi}_{m,n})}\right|}{\sqrt{\sum_{m=0}^{M}\sum_{n=0}^{N} |X_{m,n}|^2}\sqrt{\sum_{m=0}^{M}\sum_{n=0}^{N} |\hat{X}^{*}_{m,n}|^2}}$$

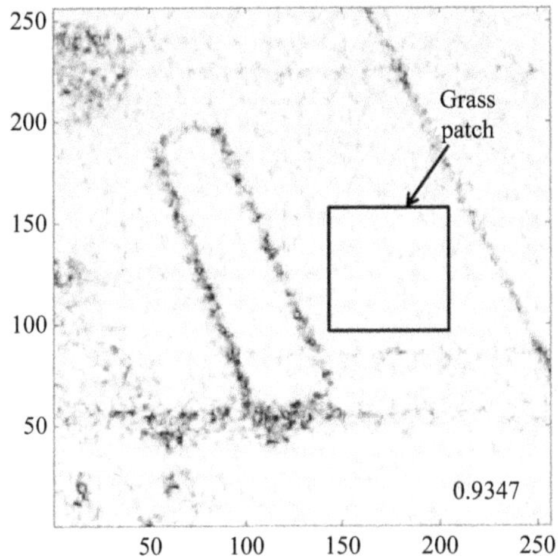

Figure 9.7 CCD image of 'racetrack' area; average image coherence = 0.9347; a grass patch is outlined

Next we make the assumption that the magnitudes of the complex reference and test pixels are equal, implying that the coherence between the images will depend only on the phase differences between pixels. With this simplifying assumption, the following result is obtained:

$$\text{if } |X_{m,n}| = |\hat{X}_{m,n}| = |X|\ \forall\, m,n \text{ then } \gamma = \left|\sum_{m=0}^{M}\sum_{n=0}^{N} e^{j(\phi_{m,n}-\hat{\phi}_{m,n})}\right|$$

We are interested in comparing the performance of the CCD algorithm using the original complex image data versus using the phase-only data; in this study we initially focus on a sub-image of the grass area including the 'racetrack feature'. Fig. 9.7 shows a 256 × 256 pixel coherence image containing the racetrack with a selected patch of grass outlined, note that the average coherence of this 256 × 256 pixel sub-image is 0.9347.

Figs. 9.8 and 9.9 present a comparison of the grass patch CCD images calculated using the original complex SAR data versus using the (amplitude-normalised) phase-only SAR data. This small patch of grass has average coherence = 0.9685 using the original complex SAR data, whereas the average coherence = 0.9361 using phase-only data. Thus, this example seems to indicate that both amplitude and phase (i.e. the complex pixel data) should be used in image exploitation using the CCD algorithm; there appears to be a loss in the level of coherence using phase-only images.

Fig. 9.10 validates the conjecture that the best CCD image is obtained using both the amplitude and phase of the data in forming the change image; the figure indicates that the average coherence using complex data, which was 0.9347 (see Fig. 9.7), has been reduced to 0.8948 using phase-only data and the change image in Fig. 9.10 also shows that a larger number of low-coherence pixels have been obtained using phase-only data.

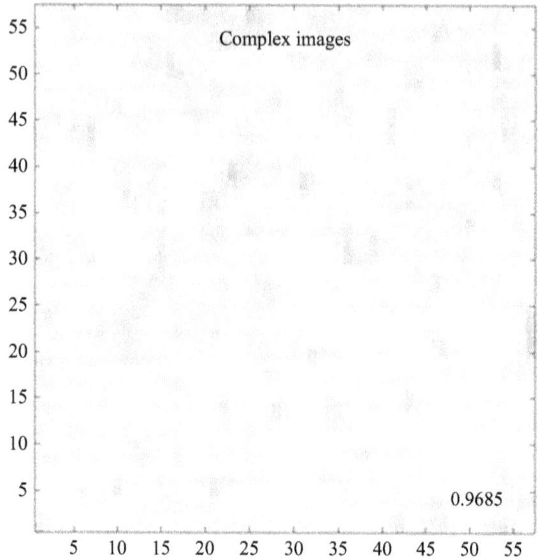

Figure 9.8 SAR CCD image of grass patch

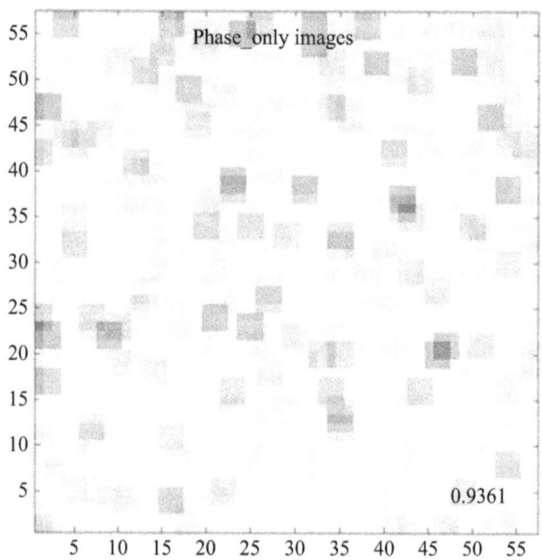

Figure 9.9 CCD image of grass patch from phase-only images

9.3 Results using the 'universal image quality index'

This section presents a summary of some interesting results that were obtained using an approach developed in References 5 and 6; these authors have proposed an image quality metric called the 'universal image quality index' and they have

Table 9.2 Definition of statistics used in calculating the 'universal image quality index'

$\text{Image } X = \begin{pmatrix} x_1 & x_2 & x_3 \\ x_4 & x_5 & x_6 \\ x_7 & x_8 & x_9 \end{pmatrix}$	$\text{Image } Y = \begin{pmatrix} y_1 & y_2 & y_3 \\ y_4 & y_5 & y_6 \\ y_7 & y_8 & y_9 \end{pmatrix}$
$\bar{x} = \frac{1}{N}\sum_{i=1}^{N} x_i$	$\bar{y} = \frac{1}{N}\sum_{i=1}^{N} y_i$
$\sigma_x^2 = \frac{1}{(N-1)}\sum_{i=1}^{N} (x_i - \bar{x})^2$	$\sigma_y^2 = \frac{1}{(N-1)}\sum_{i=1}^{N} (y_i - \bar{y})^2$
$\sigma_{xy} = \frac{1}{(N-1)}\sum_{i=1}^{N} (x_i - \bar{x})(y_i - \bar{y})$	

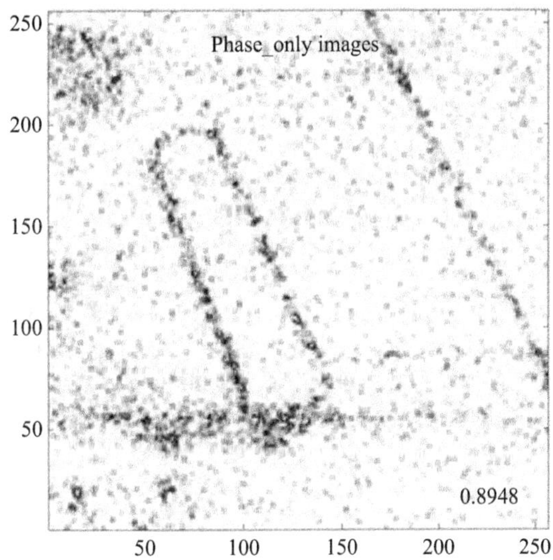

Figure 9.10 CCD image 'racetrack', phase-only data

demonstrated the application of their new metric to photographs such as the well-known 'Lena' and others. Although our SAR images comprise complex pixel values, it was of interest to apply this new metric to SAR intensity images. With this goal in mind, we give a brief description of the new metric and then present results of applying the approach to the SAR imagery shown in the previously described CCD and NCCD studies.

Table 9.2 presents details of the universal image quality index. There are two intensity images, denoted as image X and image Y; in the context of SAR change

Table 9.3 Definition of the 'universal image quality index'

$Q = \left[\dfrac{\sigma_{xy}}{\sigma_x\sigma_y}\right]\left[\dfrac{2\bar{x}\bar{y}}{(\bar{x})^2+(\bar{y})^2}\right]\left[\dfrac{2\sigma_x\sigma_y}{\sigma_x^2+\sigma_y^2}\right] \in [-1,+1]$ $Q = Q_1 \times Q_2 \times Q_3$	
$Q_1 = \left[\dfrac{\sigma_{xy}}{\sigma_x\sigma_y}\right] \in [-1,+1]$	Q_1 : A measure of image structural similarity
$Q_2 = \left[\dfrac{2\bar{x}\bar{y}}{(\bar{x})^2+(\bar{y})^2}\right] \in [0,+1]$	Q_2 : A measure of how similar image means are
$Q_3 = \left[\dfrac{2\sigma_x\sigma_y}{\sigma_x^2+\sigma_y^2}\right] \in [0,+1]$	Q_3 : A measure of how similar image contrasts are

detection, X denotes the reference image (intensity image) and Y denotes the test image (intensity image). The table shows a pair of 3 × 3 clusters of intensity values to be compared, and our goal is to find the changes between the reference and test intensity images. The mean, variance and covariance of the intensity values are calculated as indicated in Table 9.2.

As presented in Table 9.3, these mean, variance and covariance values are used to form three image quality factors: Q_1, Q_2 and Q_3. Q_1 is a measure of structural similarity, Q_2 is a measure of the similarity of the means and Q_3 is a measure of the similarity of the contrasts.

The image quality index Q is calculated from 3 × 3 clusters of intensity data at each pixel location in the image, resulting in a new image denoted as the universal IQ index image; this new image is a representation of the changes that exist between the reference and test images. The SAR reference and test images shown in Figs. 9.2 and 9.3 were converted into intensity images and processed as described above. The resulting universal IQ index image we obtained is shown in Fig. 9.11; an average IQ index of 0.8283 was obtained from the image shown. The interesting observation gleaned from the image shown in Fig. 9.11 is that this change image visually appears to be a CCD image but this change image was obtained from intensity-only SAR reference and test imagery.

Further analysis of the images produced by each of the factors Q_1, Q_2 and Q_3 showed that the image produced by the factor Q_1 was the dominant image, and this factor is simply the cross-correlation of the SAR intensity images. This observation has resulted in our researching the literature on previous mathematical analyses of the cross-correlation of SAR intensity images and its relationship with the SAR coherence parameter (see References 7 and 8). Table 9.4 presents two functional relationships derived in the references. ρ_A corresponds to intensity cross-correlation without mean removal [7] and ρ_B corresponds to intensity cross-correlation with mean removal [8].

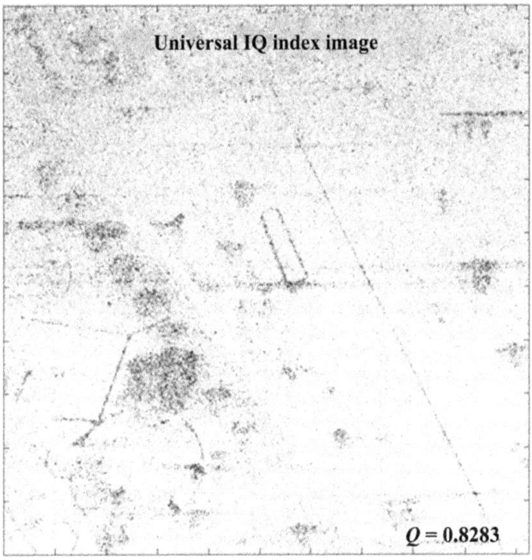

Figure 9.11 Universal image quality index image obtained using intensity images shown in Figs. 9.2 and 9.3

Additional SAR change detection studies using the intensity cross-correlation denoted as ρ_B in Table 9.4 were performed. Fig. 9.12 presents a comparison of the coherence image (a) versus the corresponding change image (b) obtained using the intensity image cross-correlation denoted as ρ_B in Table 9.4. Visually these images look quite similar and the absolute value of the difference between these images is presented in Fig. 9.13. The difference error image shows reasonably small differences between the actual coherence values γ and the coherence estimates $\gamma_B = \sqrt{\rho_B}$.

9.4 Performance comparison of change detection algorithms

In the previous sections we presented some preliminary comparisons of the CCD versus NCCD change detection algorithms. In this section we will investigate the detection performance of the maximum likelihood estimate (MLE) of the SAR coherence parameter. We will quantify and compare the change detection performance of the MLE versus the CCD. In Table 9.5 we present the definitions of these SAR coherent change detection algorithms. We will also compare the detection performance of these coherent change detection algorithms with the baseline NCCD algorithm defined previously in Table 9.1.

In our previous change detection studies we found that the MLE version of the coherent change detection algorithm gave better detection performance results than the (complex cross-correlation) CCD version of the algorithm. These previously obtained results were in agreement with a paper presented by Miriam Cha of MIT

Table 9.4 Coherence relations versus cross-correlation of SAR intensity images

Intensity correlation	Coherence relationship	
ρ_A	$\gamma_A = \sqrt{2\rho_A - 1}\,;$	$\rho_A \geq 0.5$
	$\gamma_A = \quad 0\,;$	$\rho_A < 0.5$
ρ_B	$\gamma_B = \sqrt{\rho_B}\,;$	$0 < \rho_B \leq 1.0$

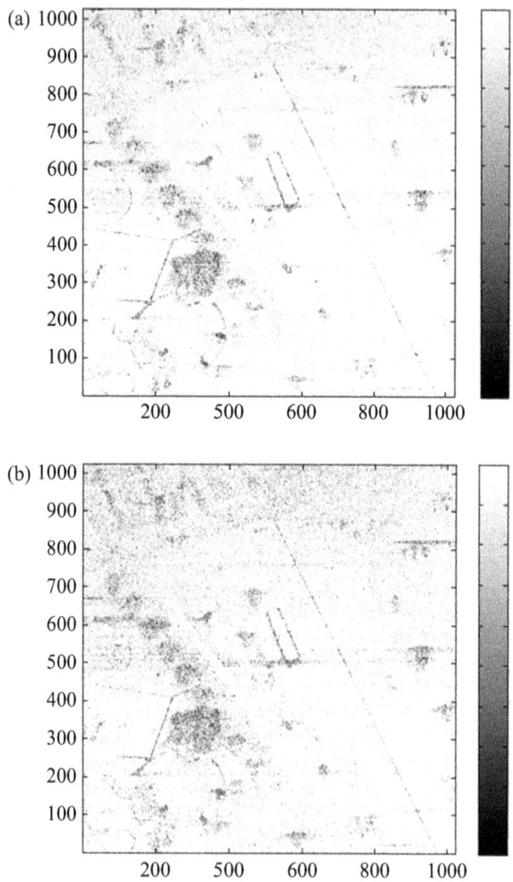

Figure 9.12 Coherence image versus approximation

Table 9.5 Coherent change algorithms

$$CCD = \frac{\left|\sum_{k=1}^{N} x_k^* x_{k+N}\right|}{\sqrt{\left(\sum_{k=1}^{N} |x_k|^2\right)\left(\sum_{k=1}^{N} |x_{k+N}|^2\right)}}$$

$$MLE = \frac{\left|\sum_{k=1}^{N} x_k^* x_{k+N}\right|}{\frac{1}{2}\left(\sum_{k=1}^{N} |x_k|^2 + \sum_{k=1}^{N} |x_{k+N}|^2\right)}$$

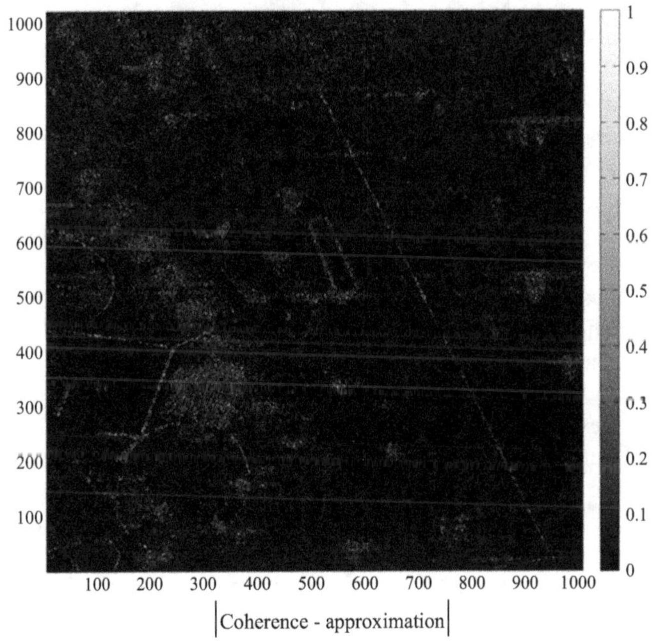

Figure 9.13 Magnitude of difference image

Lincoln Laboratory at the IEEE Statistical Signal Processing Workshop, Ann Arbor, Michigan [9]. In Miriam Cha's paper it was conjectured, based on theoretical analyses of the MLE and CCD coherent change detection algorithms, that the MLE version should provide better coherent change detection performance (i.e. PD versus PFA ROCs) than the CCD version if the reference and test images have approximately equal underlying variances. Clearly, this is the case for the

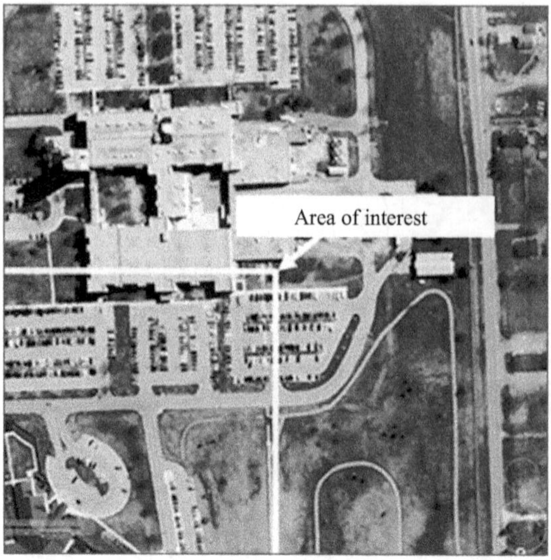

Figure 9.14 Aerial photograph of the 'area of interest' selected for CCD studies

accurately calibrated SAR imagery used in our studies [10]. In this section we summarise our studies of these two coherent change detection algorithms. Our goal is to determine the sensitivity of the MLE algorithm when the test image variance is not comparable to the reference image variance. We aim to show that the MLE outperforms the CCD over some range of calibration gain offsets between the reference and test images; thus, we aim to verify the conjecture of Miriam Cha.

Fig. 9.14 shows an aerial photo of the 'area of interest' selected for our SAR change detection studies. The area of interest comprises several parking lots, which are occupied by numerous parked (i.e. stationary) vehicles. Analysis of the SAR reference and test images of this area was found to contain a total of 33 vehicles that changed during the time interval between the gathering of the reference and test images. The vehicles in this set of 33 were either 'arrivals' or departures' that occurred during the time interval between the gathering of the reference and test images.

Fig. 9.15 presents the SAR reference and test images used in this change detection study. The locations of the change-detected vehicles are shown as circles superimposed on the images. There are a total of 33 vehicle detections in these SAR images; both arrivals and departures are contained in the 33 circles. These change detections were verified by visually flickering between the reference and test SAR images.

Fig. 9.16(a) shows a binary image of the locations of the change-detected vehicles. This binary image is used as ground truth for the targets in these CCD/MLE/NCCD change detection studies. We use the ground truth image to score the performance of the coherent and NCCD algorithms evaluated in the studies. Detections that occur in the black areas are scored as target detections. Detections

(a)

(b)

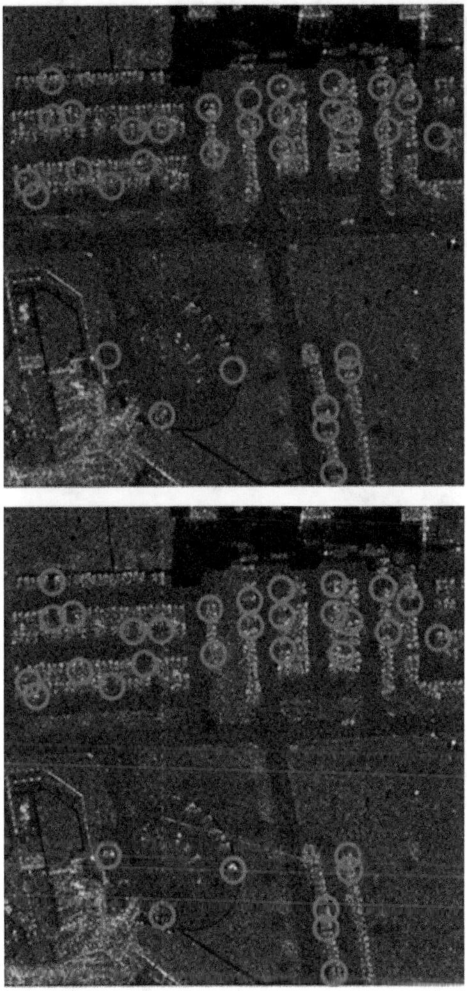

Figure 9.15 SAR reference image (a) and SAR test image (b)

that occur in the white background area are scored as false alarms (FAs). In Fig. 9.16(b) we show a 'don't care' band constructed around each of the targets to prevent clutter FAs from being declared targets.

Table 9.5 gives mathematical equations defining the SAR coherent change detection algorithms to be evaluated and compared in these studies. Two versions of the coherent change detection algorithm are given; these are denoted as MLE and CCD, depending on the algorithm normalisation used. Note that the MLE normalises the numerator by a sum of the reference and test pixel powers, whereas the CCD normalises the numerator by a product of the reference and test powers. The CCD algorithm is the well-known complex cross-correlation algorithm used for calculating (estimating) the coherence between complex-valued reference and test

(a)

(b)

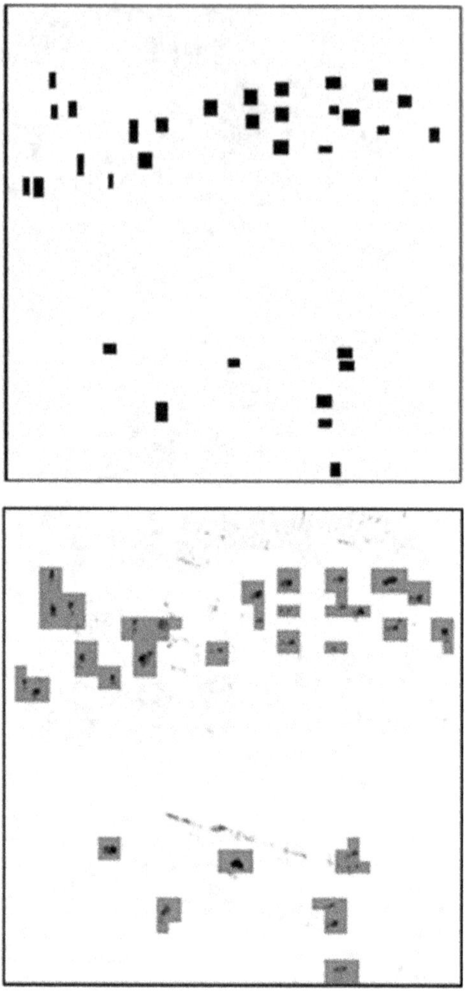

Figure 9.16 Locations of CD targets (a) and masked locations of CD Targets (b)

images; we refer to this algorithm as the 'CCD'. Since the MLE algorithm is actually the MLE of the coherence parameter, we refer to this algorithm as the 'MLE'. The non-coherent SAR change detection algorithm evaluated in these studies was given in Table 9.1; we refer to this algorithm as the 'NCCD'.

A comparison of the performance of these SAR change detection algorithms is given in Figs. 9.17–9.20; these figures show PD versus PFA obtained via change detection processing using CD-algorithm box-sizes 3×3, 5×5, 7×7 and 9×9, respectively. Fig. 9.17 shows the PD/PFA ROC curves obtained using a 3×3 CD-algorithm box size; the curves show that coherent change detection using the MLE algorithm achieves significantly better detection performance than the classical

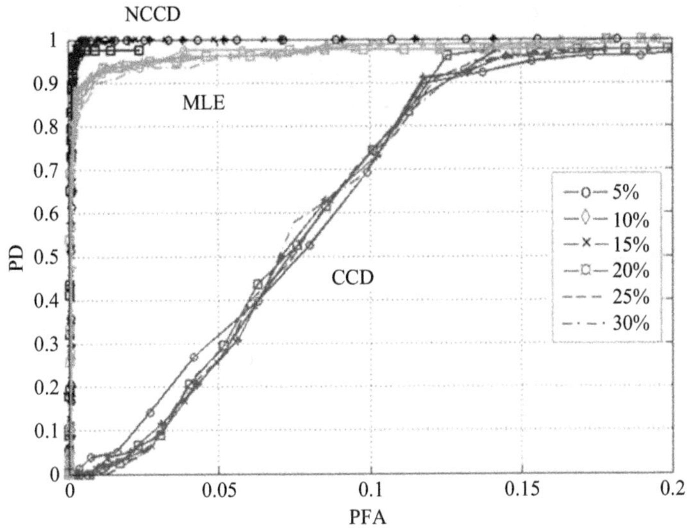

Figure 9.17 Change detection performance ROCs; CD algorithm box size = 3 × 3

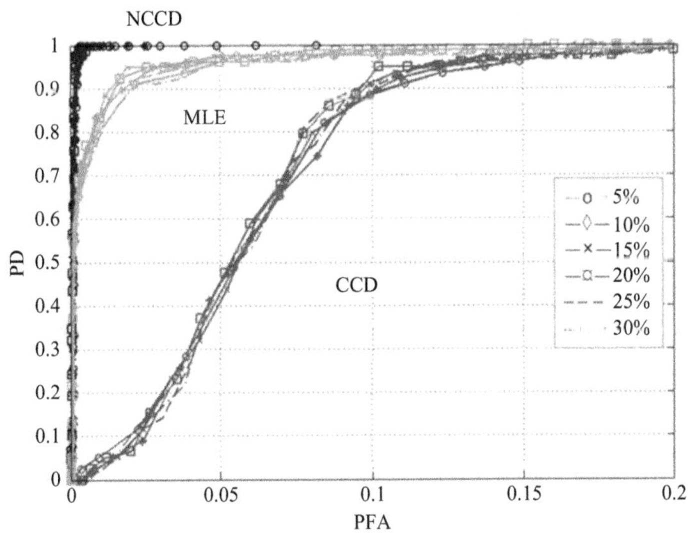

Figure 9.18 Change detection performance ROCs; CD algorithm box size = 5 × 5

CCD algorithm. At 70% PD, CCD gave ≈ 700 target-sized FAs, whereas the MLE gave ≈ 0 FAs. We also observe that the non-coherent NCCD change detection provides better performance than both of the coherent change detection algorithms. We evaluated detection performance versus the percent (%) of detected pixels in a

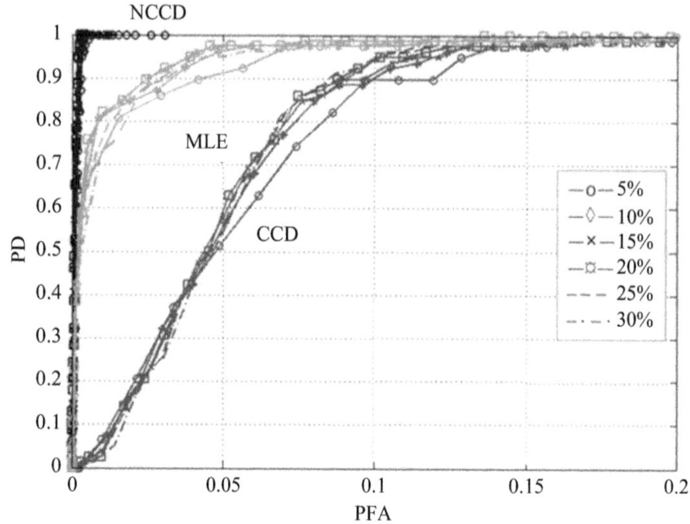

Figure 9.19 Change detection performance ROCs; CD algorithm box size = 7 × 7

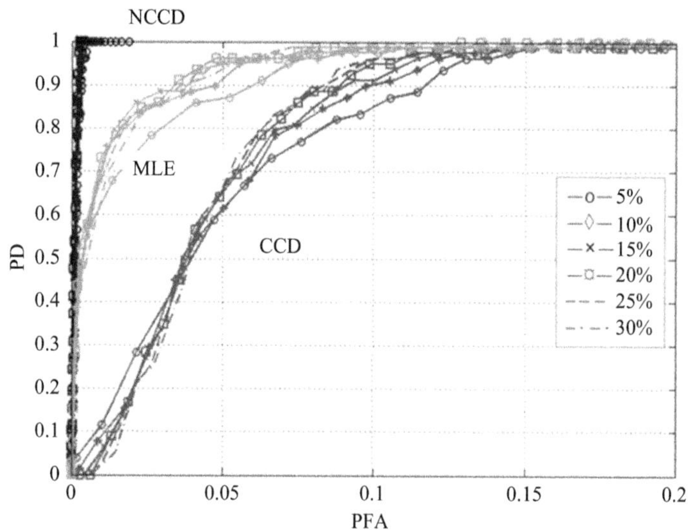

Figure 9.20 Change detection performance ROCs; CD algorithm box size = 9 × 9

target-sized box, and we found that detection performance was not very sensitive to this parameter; the ROC curves obtained for 5–30% are tightly clustered.

Fig. 9.18 summarises change detection performance using a CD-algorithm box size of 5 × 5. The ROCs shown are similar to those shown in Fig. 9.17. Fig. 9.18

shows that NCCD performance is definitely improved using the larger 5×5 box size – and CCD performance is also improved with the larger box size. The MLE algorithm shows slightly degraded performance with the larger box size.

Fig. 9.19 summarises CCD, MLE and NCCD algorithm performance using a CD-algorithm box size of 7×7 to construct the change images. The ROCs shown are similar to those shown in the previous figures. NCCD performance is again somewhat improved using a larger (7×7) box size. The CCD performance is also improved with the larger box size; for example, PD $> 90\%$ is achieved with PFA $\approx$ 0.1 (~700 FAs). The MLE algorithm has better ROC performance than the classical CCD algorithm, but again NCCD gave the best performance.

Fig. 9.20 summarises CCD, MLE and NCCD performance using a 9×9 box size. Coherent change detection using the MLE provides significantly better detection performance than the classical CCD algorithm. At 60% PD, the CCD gave approximately '350' target-sized FAs, whereas MLE gave many fewer FAs. Again we observe that the NCCD algorithm gave better performance than both coherent change detection algorithms.

In this paragraph we briefly summarise the findings of our change detection algorithm comparison studies described above. For the target and clutter data used in our studies, we found that the NCCD algorithm achieved the best overall detection performance; this was true for all box-sizes tested, and larger box-sizes gave marginally improved ROCs. PD/PFA curves were not very sensitive to the detector 'fill' parameter from 5% to 30%. The MLE version of the coherent change detector gave better detection performance than the CCD version of the algorithm, and the smaller box size (3×3) gave the best MLE performance.

9.4.1 Visual comparisons of the MLE and CCD algorithms

In this section we will evaluate and visually compare change images from the CCD and MLE versions of the coherent change algorithm with the test image scaled in amplitude relative to the given reference image. In these initial studies we scaled the test image by 0 dB, +3 dB and +6 dB relative to the original reference image. Fig. 9.21 compares CCD and MLE change images obtained using the Gotcha SAR reference and test images (with $K = 1$, i.e. no test image gain offset). Using Gotcha SAR imagery [10] similar to that used in our studies, MIT Lincoln Laboratory also showed that an MLE change image has higher contrast than the corresponding CCD change image; this phenomenon was demonstrated via change images similar to those shown in Fig. 9.21. Note that the average coherence values of the MLE and CCD change images shown in Fig. 9.21 are 0.8747 versus 0.9007, respectively; nevertheless, we show that the change vehicles in the MLE image are more easily detected than the change vehicles in the CCD image.

Fig. 9.22 shows additional examples of MLE change images obtained using amplitude scaled test images; these change images were obtained by comparing the reference image with a scaled test image (K*(test image)), where the scale factor K was either sqrt(2) or 2. The MLE baseline ($K = 1$) average coherence of 0.8747 has been reduced from this value to 0.8316 and 0.7163 for gain offsets of +3 dB and

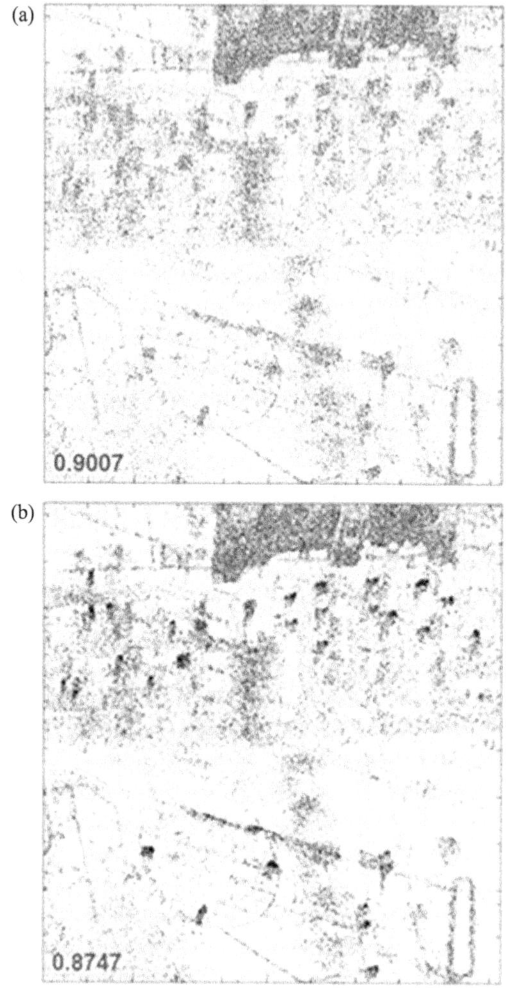

Figure 9.21 Baseline performance: CCD image (a) and MLE image (b)

+6 dB, respectively. Although these change images have less average coherence, it remains to be determined what these test image gain offsets will do to the detectability of the change vehicles in these images.

We continued these studies in order to quantify the actual change detection performance (PD versus PFA) achieved as a function of the test image gain offset parameter K. First we determined the average coherence values obtained from the CCD and MLE algorithms versus the test image gain offset for offsets as large as 10 dB; Fig. 9.23 presents these results. The data given in that figure indicates that the CCD algorithm is unaffected by the gain imbalance, whereas the average coherence estimated from the MLE algorithm is reduced in magnitude as the gain imbalance is increased. Thus, the PD/PFA ROC curves obtained using the CCD

(a)

(b)

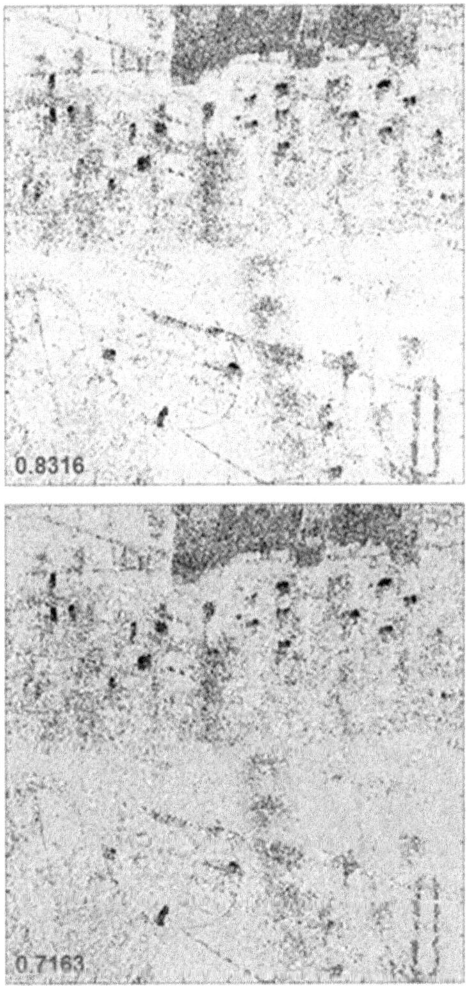

Figure 9.22 MLE image, K = sqrt(2) (a); MLE image, K = 2 (b)

version of the coherent change algorithm will be unaffected by the gain offset, and the corresponding change images for the CCD algorithm will be identical to Fig. 9.21(a).

We evaluated the change detection performance ROCs for this range of gain offsets; these ROCs are shown in Fig. 9.24. The curves in Fig. 9.24 show that MLE change detection performance is superior to CCD change detection performance for PFA < 0.04: this is true over the range of gain offsets simulated. It is of interest to visually compare the FAs and missed targets when an identical number of targets are detected by each algorithm. To this end, we have selected detection thresholds for the CCD and MLE algorithms, resulting in an identical number of targets

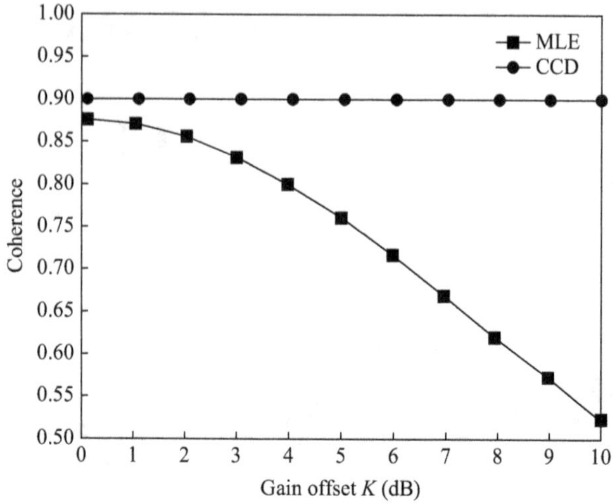

Figure 9.23 Average change image coherence versus test image gain offset K in dB

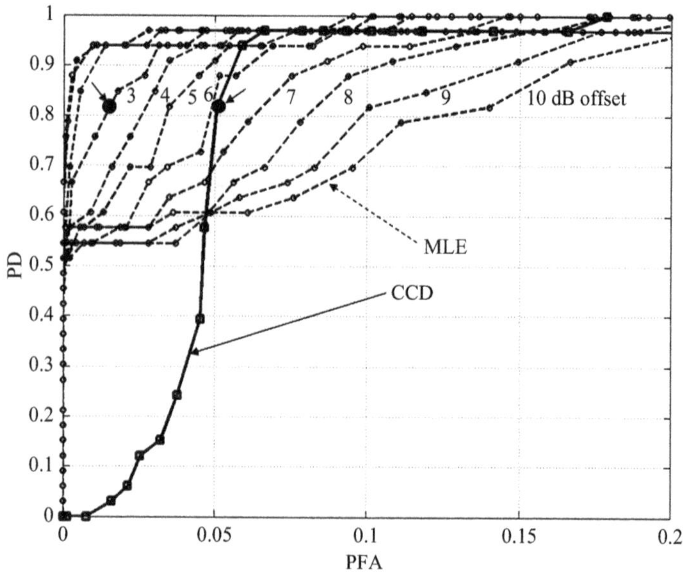

Figure 9.24 Coherent change detection ROCs for baseline CCD versus MLE. Note the CCD and MLE (3 dB) operating points at PD = 0.82

detected (see operating points at PD = 0.82 in Fig. 9.24). The corresponding PFA values at these operating points are 0.05 versus 0.014 for the CCD and MLE algorithms, respectively. The MLE PD/PFA curve used in this comparison is the 3 dB gain offset case.

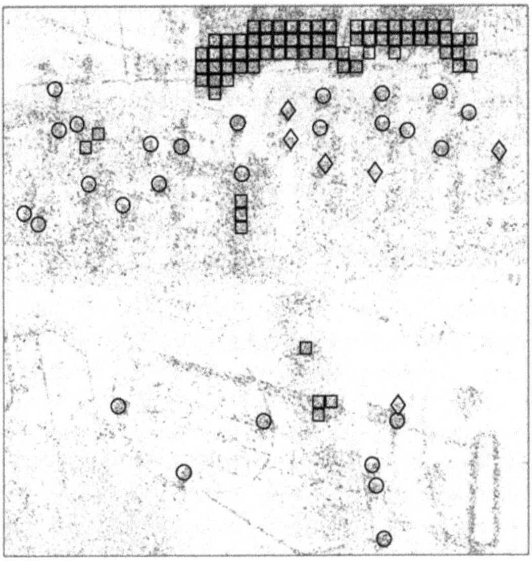

Figure 9.25 CCD change image, PD = 0.82, PFA = 0.05, missed targets = 6; threshold = 0.488; and detection fill percent = 15%

Fig. 9.25 shows the CCD change image at the selected operating point. The detection threshold used to obtain PD = 0.82 was 0.488. In this figure, detected targets are overlaid with circles, false detections are overlaid with squares and missed targets are overlaid with diamonds. At PD = 0.82 there are 6 missed targets (diamonds) and 73 false detections (squares). Note that there are a total of 64 false detections in the shadow region below the large building, whereas there are only 9 false detections in the remaining areas of the image. In the next section we will consider the problem of mitigating false detections in SAR image shadow regions and we will present a simple, robust algorithm for performing this task.

Fig. 9.26 shows the corresponding MLE change image at the selected PD = 0.82 operating point. For this change image the detection threshold used to obtain PD = 0.82 is 0.273. Note that the detection overlays on the images illustrate that a different set of targets were missed by the CD algorithms, and hence, each algorithm detects a somewhat different set of the targets. It is also important to point out that although both algorithms obtain false detections in the shadow regions below the large building, the CCD version of the algorithm obtains many more false detections in these shadow regions. In the next section we will develop and evaluate the performance of a simple approach to locate the shadow regions (tree and building shadows) and remove false detections from the SAR change image. Our algorithm for mitigating false detections in these shadow regions will, of course, improve the performance of both algorithms.

Table 9.6 presents a detection performance comparison of the CCD and MLE algorithms before and after the building shadow FAs are excluded from the FA calculation. This comparison shows that a significant improvement in coherent

Table 9.6 Coherent change detection statistics, MLE versus CCD at PD = 0.8182

	With building FAs		Without building FAs	
	PD	**FAs**	**PD**	**FAs**
MLE	27/33	20	27/33	0
CCD	27/33	73	27/33	9

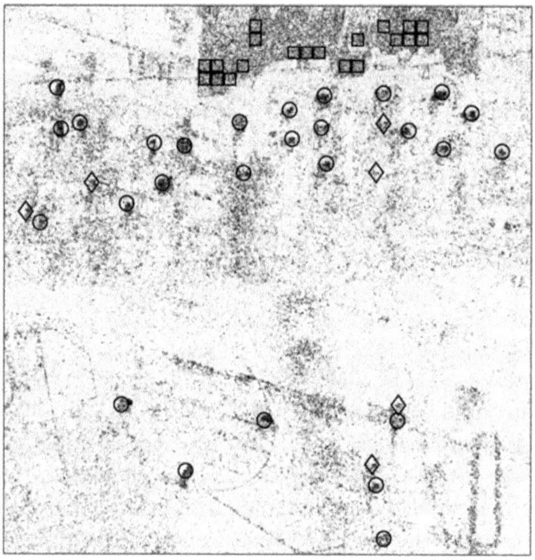

Figure 9.26 MLE change image, PD = 0.82, PFA = 0.014, missed targets = 6; threshold = 0.273, detection fill percent = 15%

change detection performance could be achieved by mitigating false detections in shadow regions of SAR change imagery.

9.4.2 Coherent change detection performance with shadow regions masked

The MLE of the SAR coherence parameter was derived by Jakowatz et al. [1]. In the previous subsection we showed that using the MLE algorithm to estimate SAR change image coherence could give better target detection performance than using the complex correlation coefficient, CCD. Furthermore, in Figs. 9.25 and 9.26 we observed that many false detections can occur in low-coherence building (and tree) shadows. Low-RCS areas such as building shadows and tree shadows, as well as flat asphalt roads, have low coherence due to random phase returns. Since an X-band SAR cannot detect target returns from targets located and masked in shadow areas, then CCD and MLE performance ROCs could be significantly improved if such areas were masked as 'don't care' areas before performing the detection operation.

One simple approach for improving the ROC curves is to post-process the standard CCD and MLE images by setting the coherence of the areas that correspond to low-RCS areas *in both the test and reference images* to unity. Target detection is then performed on the change images with low RCS areas masked. The low RCS areas can easily be detected as follows. The nth pixel in the CCD image is declared as belonging to the non-interesting, low-RCS area if the average power of the coherent sum and the coherent difference between the corresponding neighbourhoods in the test and reference images is below a selected threshold T:

$$\frac{1}{2}\left(\frac{1}{N}\sum_{k\in\Delta_n}|f_k+g_k|^2+\frac{1}{N}\sum_{k\in\Delta_n}|f_k-g_k|^2\right)<T$$

where:

$N =$ The number of pixels in the local neighbourhood Δ_n of the nth pixel
f_k and $g_k =$ The kth pixel of the reference and test data in the nth pixel neighbourhood Δ_n

This subsection applies the above algorithm for detecting low-RCS areas in the SAR scene and removing false detections from these regions of the scene. The gain in detection performance achieved will be quantified via ROC curves. We first show the SAR reference and test images used in these studies (Figs. 9.27 and 9.28).

Figure 9.27 SAR reference image

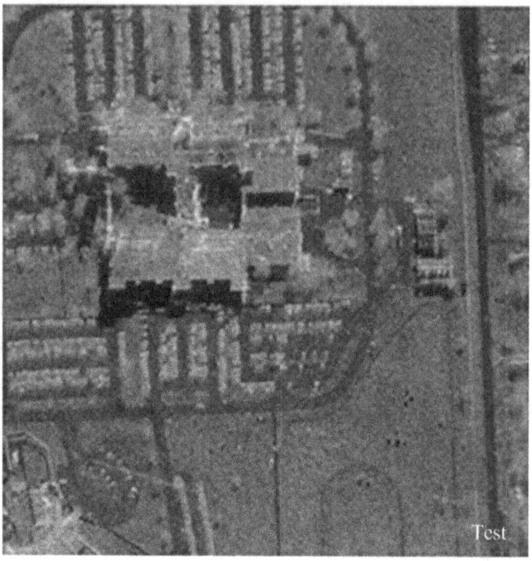

Figure 9.28 SAR test image

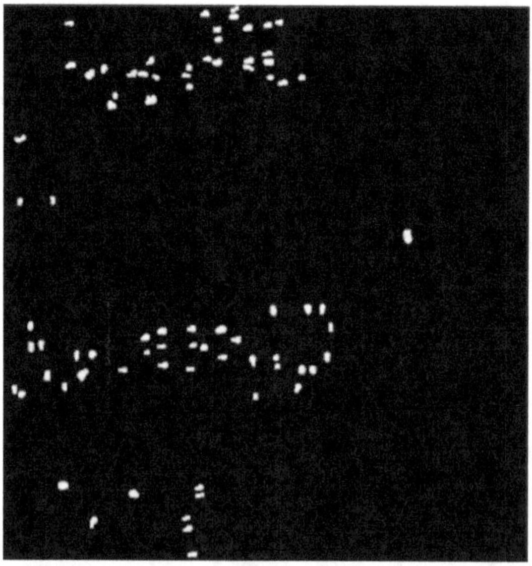

Figure 9.29 Ground truth overlay

We also show the ground truth overlay and the low-RCS mask (Figs. 9.29 and 9.30).

Coherent change detection performance was determined using the ground truth overlay shown in Fig. 9.29. Fig. 9.31 shows the original baseline ($K = 1$) MLE

Figure 9.30 Example low-RCS image

Figure 9.31 MLE, no shadow removal

change image, and Fig. 9.32 shows the MLE image after applying shadow removal using the low-RCS image shown in Fig. 9.30. Note that after shadow removal, the average MLE change image coherence has increased from 0.8325 to 0.9108.

The ROC curves in Fig. 9.33 show that a large gain in coherent change detection performance is achieved after the shadow areas in the SAR image are

Figure 9.32 MLE, after shadow removal

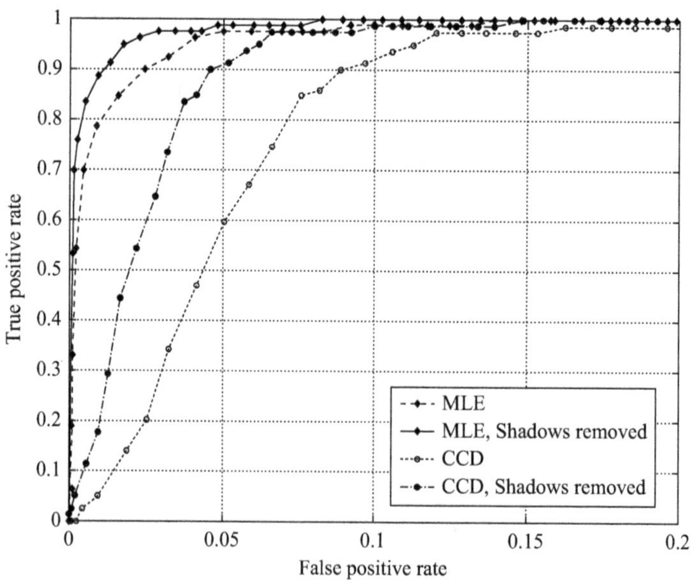

Figure 9.33 Coherent change detection ROCs (true positive rate versus false positive rate)

detected and denoted as 'don't care' areas of the scene. From the curves we see that after shadow removal, CCD performance is significantly improved relative to the original CCD performance; however, the CCD algorithm performance after shadow removal is not as good as the original MLE performance and MLE performance after shadow removal is the best overall change detection performance result.

9.5 Summary and conclusions

This chapter has reviewed some of the basic properties of SAR change detection. An example was presented comparing CCD and NCCD performance using a scene containing various types of changes that occurred between the SAR reference image and test image pair. Another example was presented comparing CCD performance using phase-only SAR images versus using the complex (amplitude and phase) SAR images. For the images used in this study we found that the coherence levels obtained from the complex imagery were somewhat better than the coherence levels obtained from the phase-only imagery.

A new image quality metric [5,6], the 'universal image quality index' was investigated. Applying this image quality metric to SAR intensity images (reference versus test intensity images) we observed that the universal IQ image was visually quite similar to the SAR CCD coherence image. We then investigated the relationship between the coherence calculated using complex image data and the correlation calculated using the corresponding intensity images. We demonstrated that the magnitude of the differences between the actual coherence image and the estimated coherence image obtained from the intensity-derived correlation image were quite small; however, further analysis is needed to quantify the errors versus the registration accuracy required between reference and test images.

We also compared the coherent change detection performance obtained using the MLE of the coherence between the reference and test complex SAR images versus using the complex correlation coefficient estimate. We showed, using an example change detection image pair from the publicly released Gotcha SAR imagery [10], that the MLE algorithm gave better PD/PFA detection ROCs than the CCD algorithm. Finally, we presented a simple, robust algorithm for mitigation of false changes in shadow areas (building and tree shadows), thereby improving the performance of both coherent change detection algorithms (MLE and CCD).

References

[1] C. Jakowatz, D. E. Wahl, P. H. Eichel, D. C. Ghiglia and P. A. Thompson, *Spotlight-Mode Synthetic Aperture Radar: A Signal Processing Approach*, Springer, New York, NY, pp. 330–340, 1996.

[2] M. Preiss and N. Stacy, *Coherent Change Detection: Theoretical Description and Experimental Results*, Intelligence, Surveillance and Reconnaissance

Division, Defense Science and Technology Organisation, Salisbury, Australia, DSTO–TR–1851.

[3] L. Novak, 'Coherent Change Detection for Multi-polarization SAR', *Asilomar Conference on Circuits, Systems, and Computers*, Pacific Grove, CA, October 2005.

[4] L. Novak, 'Change Detection for Multi-polarization, Multi-pass SAR', *SPIE Conference on Algorithms for Synthetic Aperture Radar Imagery*, Orlando, FL, March 2005.

[5] Z. Wang and A. Bovic, 'A Universal Image Quality Index', *IEEE Signal Process. Lett.* Vol. 9, No. 3, pp. 81–84, March 2002.

[6] Z. Wang and A. Bovic, 'Mean Squared Error: Love It or Leave It', *IEEE Signal Process. Mag.* Vol. 26, No. 1, pp. 98–117, January 2009.

[7] A. Guarnieri and C. Prati, 'SAR Interferometry: A "Quick and Dirty" Coherence Estimator for Data Browsing', *IEEE Trans. G.R.S.* Vol. 35, No. 3, pp. 660–669, May 1997.

[8] R. Touzi, A. Lopes, J. Bruniquel and P.W. Vachon 'Coherence Estimation for SAR Imagery', *IEEE Trans. Geoscience and Remote Sensing*, Vol. 37, No. 1, pp. 135–149, January 1999.

[9] M. Cha, R. Phillips and P. J. Wolfe 'Test Statistics for Synthetic Aperture Radar Coherent Change Detection', *IEEE Statistical Signal Processing Workshop*, Ann Arbor, MI, 5–8 August 2012.

[10] S.M. Scarborough, L. Gorham, M. J. Minardi, U.K. Majumder, M.G. Judge, L. Moore, L. M. Novak, S. Jaroszewksi, L. Spoldi and A. Pieramico 'A Challenge Problem for SAR Change Detection and Data Compression', *SPIE Conference on Algorithms for Synthetic Aperture Radar Imagery*, Orlando, FL, April 2010.

Chapter 10

Future challenges

David Blacknell and Hugh Griffiths

10.1 Introduction

The idea of automatic target recognition often conjures up visions of a completely general system that is able to classify all manner of different vehicle types in the most difficult of clutter environments. This then naturally leads to some scepticism that reliable radar ATR can ever be achieved. However, in reality the concept of ATR actually describes a continuum of problems of varying degrees of difficulty from very constrained scenarios to a completely general recognition system.

Some factors that influence the degree of difficulty of the ATR problem are listed below:

- number of target classes (from one to many)
- complexity of clutter (from open fields to dense urban environments)
- level of decision (from detection through recognition to identification and technical analysis)
- probability of false alarm (from low to extremely low)
- probability of detection (from high to extremely high)
- available training data (from previously gathered airborne SAR imagery and detailed CAD drawings to a handful of photographs)
- radar mode (from multiple, high resolution images to a single, low resolution range profile)

It is important to recognise that an ATR system does not have to address the most difficult scenarios to provide valuable military capability. Indeed, operational radar ATR systems already exist that have solved some of the problems lying on this continuum of ATR difficulty.

The aim of ongoing radar ATR research is to push the boundaries in terms of difficulty and complexity level to provide greater capability in future radar systems. Some of the key future challenges, which must be faced in order to achieve this, will now be discussed.

10.2 Future challenges

10.2.1 Target variability and practical databases

An enduring challenge with radar ATR is the immense variability in the appearance of targets from a given class as a result of within-class variations, radar parameters and imaging geometry. The fundamental problem is that particular physical structures, e.g. retro-reflective metallic corners, will give rise to strong radar returns. However, the occurrence of such structures for a given target class will vary drastically depending on the factors identified above. For example a tank observed from one aspect angle with the barrel forward, hatches closed and fuel drums attached at the back will produce a very different image to the same tank observed from a different aspect angle with the barrel angled, hatches open and no fuel drums. Indeed, once one takes into account (i) the number of possible target types, (ii) the number of degrees of freedom in the target positioning, (iii) the radar squint and depression angles, (iv) articulations of the various moving parts, (v) possible obscuration, (vi) target variants and (vii) target configurations, the number of possible variations is easily numbered in billions as graphically illustrated by DARPA during their MSTAR programme [1].

As has been discussed in Chapter 2, this extreme variation of target appearance necessitates the use of extensive databases of example imagery of the targets of interest. Options for populating such databases include the use of airborne SAR imagery, turntable ISAR imagery, scale models illuminated using appropriately scaled radar frequencies and electromagnetic prediction codes. These options tend to have an inverse relationship between reality of representation and cost. For example airborne SAR imagery of the targets of interest situated in a representative clutter background will give greatest realism but there will be a huge cost involved in gathering representative imagery for all degrees of freedom. Conversely, electromagnetic prediction codes provide a relatively cheap way of generating imagery but the degree of realism is limited because such simulations obtained from CAD models tend to be too 'clean'.

Of course, the true problem is not achieving the greatest realism that can be afforded but is to establish the degree of realism that is required to support the particular ATR problem under consideration. Furthermore, an ATR system that relies on a database of extremely accurate images of all the targets of interest will not be particularly robust as any slight, unanticipated variation will not be recognised. Indeed, a huge amount of effort has been expended over the years attempting to produce electromagnetic prediction codes that match real imagery to a high degree of accuracy. However, from an ATR perspective it could be argued that this is wasted effort as a robust, affordable ATR system must be able to perform with a database that inevitably is an imperfect representation of the real world and all its inherent variability.

Thus the requirement is for practical and affordable databases matched to the ATR problem under consideration. There are various strategies for making the best use of available data sources to generate databases. For example it is possible to form a mixed database containing a limited number of airborne SAR images at key points in the parameter space augmented by imagery from other sources as appropriate. The key question is how the parameter space of target/scenario variability should be sampled and at which points is the greatest realism required in the imagery. For example, thinking purely of aspect angle variation, does the ATR system require a database containing an image at every degree or at every ten degrees? Also, the most rapid change tends to be when a target is at broadside or end on when the image will contain large flashes, so is it at these points that the greatest realism is required? The future challenge is thus to model the sensitivity of ATR performance to the characteristics of the database in order to allow practical and affordable databases to be designed.

10.2.2 Complex clutter environments

ATR development, in common with many problems, quite naturally tends to proceed from less complex scenarios to more complex scenarios. For example, for the ground target case, much development has taken place assuming that the targets are seen against an essentially featureless background. This assumption is valid for some operational scenarios and provides valuable military capability. However, in progressing towards further levels of complexity and utility, a key consideration is the situation when the targets are located in more complex clutter environments, e.g. heavy scrubland or urban areas.

Some progress can be made by the use of advanced radar modes. For example Fig. 10.1 shows a single channel SAR image (a) of an area of scrubland containing some vehicles. These are difficult to see at first glance and would be challenging to an operator having to examine huge amounts of SAR imagery. However, the use of polarimetric decompositions enhances the contrast between the targets and the background making them more visible and hence more easily detected.

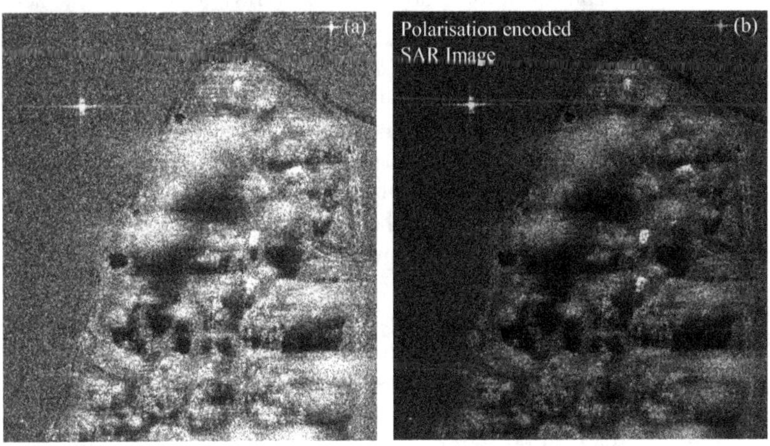

Figure 10.1 Use of polarimetry to enhance contrast of target signatures against scrubland background

Fundamentally, hard targets contain a lot of dihedral scattering events and so decomposition into odd and even scattering events [2] will highlight these against the mixed polarisation of the scrubland.

Moving towards the challenge of more complex clutter and the urban environment, approaches based on image segmentation to delineate and remove the unwanted clutter objects are one way forward [3]. However, increasingly in military applications there are many other sources of information available in addition to the radar data. Thus an alternative approach is to use collateral information in the form of maps or other sensor information to provide a template of the buildings in the scene. These can then be masked out so that the search for targets is confined to only those regions where it is known that valid targets can be found.

The development of robust ATR systems for use in complex clutter environments is an ongoing challenge and leads on to the idea of using contextual information.

10.2.3 Use of contextual information

The standard approach to ATR is to extract an image chip containing the target and perform the classification purely on the basis of this chip. However, certainly in the ground target case, there is a substantial amount of contextual information that an operator would use in making a target declaration, which is ignored in this 'tunnel-vision' approach. For example the nearby lines of communication, the type of terrain and the military doctrine, which govern deployment of targets, will influence the probability of a target of a particular type being present.

A simple illustration of this is given in Fig. 10.2. The target recognition process is supported by SAR imagery but also by imagery from other sensors, e.g. electro-optic imagery. This imagery provides the target chips that will be analysed but also provides a means of making an up-to-date assessment of the whole scene in which the targets reside. This whole scene analysis can be used to augment information from geographical information systems (GIS) to provide as accurate representation of the scene context as possible. Another source of contextual information comes from an understanding of military behaviour in terms of both established behaviour characterised by military doctrine and up-to-date situational awareness provided by operational intelligence. An ATR system should make use of all these sources of contextual information to make the best possible decision regarding target declaration. This type of approach is an example of knowledge-based radar signal processing in which the analysis uses not only the information contained in the radar signal but also knowledge obtained from external sources.

An example of an approach to using contextual information has been provided in Chapter 2 specifically addressing the target detection problem. Simple forms of knowledge were used such as the tendency for vehicles to avoid certain terrain types, to be parked up near hedge boundaries and to move around in groups rather than in isolation. A simulated example showed how use of such knowledge could significantly improve the probability of detection for a given false alarm rate.

The challenge for the future is to develop knowledge-based ATR, which makes best use of the myriad of contextual information sources that are becoming

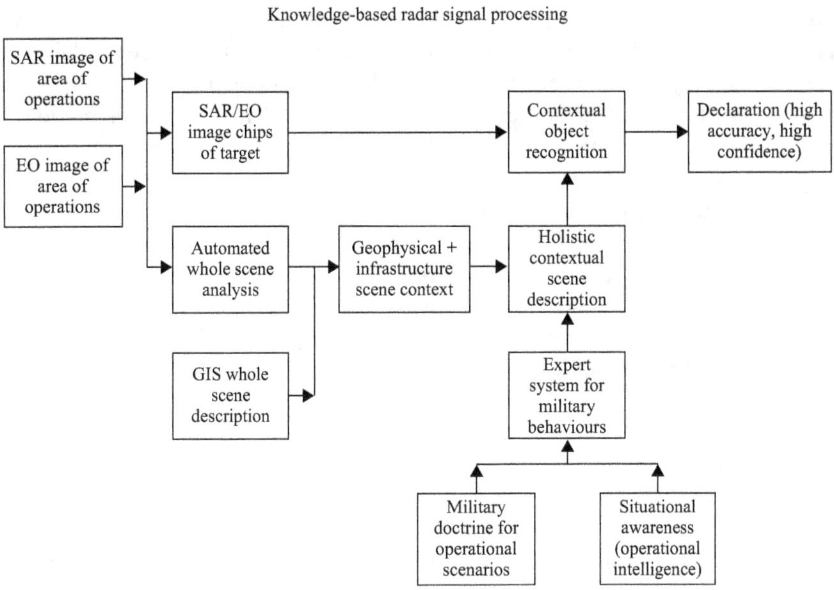

Figure 10.2 Knowledge-based radar signal processing based on contextual information

available. An associated challenge is how to maintain rigorous control and quantification of performance when the reliability or accuracy of the various data sources may be variable or unknown. This is set to remain an enduring challenge as greater and greater connectivity between information sources on the battlefield emerges.

10.2.4 Performance assessment and prediction

A crucial aspect of achieving an operational ATR capability is the process of validating performance against the required specification. This is an area with multiple challenges. First, being able to articulate the required operational performance in terms that can be meaningfully assessed is not straightforward. Typically the metrics used in discussions of performance in the literature will not ideally match the key performance metrics for operations. This has been illustrated to some extent in Chapter 2 where it has been seen that factors such as the probability of critical error and the ground order of battle can significantly influence performance assessment. However, more generally, the challenge is to specify requirements for performance that can be measured against necessarily limited validation and verification trial data obtained in non-operational circumstances that will provide some level of confidence in obtaining similar levels of performance in operational scenarios. This is an extremely difficult and challenging problem, yet it is of crucial importance.

A second issue with performance assessment is that test data will not always be available. It may be that it is prohibitively expensive to gather sufficient data to populate the region of operational conditions. Alternatively, in the design stage it may be wished to understand the trade-off in radar parameter choices in terms of eventual ATR performance prior to building any demonstrator system. The resulting challenge is then to develop a theory of ATR that allows prediction of performance as a function of the variable that defines the radar system, the target set, the operational scenario and the environmental conditions.

As a simple example, it may be desired to understand the performance trade-off between resolution and signal-to-noise ratio (SNR) in an ATR system. In Reference 4, a simple theory of ATR performance is developed for an ATR system that classifies purely on the basis of properties of the target shadow. Using this theory, it is possible to predict the probability of correct classification as a function of the contrast of the shadow (analogous to SNR) in comparison to the surrounding clutter and the resolution. This results in the performance curve are shown in Fig. 10.3(a). It can be seen that there is an area of poor performance, a region where performance increases rapidly and then an area where the performance reaches a plateau. Once in this plateau region, it is clear that improvements in shadow contrast (SNR) and resolution will be very little in terms of improved ATR performance. The key region is where the performance rises rapidly as changes in the radar performance can have a large impact on ATR performance. Furthermore, contours on this surface at a given probability of correct classification can be used to determine the trade-off between shadow contrast (SNR) and resolution.

This is a simulated example but the performance curve in Fig. 10.3(b) shows results for the classification of radar range profiles as a function of SNR and bandwidth (resolution) [5]. It can be seen that this example from real data exhibits the same characteristics as were observed in the simulated example. While the simulated example was not intended to model the real example in this case, the qualitative comparison gives some confidence that the ATR theory developed in Reference 4 is capturing the correct behaviour.

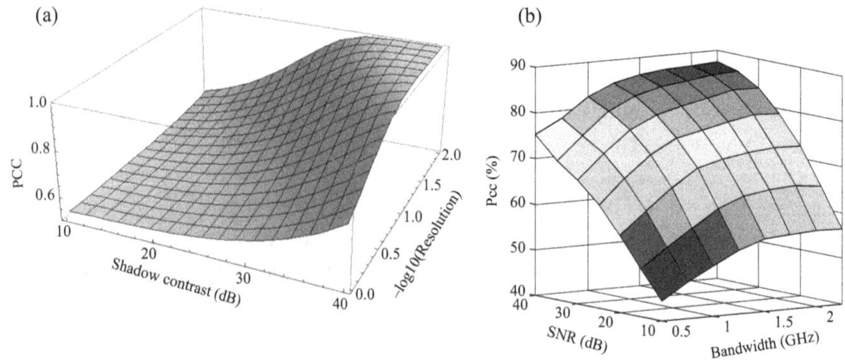

Figure 10.3 Theoretical prediction (a) and measurement on real data (b) of ATR performance as a function of parameters related to SNR and resolution

The challenge of developing a completely general theory for ATR performance is immense and is at a very early stage in development. However, such a theory would provide huge cost benefits in terms of ATR system design and development and so this remains a key challenge for the future.

10.2.5 Deception and countermeasures

As a final, brief comment, we can look at the problem from the opposite point of view – in other words, how to make our own targets more difficult to recognise by an adversary. Earlier chapters have shown the importance of identifying the most robust set of features to use as the basis for classification. It follows, then, that the best way of confusing an enemy's target recognition processing will be to understand the set of features of the radar signature of our own assets, and to set about modifying them in some suitable way. A full treatment of this is well beyond the scope of this book, but the principle may offer some interesting ideas for further work.

References

[1] R. Hummel, 'Model-based ATR Using Synthetic Aperture Radar', *IEEE International Radar Conference*, pp. 856–886, Washington, DC, 2000.

[2] D. E. Kreithen, S. D. Halversen and G. J. Owirka, 'Discriminating Targets from Clutter', *Lincoln Lab. J.*, Vol. 6, No. 1, 1993.

[3] C. J. Oliver and S. Quegan, *Understanding Synthetic Aperture Radar Images*, SciTech Publishing Inc., Raleigh, NC, 2004.

[4] D. Blacknell, 'Shadow-based SAR ATR Performance Prediction', *Proceedings of SPIE 7337, Algorithms for Synthetic Aperture Radar Imagery XVI*, Orlando, FLA, 2009.

[5] E. Cerrone, Research Thesis, University College London, 2008.

Index

www.ingramcontent.com/pod-product-compliance
Lightning Source LLC
LaVergne TN
LVHW011749200926
848484LV00003B/657

* 9 7 8 1 8 4 9 1 9 6 8 5 7 *